# Indian Institute of Metals Series

**About the Book Series**:

The study of metallurgy and materials science is vital for developing advanced materials for diverse applications. In the last decade, the progress in this field has been rapid and extensive, giving us a new array of materials, with a wide range of applications, and a variety of possibilities for design of new materials, processing and characterizing the materials. In order to make this growing volume of knowledge available, an initiative to publish a series of books in Metallurgy and Materials Science was taken during the Diamond Jubilee year of the Indian Institute of Metals (IIM) in the year 2006, and has been published in partnership with Springer since 2016.

This book series publishes different categories of publications: textbooks to satisfy the requirements of students and beginners in the field, monographs on select topics by experts in the field, professional books to cater to the needs of practising engineers, and proceedings of select international conferences organized by IIM after mandatory peer review. The series publishes across all areas of materials sciences and metallurgy. An panel of eminent international and national experts serves as the advisory body in overseeing the selection of topics, important areas to be covered, and the selection of contributing authors.

Eswara Prasad Namburi · R. J. H. Wanhill ·
Dipak Kumar Setua
Editors

# Novel Defence Functional and Engineering Materials (NDFEM) Volume 1

## Functional Materials for Defence Applications

Metallurgy
Materials Engineering

Springer

*Editors*
Eswara Prasad Namburi
Defence Materials and Stores Research and Development Establishment (DMSRDE)
Defence Research and Development Organisation (DRDO)
Kanpur, Uttar Pradesh, India

R. J. H. Wanhill
Emeritus Principal Research Scientist, Aerospace Vehicles Division
Royal Netherlands Aerospace Centre
Amsterdam, Flevoland, the Netherlands

Dipak Kumar Setua
ACRHEM/UoH, Defence Materials and Stores Research and Development Establishment (DMSRDE)
Defence Research and Development Organisation (DRDO)
Kanpur, Uttar Pradesh, India

ISSN 2509-6400 ISSN 2509-6419 (electronic)
Indian Institute of Metals Series
ISBN 978-981-99-9790-9 ISBN 978-981-99-9791-6 (eBook)
https://doi.org/10.1007/978-981-99-9791-6



This Springer imprint is published by the registered company Springer Nature Singapore Pte Ltd.
The registered company address is: 152 Beach Road, #21-01/04 Gateway East, Singapore 189721, Singapore

Paper in this product is recyclable.

*These book volumes are respectfully dedicated to all our Indian Defence Functional Materials Scientists, Technologists, and Industrials Associates, and also the Defence Personnel, who have been collaborating with the authors and editors of these volumes; and our partners and families for their continuous support, notably during 2020–2023, whose constant encouragement and kind support have been the main sources of motivation for maintaining this task until its completion.*

*—Eswara Prasad Namburi, R. J. H. Wanhill and Dipak Kumar Setua*

# Series Editor's Preface

The *Indian Institute of Metals Series* is an institutional partnership series focusing on metallurgy and materials science and engineering.

## About the Indian Institute of Metals

The Indian Institute of Metals (IIM) is a premier professional body (since 1947) representing an eminent and dynamic group of metallurgists and materials scientists and engineers from R&D institutions, academia, and industry, mostly from India. It is a registered professional institute with the primary objective of promoting and advancing the study and practice of the science and technology of metals, alloys, and novel materials. The institute is actively engaged in promoting academia–research and institute–industry interactions.

## Genesis and History of the Series

The study of metallurgy and materials science and engineering is vital for developing advanced materials for diverse applications. In the last decade, progress in this field has been rapid and extensive, giving us a new array of materials, with a wide range of applications and a variety of possibilities for processing and characterizing the materials. In order to make this growing volume of knowledge available, an initiative to publish a series of books in metallurgy and materials science and engineering was taken during the Diamond Jubilee year of the Indian Institute of Metals (IIM) in the year 2006. IIM entered into a partnership with Universities Press, Hyderabad, in 2006, and from 2016 the book series is under MoU with M/s Springer Nature, and as part of the IIM book series, a total of 24 books were published till 2023. The books were authored by eminent professionals in academia, industry, and R&D with outstanding background in their respective domains, thus generating unique resources of validated expertise of interest in metallurgy. The international character of the authors' and editors has enabled the books to command national and global readership. This book series includes different categories of publications: textbooks to satisfy the requirements of undergraduates and beginners in the field, monographs on select topics by experts in the field, and proceedings of select international conferences organized by IIM, after mandatory peer review. An eminent panel of experts constitutes the advisory body in overseeing the selection of topics, important areas to be covered in the books and the selection of contributing authors.

## About "Novel Defence Functional and Engineering Materials"

This book on "Novel Defence Functional and Engineering Materials" with Volume 1: Defence Functional Materials, and Volume 2: Defence Engineering Materials, will be of great interest to the scientists and researchers who are involved in developing new and innovative functional polymeric materials, synthetic and special fuels and fluids, and composite systems for the development of advanced systems like hypersonic vehicles, fighter aircraft, e.g., LCA Mk II, unmanned air vehicle systems, a variety of Defence platforms, underwater missiles and torpedoes, radar detection and deception systems, strategic applications for future MBTs, in addition to advanced futuristic technologies for structural sensors, stealth and environmental applications. Both volumes of the book cover three important categories of functional materials: viz., polymers, elastomers and textiles; high temperature ceramics; and nanomaterials, particularly the design and development of polymer matrix composites, inorganic–organic hybrids, active and responsive smart materials, magneto-rheological fluids, gels and lubricant fluids, polymeric systems and coatings for stealth and camouflage applications, with examples of a few critical products. To some extent theoretical aspects are dealt with, where appropriate, to provide core knowledge on phases and microstructures; in addition, the use of sophisticated characterization techniques, like transmission and scanning electron microscopy equipped with various analyzers; X-ray diffraction; nanoindentation; atomic force microscopy; FTIR and Raman Spectroscopy; vector network and particle size analyzers; and a variety of thermal and magnetic property measurement instruments have been brought out, including a few improvised analytical techniques.

Both volumes 1 and 2 contain 10 chapters each by well-known experts in the field from India and abroad, and finely edited and delivered by the editors, Dr. Eswara Prasad Namburi, Dr. R. J. H. Wanhill and Dr. Dipak Kumar Setua. They have made excellent efforts to coordinate with specialists in the respective fields of polymers, elastomers and textiles; high temperature ceramics; and nanomaterials, to provide 20 articles of high relevance. Both volumes 1 and 2 of this Book will be a treasure for those who are interested in learning everything about functional and engineering materials of relevance to defense applications, and pursue a research career and study. The IIM-Springer Series gratefully acknowledge the three editors and all chapter authors for their excellent efforts in covering wide range of information on the subject matter of interest to the readers.

Dr. U. Kamachi Mudali
FIIM, FNAE, FNASI, FASM, FNACE
FAPAM, FICS, HFECSI, FIIChE FASCh.
Editor-in-Chief, Series in
Metallurgy and Materials Engineering
Vice Chancellor
Homi Bhabha National Institute
(HBNI)
Mumbai, India

# Foreword

Materials and technologies and their development play fundamental roles for the design of advanced tactical and operational systems required for Defence Services. The functional non-metallic materials that are described in this book cover a vast array of materials, including polymers, elastomers and composites; interactive and smart textiles; ceramics for high-temperature applications; protective armours; special organics for radar absorption and electromagnetic radiation shielding properties; polymer nanocomposites (PNCs) based on magnetostrictive and magneto-rheological solids, liquids and gels; and hybrid and metal–organic framework materials. These materials have been conceived and fabricated by the use of novel synthesis and processing routes and tailored to manipulate the microstructures to obtain classes of materials that possess specific properties for particular applications.

Recently, some of these materials have started to be classified on the basis of particular functions or specific tasks they perform. This is a major change from the classical categorization of materials in terms of their structure–property relationships. The change to the two major subject areas, Functional and Engineering, has become a trendsetter for research and development (R&D) in several sectors, e.g., academia, industry and strategic, especially the Defence and Aerospace sectors.

This volume "Functional Materials", will, in my opinion, be of significant interest to scientists and researchers who are either involved in the development of new functional polymeric materials and composite systems, or others who have been planning to use them in their forthcoming advanced systems like hypersonic vehicles; fighter aircraft like the light combat aircraft (LCA); unmanned aerial/aerospace vehicles (UAVs); underwater missiles; radar systems; strategic materials for main battle tanks (MBTs); and advanced futuristic technologies for structural sensors, stealth and environmental applications.

The editors, Dr. Eswara Prasad Namburi, Dr. R. J. H. Wanhill and Dr. Dipak Kumar Setua, are quite well known, both nationally and internationally, and have made significant contributions in their respective scientific fields. They have done an admirable task in making an impressive configuration for the two volumes, containing a total of 20 book chapters. Each of these 20 chapters draws on the expert knowledge and contributions of the scientists who have first-hand knowledge on the subject. I also understand that the entire process of reviewing; presentation of the subject materials; verifying and thus authenticating the contents; and also rephrasing texts to improve the readability, was accomplished by Dr. Wanhill. My best compliments to him and to Dr. Prasad and Dr. Setua for ensuring the quality of this special book publication effort.

The array of materials and their processing methodologies for creating different functional materials; and their simulation and modelling methodologies, have been adopted to evolve some new generation functional materials of potential commercial interest; as well as synergy with the ultimate goal of device-making particularly for the Defence R&D Organization (DRDO). Further, the problems associated with applications related to nanotechnology, optics, electronics, stealth and camouflage technologies and devices, sensors and information technology have especially been investigated or refined further. The outcomes and reproducibility of in-house studies are likely to enable allied technologies for extraordinary molecules with manyfold structural complexity and intricate chemical reactivity; supra-molecular assemblies; embedded electronics in material structures that can sense, actuate and communicate with the systems and surroundings.

The functional molecules can be associated with conventional materials and structures and could be organized into assemblies with specific architectures, stoichiometries and dimensionalities. The full potential of functional materials and composites will be realized in applications that require combinations of properties, where the overall performance outweighs one particular property; and the use of existing process technologies and manufacturing to optimize the costs versus performance of new devices. The exciting journey of functional materials is still in its infancy. Success during this demanding and challenging journey will be one of the most crucial contributions to a stable, rewarding and prosperous future.

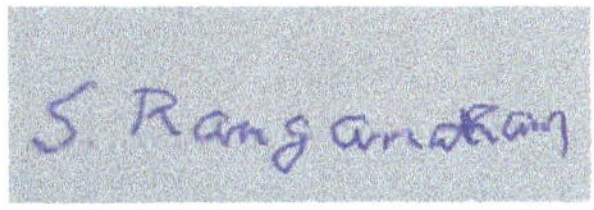

Prof. Srinivasa Ranganathan
FASc, FNASc, FNAE, FTWAS
Emeritus Professor
Indian Institute of Science (IISc)
Bengaluru, India

# Preface

The designation "functional materials" covers an exceptionally wide range of materials: polymers, elastomers, composites and polymer nanocomposites; nanomaterials with 0D, 1D, 2D and 3D structures; inorganic–organic and hybrid framework materials; interactive and smart textiles; ceramics for high-temperature applications and protective armours; special organics for radar absorption, camouflage and electromagnetic radiation shielding; and super-specialized coatings for sensing and deception. There are also applications as nano-fluids, ionic liquids and hydrogels, and precursor materials in the development of advanced semiconductors, ultrahigh temperature ceramic components, super hydrophobic and oleophobic textiles, and magnetostrictive and magneto-rheological solids.

The performances of functional materials are based on understanding the solid state physical and chemical principles and phenomena, atomic orbital and bond structures and the material microstructures. This necessarily wide-ranging and detailed approach aims to explain both the micro-scale phenomena as well as the macroscopic behaviour, with the additional goal of developing novel functionalities and evaluating their potential for applications. For these reasons, and because functional materials are recognized as key research areas for future expansion and growth, they have become a "hot topic" for academic and industrial R&D, both nationally and globally, and also for the strategic sectors like Defence and Aerospace. The special books on Novel Defence Functional and Engineering Materials (NDFEM): Volume 1. Functional Materials, and Volume 2. Engineering Materials will be of significant interest to scientists and researchers who are (i) involved in developing new functional polymeric materials and composite systems for forthcoming development of advanced systems like hypersonic vehicles, fighter aircraft, e.g., LCA-II, unmanned air vehicle systems, a variety of defence platforms, underwater missiles and torpedoes, radar detection and deception systems, and strategic applications for future MBTs and (ii) advanced futuristic technologies for structural sensors, stealth and environmental applications.

Owing to the necessary detailed understanding of functional materials, the conventional trend is to classify them based on their origins, the nature of physical–chemical bonding, or the processing techniques for their preparation. However, recent trends have shifted to classifications based on mechanical and structural load-bearing properties *in combination with* their particular functional properties. This is a major shift from the traditional classification in terms of structure–property relationships.

The list of functional materials is long and to review them all in one go is a demanding task. Textbooks related to the subject have not kept pace with the technological needs and state-of-the-art advancement in this field. While Materials Science textbooks deal with the properties of functional materials, the Electronic Materials textbooks focus on the advent of semiconductors and smart materials and structures. Therefore, our focus in these two volumes will consider various kinds of non-metallic functional materials and particularly their applications to important and strategic sectors like Aerospace and Defence.

A broad overview of the status and technological gap has been attempted by selecting some topics for detailed treatment of their critical aspects. The chapters in these books mostly cover three important categories of functional materials: viz., polymers, elastomers and textiles; high-temperature ceramics; and nanomaterials. The scope also includes a variety of specific functional materials and some assemblies and sub-assemblies. It provides details on the design and development of polymer matrix composites, inorganic–organic hybrids, active and responsive smart materials, magneto-rheological solids, gels and lubricant fluids, polymeric systems and coatings for stealth and camouflage applications, with examples of a few critical products. Theoretical aspects are dealt with, where appropriate, to provide core knowledge on phases and microstructures; the use of sophisticated characterization techniques, e.g., transmission and scanning electron microscopy equipped with various analysers; X-ray diffraction; nanoindentation; atomic force microscopy; FTIR and Raman Spectroscopy; vector network and particle size analysers; and a variety of thermal and magnetic property measurement instruments. Improvised analytical techniques have also been mentioned.

The nature and contents of these special volumes are most suitable as ready reference material for working engineers, faculty members and research scholars working on special functional materials both in the science and engineering of chemistry, materials science, materials technology, industrial engineering, chemical engineering and manufacturing engineering.

Eswara Prasad Namburi

Kanpur, India
May 2023

R. J. H. Wanhill

Dipak Kumar Setua

# Acknowledgements

When I met Dr. Baldev Raj, the then Editor-in-Chief of IIM-Springer Book Series in his Director's office at NIAS, Bengaluru soon after we published the two-volume book on Aerospace Materials and Materials Technologies, he complemented our near 3-year effort to fructify that book project. [As he liked the book contents very much, he predicted they will do well both in stands (in sales) as well as Vade Mecums (in Libraries). His predictions came true as these books did extremely well on both fronts.] He then inquired whether we can bring out similar volumes for Functional Materials, as I just started leading a premier Functional Materials Laboratory of DRDO, the DMSRDE, Kanpur. I immediately said emphatically "NO" as I or my associates till then never worked on functional materials. But, he was persistent; and, after an year or so, I conceded and consented to do another two-volume book series—another set of Vade Mecums on Functional and Engineering Materials. Since this is a very vast subject, we agreed to cover only Defence-related materials, in which my colleagues at DMSRDE and I started having a good insight and we were greatly strengthened in our cause when Dr. Dipak Kumar Setua joined hands as he was a known international expert in the area of polymers and their derivatives. Soon, the contents took shape and all the preliminary drafts were sent for review and basic approval to Dr. R. J. H. Wanhill, by then he was a very close associate and a co-editor for three of my earlier 16 books that I edited. He liked the idea and the outline, but rendered his apologies for not being able to do the honors (partly because the contents were not in the best of the shape for any international book publications). Later, he too had to agree to Dr. Baldev Raj's request and put in enormous efforts to bring the contents to good standards of readability and presentability. A few special efforts by both Dr. Russell and Dr. Setua improved the quality of results and further due to their undeterred insistences, a few initial drafts took totally new shape with large omittances and many new additions. Some of the authors of this book series are greatly benefited and their own contributions elsewhere too have become widely accepted. All this has happened as this project was executed to near perfection for more five years of hardship and total dedication.

Special mention should be made to the perseverance of all the stakeholders not to lose the sight of ultimate goal (even in the worst situations of the COVID-19 pandemic that prevailed during 2019 all the way up to 2022), i.e., to complete the book project without lowering the standards, and also to cover most of the topics that are relevant not only to Indian defence, but also to defence applications for the world over. Hence, my heart goes out to all the authors and my two co-editors and I would like to express my sincere gratitude to all these contributors; I am indebted to each one of them: *Bablu Mordina, Dipak Kumar Setua, Swati Chopra, Vishal Das, Ajit S. Singh, Arvind Kr. Pandey, Debmalya Roy, Alok Kr. Srivastava, Kingsuk Mukhopadhya, Jyoti P. Singh, Ashok Kr. Gautam, Jyoti Srivastava, Tandra Nandi, Harish Kumar, Sandeep Kumar, Ajay Katiyar, Nizmuddin Khan, Rakesh Kr. Gupta, Raghwesh Kr. Mishra, Suresh Kumar, Ashok Ranjan, Sathosh Kr. Tripathi, Himanshi Chaurasia, Shilendra Kumar, Sunil Kumar, Dibyendu S. Bag, Akansha Dixit, Satyen Saha.*

Dr. Russell Wanhill's efforts need a very special mention, especially for these book volumes. This is because in case of many book chapters, we had minimum three or four versions, and further some of these versions were differently modified by either Dr. Setua or our internal DMSRDE experts, causing utter chaos and confusion. Dr. Wanhill not only kept track of these versions and progressively worked on all the drafts, and he singlehandedly and painstakingly shaped the entire book contents. No works will suffice to thank him enough and these two volumes took final shape only due to his unparalleled efforts. I also take this opportunity to thank Dr. Setua, who went through lots of health problems, still never excused himself from this daunting task. I owe a lot to him and also learnt a lot in this process with regard to all these special defence functional and engineering materials, which were till then foreign to me. Some of my colleagues who authored many special and classified subjects took special care to work diligently not to reveal any classified information, but, at the same time, did justice to cover amply on all the important aspects of each of the materials and material systems. I thank all of them profoundly.

As I said the book volumes underwent several interactions, including one to take care of stringent plagiarism limits of Springer. Mr. Devroop Arya and Mrs. Bhavana Srivastava of DMSRDE as well as Mr. S. Shashinath of DMRL, DRDO, Hyderabad helped immensely with minimum 2–3 rounds of checks. Mr. Devroop, Mrs. Pratibha Singh, and Mr. Surendra Kr. Yadav of DMSRDE further helped in file maintenance as well as several rounds of formatting. A few colleagues from the various materials directorates of DMSRDE too have helped and my special thanks are due to them. I also thank several senior and yesteryear DMSRDE scientists and technologists and former directors—Prof. G. N. Mathur, Dr. K. U. Bhasker Rao, Dr. A. K. Saxena and the present director, Dr. Mayank Dwivedi. Besides these dignitaries, my heart goes out to the colleagues and their family members whom we lost dearly due to COVID-19 pandemic, especially Dr. Mahender Prasad; Shri Avinash Pankaj of DMSRDE; and Shri Alok Mal, Director, DHR, DRDO Hqs (Formerly with DMSRDE).

I consider it is an honor that none other than Prof. Srinivasa Ranganathan, a doyen of Materials Research in India has agreed to write the Foreword for the first of the two volumes. I am most indebted to Dr. Ranganathan. Dr. Samir V. Kamat who recently took charge as Chairman, DRDO has obliged my request and wrote the Foreword for the second volume despite his many pressing engagements within a very short time and I consider this as a special favor that he has done to me personally. I should also place on record my sincere thanks to Dr. K. Bhanu Sanakara Rao, Dr. G. Malakondaiah, Prof. V. V. Kutumbarao, Prof. J. Viplava Kumar, Prof. Rajiv Prakash, Prof. Sandeep Verma, Prof. B. S. Murty, Prof. I. Manna, Prof. D. Banerjee, Prof. Shrikant Lele, Dr. G. Sundararajan, Prof. C. Suryanarayana, Prof. K. A. Padmanabhan, Dr. Shantanu Chakrabarti, Prof. C. Ravi Ravindran, Dr. Kamachi U. Mudali, Dr. Amol A. Gokhale, Dr. G. Satheesh Reddy, Dr. Kota Harinarayana, Dr. V. K. Saraswat, and Prof. P. Rama Rao for their constant encouragement and exemplary support. Special mention should be made to the spontaneous encouragement given by Prof. Dr. Eduard Arzt and Prof. Tobias Kraus, though I could not involve them much directly in this mammoth publishing effort. The final contribution to manuscript preparations and support was provided by a few intimate dignitaries, officials, and colleagues at BEPL/NH, located at the Mind Space, HITEC City, Hyderabad, where I was on a 6-month sabbatical as Expert Advisor. I profoundly thank all of them.

I consider it my honor that Springer Nature Publications agreed to IIM Book Publication Committees' recommendations and accepted to publish these book volumes. I place on record my profound gratitude to Mr. William Achauer, Director and Mr. Anil Chandy, Managing Director, Springer Nature Singapore Private Limited for official contracting; Mrs. Swati Maharishi, Editorial Director, Springer, New Delhi for her special, continued, and constant support; and Mr. Ashok Kumar, Senior Production Executive, Springer, New Delhi for the kind follow-up. Our special thanks are due to Mrs. Ramya Somasundaram, Project Coordinator, Book Production, Springer Nature for valuable support in printing these volumes, and more importantly for bearing with us with many a number of special requests. At this juncture I should acknowledge gratefully the initiative taken by the IIM, particularly by Late Dr. Baldev Raj, without whom I would not have ventured to this project and without his invisible support would not have completed this book project.

Eswara Prasad Namburi

# Current Series Information

To increase the readership and to ensure wide dissemination among global readers, this new chapter of the series has been initiated with Springer in the year 2016. The goal is to continue publishing high-value content on metallurgy and materials science and engineering, focusing on current trends and applications. So far, eleven important books on state of the art in metallurgy and materials science and engineering have been published and three books were released during IIM-ATM 2022 at Hyderabad. Readers who are interested in writing books for the Series may contact the Series Editor-in-Chief, Dr. U. Kamachi Mudali, Former President of IIM and Vice Chancellor of Homi Bhabha National Institute (HBNI), Mumbai at ukmudali1@gmail.com, vicechancellor@hbni.ac.in or the Springer Editorial Director, Ms. Swati Meherishi at swati.meherishi@springer.com.

## Editorial Advisory Board

(2022–2024)

## Editor-in-Chief

# Contents

# Editors and Contributors

## About the Editors

**Dr. Eswara Prasad Namburi** joined DRDO in 1985 after obtaining a B.Tech. and Ph.D. in Metallurgical Engineering from the Indian Institute of Technology, Banaras Hindu University. He retired in September 2022 after serving the Indian Ministry of Defence for more than three decades as an Outstanding Scientist of DRDO. Dr. Prasad has contributed immensely to the development of several defence systems, including fighter aircraft structures, aero engines, missiles, naval propulsion, and personal protection systems. He has been instrumental in awarding 14 defence production technologies to over 40 Indian industries. Dr. Eswara Prasad is internationally known for his research on fatigue, fracture and life extension behaviour of several aero, structural and functional materials; has, over 750 wide-ranging publications that include 12 government white papers; 17 international edited books/conference proceedings, a monograph on Aluminium-Lithium Alloys: Processing, Properties and Applications and a 2-volume Reference Book (A Vade Mecum) on Aerospace Materials and Material Technologies. He has been Chairman, President or Member of 10+ national committees and bodies and received 15+ national and international awards and 70+ honours and recognitions. These include fellowships of six national and international science and engineering academies/institutions, notably Founder Fellow of the Indian Structural Integrity Society (FInSIS) and Fellow of the Asia Pacific Academy of Sciences (FAPAM).

**Dr. R. J. H. Wanhill** is emeritus Principal Research Scientist, Aerospace Vehicles Division, Royal Netherlands Aerospace Centre (NLR), in the Netherlands. He holds two Doctorates, one from the University of Manchester (1968) and the second from the Delft University of Technology (1994). He joined the NLR in 1970, and since then has investigated fatigue and fracture of all classes of aerospace alloys. He is co-author of the book Fracture Mechanics (1984), which ran into a second edition; co-author with Simon Barter of the monograph Fatigue of Beta Processed and Beta Heat-treated Titanium Alloys (2012); co-author and co-editor for the book Aluminium-Lithium Alloys: Processing, Properties and Applications (2014); co-author and co-editor for the 2-volume series Aerospace Materials and Material Technologies (2017); co-author of the monograph Corrosion and Stress Corrosion Testing of Aerospace Vehicle Structural Alloys (2018); and co-author of the monograph Fatigue Crack growth Failure and Lifing Analyses for Metallic Aircraft Structures and Components (2019).

**Dr. Dipak Kumar Setua** obtained his M.Sc. (Chemistry) with top rank in 1981 and Ph.D. (Polymer Science and Rubber Technology) in 1985, both from the Indian Institute of Technology (IIT), Kharagpur, India. Subsequently, he joined as a research scientist and spent more than three decades at the Defence Materials and Stores R&D Establishment (DMSRDE), Kanpur, India. Furthermore, Dr. Setua served DMRSRDE and DRDO as a DRDO Fellow until 2022. He spent two years (1989–1991) at the Institute of Polymer Engineering, University of Akron, Ohio, Akron, USA, as a visiting scientist and remained Director of the Advanced Centre of Research on High Energy Materials (ACRHEM), University of Hyderabad, India, during 2013–2014. Some of the of his research include short fibre-rubber composites, polymer blends and compatibility, polymer characterization, magneto-rheological elastomer, and design and development of polymer-based composites and nanocomposites and products for personal protection, aerospace systems and bio-medical applications. Dr. Setua has published more than 100 archival papers, several patents and book chapters. He has also received many prestigious honours and awards.

## Contributors

**Dibyendu S. Bag** Defence Materials and Stores Research and Development Establishment (DMSRDE), Defence Research and Development Organisation (DRDO), Kanpur, Uttar Pradesh, India

**Himanshi Chaurasia** Formerly With Defence Materials and Stores Research and Development Establishment (DMSRDE), Defence Research and Development Organisation (DRDO), Kanpur, Uttar Pradesh, India

**Swati Chopra** Defence Materials and Stores Research and Development Establishment (DMSRDE), Defence Research and Development Organisation (DRDO), Kanpur, India

**Vishal Das** Defence Materials and Stores Research and Development Establishment (DMSRDE), Defence Research and Development Organisation (DRDO), Kanpur, India

**Akansha Dixit** Indian Institute of Technology, Kanpur, Uttar Pradesh, India

**Ashok Kr. Gautam** Defence Materials and Stores Research and Development Establishment (DMSRDE), Defence Research and Development Organisation (DRDO), Kanpur, Uttar Pradesh, India

**Rakesh Kr. Gupta** Defence Materials and Stores Research and Development Establishment (DMSRDE), Defence Research and Development Organisation (DRDO), Kanpur, Uttar Pradesh, India

**Ajay Katiyar** Defence Materials and Stores Research and Development Establishment (DMSRDE), Defence Research and Development Organisation ((DRDO), Kanpur, Uttar Pradesh, India

**Nizamuddin Khan** Defence Materials and Stores Research and Development Establishment (DMSRDE), Defence Research and Development Organisation ((DRDO), Kanpur, Uttar Pradesh, India

**Sandeep Kumar** Defence Materials and Stores Research and Development Establishment (DMSRDE), Defence Research and Development Organisation ((DRDO), Kanpur, Uttar Pradesh, India

**Shilendra Kumar** Defence Materials and Stores Research and Development Establishment (DMSRDE), Defence Research and Development Organisation (DRDO), Kanpur, Uttar Pradesh, India

**Sunil Kumar** Defence Materials and Stores Research and Development Establishment (DMSRDE), Defence Research and Development Organisation (DRDO), Kanpur, Uttar Pradesh, India

**Suresh Kumar** Defence Materials and Stores Research and Development Establishment (DMSRDE), Defence Research and Development Organisation (DRDO), Kanpur, Uttar Pradesh, India

**Raghwesh Mishra** Defence Materials and Stores Research and Development Establishment (DMSRDE), Defence Research and Development Organisation (DRDO), Kanpur, Uttar Pradesh, India

**Bablu Mordina** Defence Materials and Stores Research and Development Establishment (DMSRDE), Defence Research and Development Organisation (DRDO), Kanpur, Uttar Pradesh, India

**Kingsuk Mukhopadhyay** Defence Materials and Stores Research and Development Establishment (DMSRDE), Defence Research and Development Organisation (DRDO), Kanpur, Uttar Pradesh, India

**Eswara Prasad Namburi** Former Outstanding Scientist and Ex-Director, Defence Materials and Stores Research and Development Establishment (DMSRDE), Defence Research and Development Organisation (DRDO), Kanpur, Uttar Pradesh, India

**Tandra Nandi** Formerly With Defence Materials and Stores Research and Development Establishment (DMSRDE), Defence Research and Development Organisation (DRDO), Kanpur, Uttar Pradesh, India

**Arvind Kr. Pandey** Defence Materials and Stores Research and Development Establishment (DMSRDE), Defence Research and Development Organisation (DRDO), Kanpur, India

**Ashok Ranjan** Defence Materials and Stores Research and Development Establishment (DMSRDE), Defence Research and Development Organisation (DRDO), Kanpur, Uttar Pradesh, India

**Debmalya Roy** Defence Materials and Stores Research and Development Establishment (DMSRDE), Defence Research and Development Organisation (DRDO), Kanpur, Uttar Pradesh, India

**Satyen Saha** Department of Chemistry, Institute of Science, Banaras Hindu University, Varanasi, India

**Dipak Kumar Setua** Defence Materials and Stores Research and Development Establishment (DMSRDE), Defence Research and Development Organisation (DRDO), Kanpur, Uttar Pradesh, India;
ACRHEM/UoH, DRDO, Hyderabad, India

**Ajit S. Singh** Defence Materials and Stores Research and Development Establishment (DMSRDE), Defence Research and Development Organisation (DRDO), Kanpur, India

**Jyoti P. Singh** Defence Materials and Stores Research and Development Establishment (DMSRDE), Defence Research and Development Organisation (DRDO), Kanpur, Uttar Pradesh, India

**Alok Kr. Srivastava** (On Deputation From DMSRDE, DRDO), National Test House, Min. Consumer Affairs, GoI, Kolkata, West Bengal, India

**Jyoti Srivastava** Defence Materials and Stores Research and Development Establishment (DMSRDE), Defence Research and Development Organisation (DRDO), Kanpur, Uttar Pradesh, India

**Santosh Kr. Tripathi** Defence Materials and Stores Research and Development Establishment (DMSRDE), Defence Research and Development Organisation (DRDO), Kanpur, Uttar Pradesh, India

# Chapter 1
# Polymer Materials for Defence & Aerospace Applications

**Bablu Mordina, Dipak Kumar Setua, and Eswara Prasad Namburi**

**Abstract** This chapter gives an overview of the synthesis of different magnetic materials (including particles, fibres and rods) and their incorporation into PNCs (polymer nanocomposites) to obtain so-called MREs (magneto-rheological elastomers). Characterization of the structural and magnetic properties of these materials and evaluation of their magneto-rheological properties are discussed. MREs are 'smart' materials with numerous functional applications, examples of which are given towards the end of the chapter.

## 1.1 Introduction

Development of new materials and technologies has become a foremost necessity during the recent decades of rapid technological growth. In this context, smart materials have attracted much attention owing to their versatile and tailorable functional properties. These materials have unique combinations of physical and chemical properties that are influenced by external stimuli, but in controlled and reversible ways. The external stimuli may be magnetic or electric fields, mechanical stresses, temperature, moisture, pH, etc.

Magneto-rheological (MR) materials are an important class of smart materials. These show a continuous, reversible and rapid change in their rheological properties when the applied magnetic field intensity is varied. The development of MR materials

B. Mordina · D. K. Setua (✉)
Defence Materials and Stores Research and Development Establishment (DMSRDE), Defence Research and Development Organisation (DRDO), Kanpur, Uttar Pradesh, India
e-mail: dksetua@rediffmail.com

D. K. Setua
ACRHEM/UoH, DRDO, Hyderabad, India

E. P. Namburi
Former Outstanding Scientist and Ex-Director, Defence Materials and Stores Research and Development Establishment (DMSRDE), Defence Research and Development Organisation (DRDO), Kanpur, Uttar Pradesh, India

 
E. P. Namburi et al. (eds.), *Novel Defence Functional and Engineering Materials (NDFEM) Volume 1*, Indian Institute of Metals Series,
https://doi.org/10.1007/978-981-99-9791-6_1

was initiated by three major companies: Toyota Central Research and Development Laboratories Inc., Lord Corporation, and the Ford Research Laboratory (Shiga et al. 1995; Jolly et al. 1996a, 1996b; Ginder et al. 1999). Subsequently, extensive worldwide research has been done on different MR materials, namely, magneto-rheological fluids (MRFs), ferrofluids and magneto-rheological elastomers (MREs). These materials differ in several respects: carrier medium, types and sizes of magnetic particles, types of additives, filler loading, properties, working regime and application areas.

Table 1.1 summarizes the differences between typical MR materials. It is seen that the particle sizes cover a wide range, and the carriers may be oils, water or elastomers. The physical phenomena responsible for magnetic field-dependent sensitivity are very similar for MR fluids and elastomers, but there are some distinct differences: MR fluids operate mainly in the post-yield continuous shear or flow regime and show field-dependent yield stress; whereas an MRE always operates in the pre-yield regime and possesses a field-dependent modulus. In fact, MREs are the solid-state analogues of MR fluids: the use of a solid elastomer matrix imparts flexibility due to their soft and elastically deformable characteristics at room temperature, together with considerable reversible extensibility.

**Table 1.1** Composition and properties of typical magneto-rheological materials

| Properties | MRF | Ferrofluid | MRE |
|---|---|---|---|
| Particle types | Iron | Magnetite | Iron |
| Particle sizes | 0.1–10 μm | 2–10 nm | 10–50 μm |
| Carrier medium | Oils | Oils, water | Elastomers |
| Volume fraction of particles | 0.1–0.5 | 0.02–0.2 | 0.1–0.5 |
| Additives | Surfactants, thixotropic agents | Surfactants | None |
| Viscosity in absence of magnetic field [mPa s] | 100–1000 | 2–200 | None |
| Working region | Post-yield continuous shear or flow regime | Post-yield continuous shear or flow regime | Pre-yield regime |
| Characteristic property | Field-dependent yield stress | Field-dependent yield stress | Field dependent modulus |
| Change of properties under external magnetic field | Yield stress ($\tau_y$) can reach up to ~100 kPa | Relative viscosity $\Delta\eta/\eta - 1$ | Change in storage modulus ($\Delta G'$) ~20 kPa |
| Container requirement | Yes | Yes | Not required |
| Settling of particles | Very prone to settle | Low settling | No settling |
| Toxicity | May be toxic depending on carrier fluid | May be toxic depending on carrier fluid | Nontoxic |

All three types of MRs are interesting and important. However, only MREs are discussed in this chapter, notably the more recently developed MREs using nanomaterial and nanotechnology developments. These are functional polymer nanocomposites (PNCs) consisting of nano-sized particles/fibres/whiskers, etc., at a very low filler concentration in a variety of polymers. They are important owing to their potential for novel and strategic applications, e.g. as flexible polymer magnets and active elastomeric vibration dampers. Example applications are given in Sect. 1.9 of this chapter.

## 1.2 Classification of MREs

MREs consist of ferromagnetic or ferrimagnetic particles embedded in an elastomer matrix with a very low level of cross-link density. An external magnetic field causes dipole magnetic forces between the embedded particles, which then exhibit a field-dependent MR effect. This effect can result in very significant changes in the modulus and loss (damping) factor of the filled elastomers. In turn, these changes depend not only on the magnetic field strength but also the concentration, size, shape, orientation, and inherent properties of the magnetic particles, as well as the physico-mechanical properties of the (elastomer) matrix.

MREs can be classified in several ways, depending on the types of particles, their saturation magnetization their distribution and dispersion into an elastomeric matrix, and the properties of the matrix itself. Figure 1.1 is an example classification. Note, however, that while Fig. 1.1 is apparently comprehensive, this classification does not specifically mention the more recently developed group of PNCs.

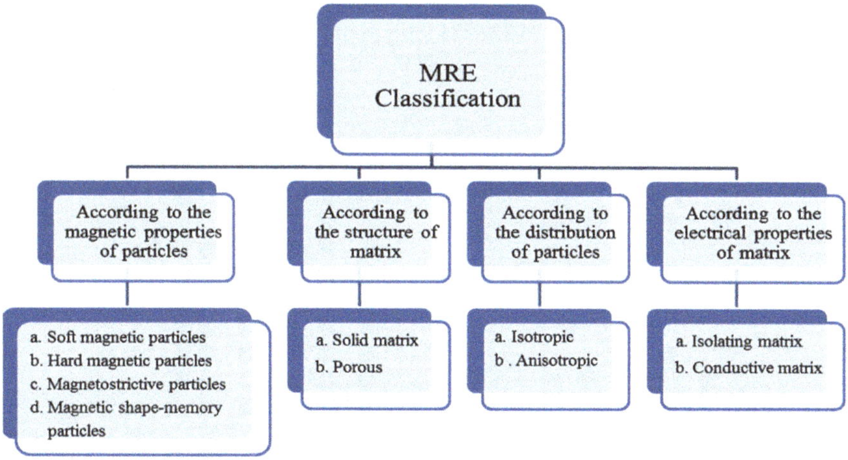


**Fig. 1.1** Classification of MREs

## 1.3 MRE Basics

### *1.3.1 MRE Compositions and Structures*

Conventional MREs consist of micron-sized magnetic particles embedded in a non-magnetic elastomeric solid matrix. Most are based on natural rubber, polyurethane, chloroprene, silicone, nitrile rubbers, polybutadiene, and polyvinyl chloride matrix materials containing soft magnetic particles of iron, cobalt, nickel, and their oxides (Vicente et al. 2002; An and Shaw 2003; Kallio et al. 2003; Lokander and Stenberg 2003a). MREs with hard magnetic materials like $BaFe_{12}O_{19}$, $SrFe_{12}O_{19}$, $Nd_2Fe_{14}B$, and $SmCo_5$ have also been investigated (Dishovsky et al. 2001; Koo et al. 2012; Stepanov et al. 2012). Of these particles, pure iron shows the highest saturation magnetization, coupled with high permeability and low remnant magnetization. This combination of magnetic properties makes the use of pure iron particles popular and commercially attractive. Carbonyl iron particles are also extensively used.

Either isotropic or anisotropic MREs can be made, depending on whether the cross-linking (curing) of the elastomer matrix is done in the absence or presence of a magnetic field. In the absence of a magnetic field the magnetic particles are distributed homogeneously in the polymer matrix and the MREs are isotropic, see Fig. 1.2a; but when curing is done in the presence of a magnetic field the particles become arranged in a chain-like columnar or complex three-dimensional network, such that the MREs are anisotropic, see Fig. 1.2b.

When cured MREs are subjected to a magnetic field the magnetic particles approach each other owing to interparticle forces, resulting in *magnetostriction*. This is manifested in anisotropic MREs by length decreases, Fig. 1.3a; and in isotropic MREs by length increases, Fig. 1.3b. Obviously, for these effects to occur, the matrix/particle interfacial bonding must be sufficiently strong. Figure 1.3c illustrates a special case of a giant magnetostriction effect in MREs containing Terfenol-D particles (Wang et al. 2003; Yin et al. 2006): applying an external magnetic field causes

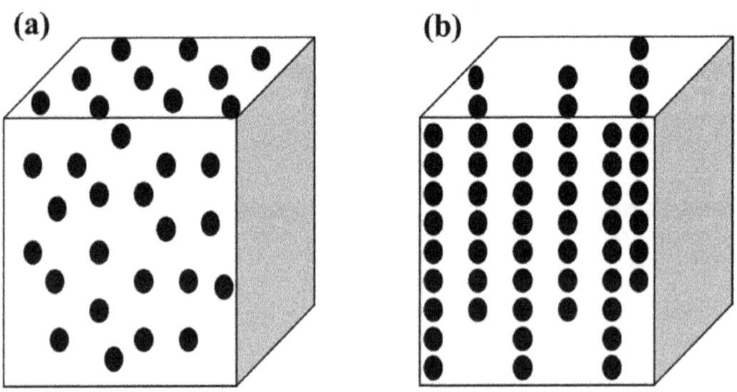


**Fig. 1.2** MRE microstructures: **a** isotropic and **b** anisotropic distribution of magnetic particles

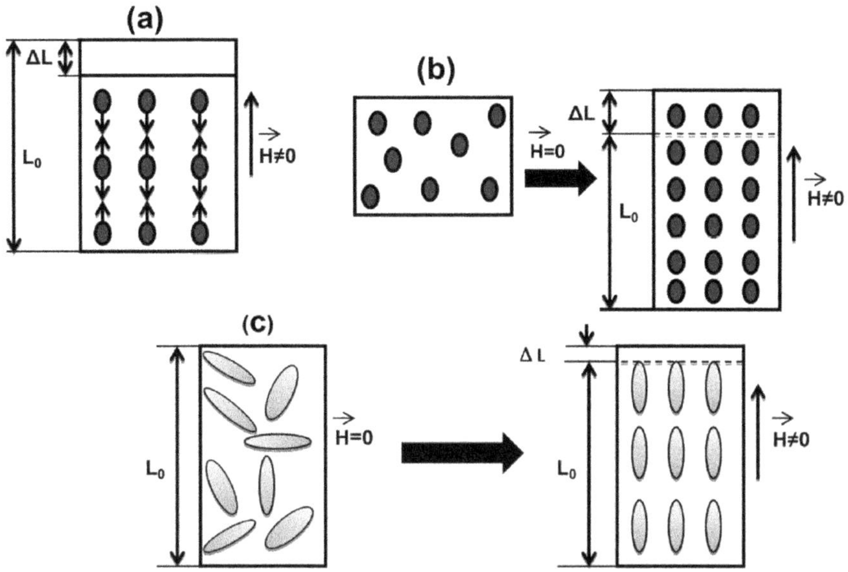


**Fig. 1.3** Magnetostriction effects in MREs under externally applied magnetic fields: **a** anisotropic MREs, **b** isotropic MREs and **c** illustration of a giant magnetostriction effect. The symbol H represents an external magnetic field

magnetic dipoles to align with the magnetic field vector and induce deformation in the matrix, thereby increasing the length.

### *1.3.2 Origin of the MR Effect and Its Representation*

Ferromagnetic particles in an elastomer matrix possess a net magnetic moment and exist in the minimum dipolar energy state. Application of a shear stress displaces the particles and hence disturbs their low dipolar energy state, which requires some additional work. The amount of additional work increases monotonically with the externally applied magnetic field strength, thereby resulting in a field-dependent shear modulus. Movement of the particles also causes deformation of the elastomer matrix and results in increases in the MRE stiffness and shear modulus.

Changes in the physico-mechanical properties of MR elastomers upon application of a magnetic field are expressed mathematically by two terms called the absolute MR effect and relative MR effect (Lokander and Stenberg 2003b). The absolute MR effect ($\Delta G$) is expressed by:

$$\Delta G = G_{\mathrm{MAX}} - G_0 \tag{1.1}$$

where $G_{MAX}$ is the maximum value of the storage shear modulus achieved when the MREs are exposed to the magnetic field, and $G_0$ is the storage shear modulus in the absence of the magnetic field.

The relative MR effect ($\Delta G_r$) is expressed by:

$$\Delta G_r = \{(G_{MAX} - G_0) \times 100\}/G_0 = (\Delta G \times 100)/G_0 \qquad (1.2)$$

The absolute and relative MR effects of a particular MRE depend on the loading on the magnetic particles, the test oscillation frequency, the amplitude of the applied strain and the strength of the external magnetic field (Lokander and Stenberg 2003b; Setua 2008).

### *1.3.3 Factors Affecting the Properties of MREs*

The mechanical properties of MREs follow the rule of mixture of matrix materials (dispersion medium) and fillers (magnetic particles). However, this rule is influenced by synergism, since there are magnetically induced filler–matrix interactions:

(1) Matrix polymers: These are inherently viscoelastic and they exhibit characteristic properties like elastic modulus and density. They demonstrate maximum increases in stress and modulus when the magnetic particles attain saturation magnetization in the external magnetic field.
(2) Magnetic particles: These have sizes generally ranging from 10 to 50 μm, see Table 1.1, which enable both isotropic and anisotropic MREs to be made. However, ferromagnetic particles with sizes less than about 1.5 μm have negligible anisotropy and uniform magnetization, such that the MREs are always isotropic.

Several other factors contribute to the overall properties of the MREs. These include particle shapes, distributions and orientations, improved particle–matrix adhesion via the use of a suitable plasticizer, and hybrid combinations of fillers. In particular, the use of nano-sized particles as fillers can greatly change the elastic, rheological and damping properties, as will be discussed later in this chapter.

The quality of MREs is generally assessed by measurement of the MR effect. This is the reversible change of a material property, in the present instance the rheological behaviour generally expressed by shear moduli (e.g. the storage ($G'$) and loss ($G''$) moduli) under an external magnetic field compared to the shear moduli in the absence of a magnetic field. Sometimes the MR effect is also expressed in terms of the elastic modulus increase during a compression test, or a change in slope of the stress–strain curve, or the damping properties.

## 1.4 Background Literature on MREs

### *1.4.1 Conventional MREs*

Most of the literature on MREs concerns conventional MREs that, as stated in Sect. 1.3.1, consist of micron-sized magnetic particles embedded in a non-magnetic elastomeric solid matrix. Furthermore, the favoured particles were mainly soft magnetic, e.g. iron, nickel, cobalt and their oxides. There is extensive literature on conventional MREs, and a selection (Lokander and Stenberg 2003a; Koo et al. 2012; Gong et al. 2007, 2012; Chen et al. 2008a; Lejon and Kari 2009; Alberdi-Muniain et al. 2012; Hu et al. 2005a, 2005b; Fuchs et al. 2007; Wang et al. 2007, 2006; Kaleta et al. 2011; Farshad and Benine 2004; Varga et al. 2006; Li et al. 2008, 2012; Dong et al. 2012; Abramchuk et al. 2006; Kallio et al. 2007; Popp et al. 2009; Du et al. 2011; Eem et al. 2011; Liao et al. 2011a; Jiang et al. 2008; Fan et al. 2010; Wu et al. 2012; Von et al. 2011, 2008) is given in the Reference list at the end of this chapter. A study of this literature showed that it is very difficult to compare the performance of different MRE systems, since they vary in types of polymer matrix, magnetic particles, filler loading, the external magnetic fields applied during fabrication or testing, and the types of properties evaluated. However, some data which may reasonably be compared for different MREs are given in Table 1.2. Additional comments and information are summarized here:

(1) Polyurethane-based MREs: Relative MR effects varied widely in magnetic particle filled (70–80 wt.%) polyurethane matrices. Note the very large maximum MR effect for iron particles in Table 2.
(2) Natural rubber MREs: Several investigations examined the damping properties (Lokander and Stenberg 2003a; Chen et al. 2008a; Lejon and Kari 2009). A vibration isolation system using iron particle filled natural rubber demonstrated 100–200 times less energy flow in the foundation than the total energy flow into the system (Alberdi-Muniain et al. 2012).
(3) Carbonyl iron filled polymer blend-based MREs: Blends of (polyurethane–silicone rubber, polyurethane–bromobutyl rubber, polyurethane–silicone–bromobutyl rubber, silicone rubber–polystyrene etc.) were investigated for dynamic mechanical, morphological, thermal, mechanical, and MR properties (Hu et al. 2005a, 2005b; Fuchs et al. 2007; Wang et al. 2007). Some results were: (i) the polyurethane–silicone rubber blend had a greater MR effect than polyurethane and silicone rubber, (ii) silicone rubber–polystyrene blends showed a maximum MR effect when the polystyrene content was 20% (Wang et al. 2007).
(4) Silicone rubber MREs: These are the most popular, owing to their design flexibility, easy processing, and high and low temperature stability. Silicone rubber MREs have been extensively studied (Koo et al. 2012; Farshad and Benine 2004; Varga et al. 2006; Li et al. 2008, 2012; Dong et al. 2012; Abramchuk et al. 2006; Kallio et al. 2007; Popp et al. 2009; Du et al. 2011; Eem et al. 2011; Liao et al.

**Table 1.2** Some comparisons of conventional MREs

| Polymer matrix | Filler type | Filler loading | Relative MR effect (%) | Refs |
|---|---|---|---|---|
| Polyurethane | Carbonyl iron | 70–80 wt% | 21–121 | Mitsumata and Ohori (2011); Wu et al. ( 2010); Xu et al. (2011) |
| | Iron particles | 80 wt% | 1225 (max) | Boczkowska et al. (2012) |
| Natural rubber | Carbonyl iron | 90 wt% | 107 | Chen et al. (2007) |
| | Carbonyl iron | 80 wt% | 133 (max) | Gong et al. (2007) |
| Silicone rubber–polystyrene (80:20) | Carbonyl iron | – | 33 | Wang et al. (2007) |
| Thermoplastic elastomer | Iron particles | – | 24–38 (isotropic)<br>25–44 (anisotropic) | Kaleta et al. (2011) |
| Silicone rubber | Carbonyl iron | 80 wt% | 878 | Gong et al. (2007) |
| | Carbonyl iron | – | Tensile and compression moduli increased by 200 and 300% | Farshad and Benine (2004) |
| | Magnetite | – | Shear modulus increased by 10,000% | Abramchuk et al. (2006) |

2011a; Von et al. 2008), including the design and performance of several devices and components (Kallio et al. 2007; Popp et al. 2009; Du et al. 2011; Li et al. 2012; Eem et al. 2011; Liao et al. 2011a).

(5) Filler surface treatments and plasticizers: Surface treatment of magnetic fillers to improve particle–matrix adhesion reduces the relative MR effect unless the surface treatment agents have a plasticizing effect on the elastomer matrix (Wang et al. 2006; Jiang et al. 2008; Fan et al. 2010; Wu et al. 2012).

### *1.4.2 Polymer Nanocomposite (PNC) MREs*

A survey of the literature for conventional MREs shows that most require high concentrations (60–80 wt%) of magnetic particle fillers. Such high particle concentrations make processing difficult owing to large increases in viscosity and may give rise to particle agglomeration, which affects the matrix elasticity and also inhibits the formation of well-arranged 3D networks, which are essential to obtaining enhanced MR effects. These actual and potential disadvantages are avoided when the required high concentrations of relatively large particles are replaced by small quantities of nano-sized particles to make polymer nanocomposite (PNC) MREs. Moreover, the use of nano-sized particles can have large effects on the rheological, elastic and damping properties of MREs.

The literature on PNC MREs is less extensive than for conventional MREs but is increasing rapidly. The most relevant investigations for this chapter are Refs. (Wu et al. 2012; Evans et al. 2012; Denver et al. 2009; Chen et al. 2008b; Tian et al. 2011, 2013; Mordina et al. 2016, 2014, 2015; Song et al. 2009a, 2009b; Park et al. 2010; Szabo et al. 1998; Zrínyi et al. 1996; An et al. 2012; Padalka et al. 2010; Yanez-Flores et al. 2007; Rwei et al. 2013; Mousavian et al. 2012; Moeen et al. 2012; Landa et al. 2013; Zhu et al. 2013; Antonel et al. 2011). Silicones are often chosen for the matrix, e.g. polydimethylsiloxane (PDMS). On the other hand, the choice of nanoparticles has been very varied, including carbon nanotubes (Wu et al. 2012; Chen et al. 2008b; Tian et al. 2011; Mordina et al. 2016), iron oxides ($Fe_2O_3$, $Fe_3O_4$) (Evans et al. 2012; Szabo et al. 1998; Zrínyi et al. 1996), Ni particles and nanowires (Denver et al. 2009; Padalka et al. 2010) and $FeCo_3$ particles (Mordina et al. 2014).

Large (beneficial) property changes may be obtained with low concentrations of nano-size particles, fibres and wires. Note, however, that these property changes are very sensitive to the nanomaterial morphologies. For example, Denver et al. (Denver et al. 2009) found that 5 wt% additions of nanoparticles or nanowires to a PDMS matrix resulted in 30% and 80% increases in elastic modulus, respectively. Similarly, adding nanowires, nanofibres and nanochains to PDMS instead of nanoparticles was beneficial to a range of magnetic properties, including larger saturation magnetization, remanent magnetization, coercivity, and magnetic anisotropy (Denver et al. 2009; Landa et al. 2013; Zhu et al. 2013). The results of adding cobalt ferrite ($CoFe_2O_4$) nanofibres instead of nanoparticles to PDMS were also remarkable. The nanofibre MRE showed 100–400% improvements in MR properties compared to the nanoparticle MRE (Mordina et al. 2015; Antonel et al. 2011).

More specific details on PNC MREs are available in the referenced literature (Wu et al. 2012; Evans et al. 2012; Denver et al. 2009; Chen et al. 2008b; Tian et al. 2011, 2013; Mordina et al. 2016, 2014, 2015; Song et al. 2009a, 2009b; Park et al. 2010; Szabo et al. 1998; Zrínyi et al. 1996; An et al. 2012; Padalka et al. 2010; Yanez-Flores et al. 2007; Rwei et al. 2013; Mousavian et al. 2012; Moeen et al. 2012; Landa et al. 2013; Zhu et al. 2013; Antonel et al. 2011). However, for a broader view, including innovative modelling approaches to predict mechanical performances, the reader is referred to the Bibliography, see Sect. 1.11 of this chapter.

## 1.5 Synthesis of Magnetic Nanoparticles

Magnetic nanoparticles can be prepared using a number of synthetic routes. The most important and widely used processes are:

- Co-precipitation
- Hydrothermal and solvothermal methods
- Thermal decomposition

- Microemulsion or nanoemulsion-mediated synthesis
- Sol–gel method
- Reduction of metal-containing compounds
- Ceramic method
- Electrospinning and co-electrospinning.

These processes are surveyed in Sects. 1.5.1–1.5.8.

### 1.5.1 Co-precipitation Method

This is the easiest technique for the preparation of iron oxide nanoparticles. Magnetite ($Fe_3O_4$) and maghemite ($\gamma$-$Fe_2O_3$) can be prepared easily from a mixture of $Fe^{2+}$/$Fe^{3+}$ salt solutions by reaction with bases (e.g. NaOH, KOH, and organic amines like propylamine) under an inert atmosphere and at elevated or room temperatures. The iron salts form hydroxide or amine complexes and are precipitated together from the solution as shown here:

$$Fe^{2+} + 2Fe^{3+} + 8OH^- \leftrightarrow Fe(OH)_2 + 2Fe(OH)_3 \rightarrow Fe_3O_4 + 4H_2O \quad (1.3)$$

Heating these precipitates at high temperature produces the magnetic nanoparticles. The size, shape and composition of the synthesized nanoparticles depend on several factors such as the nature of the salts (e.g. nitrates, chlorides, sulphates, acetates); the molar ratio of $Fe^{2+}$/$Fe^{3+}$; molar strength of the solution; type of surfactant and its concentration; pH; temperature; and stirring rate. The $Fe_3O_4$ nanoparticles are highly susceptible to oxidation and are finally converted to $\gamma$-$Fe_2O_3$ by the following reaction:

$$Fe_3O_4 + 2H^+ = \gamma - Fe_2O_3 + Fe^{2+} + H_2O \quad (1.4)$$

The main advantage of the co-precipitation technique is that it can produce large quantities of magnetic nanoparticles. However, this process has the disadvantage of producing polydisperse magnetic nanoparticles, since the growth of the magnetic particles is directed by kinetic factors. Hence, it is necessary to produce monodisperse nanoparticles in short burst nucleations together with slow controlled growth.

### 1.5.2 Hydrothermal and Solvothermal Methods

The hydrothermal process can provide a wide variety of magnetic nanoparticles. The synthesis is carried out at high pressure (~2000 psi) and temperature (~ 200 °C) within a Teflon-lined stainless steel reactor or autoclave. Wang et al. (Wang et al. 2005) synthesized a variety of nanocrystals by using a generalized hydrothermal method involving liquid–solid solution reactions at the interfaces. This method uses a metal linoleate as the solid and an ethanol–linoleic acid as the liquid phase in a water–ethanol solution. By using bimetallic precursors in a particular molar ratio, different monodispersed bimetallic oxide nanocrystals such as magnetic $MFe_2O_4$ (where M can be Ni, Co, Zn or Mn) can be synthesized via co-precipitation followed by a liquid–solid-solution phase transfer reaction and separation process.

Deng et al. (2005) synthesized monodispersed ferrite spheres in the size range of 200–800 nm from a mixture of $FeCl_3$, ethylene glycol, polyethylene glycol and sodium acetate held at 200 °C for 8–72 h. Another possibility is to synthesize monodispersed and highly crystalline $\gamma$-$Fe_2O_3$ nanocrystallites by controlled oxidation of uniform iron nanoparticles produced via solvothermal decomposition of an iron complex (Hyeon et al. 2001).

### 1.5.3 Thermal Decomposition

Monodispersed magnetic nanoparticles with controllable size and excellent crystallite morphology can be produced by thermal decomposition of organometallic precursors in high boiling-point organic solvents consisting of fatty acids (e.g. oleic acid, decanoic acid, myristic acid, lauric acid, palmitic acid, and stearic acid) or amines (e.g. hexadecylamine) as stabilising surfactants (Sun et al. 2000; Hu et al. 2006). The organometallic precursors may be metal acetylacetonates, $[M(\text{acac})_n]$, where M = Fe, Co, Ni, Mn, Cr; acac = acetylacetonate, and $n = 2$ or 3); carbonyls; or metal cupferronates $[M^xCup_x)$, where M = metal ion and Cup = *N*-nitrosophenylhydroxylamine, $C_6H_5N(NO)O$–] (Farrell et al. 2003; Rockenberger et al. 1999).

Two examples of thermal decomposition are given here:

(1) Iron pentacarbonyl $Fe(CO)_5$ is a homoleptic metal carbonyl, where carbon monoxide is the only ligand complexed with a metal. Thermal decomposition of $Fe(CO)_5$ in a mixture of octyl ether and oleic acid at 100 °C initially led to the formation of metal particles. Subsequent controlled oxidation of these particles using a mild oxidizing agent such as trimethylamine oxide, $(CH_3)_3NO$, produced monodispersed $\gamma$-$Fe_2O_3$ nanoparticles (Hyeon et al. 2001).
(2) Park et al. (2004) have reported on the large-scale synthesis of monodispersed iron oxide nanoparticles from the thermal decomposition of an iron (III) oleate precursor. This synthesis process reacted inexpensive and nontoxic ferric chloride with sodium oleate for in situ generation of the iron oleate complex. This complex was then decomposed at 240–320 °C in the presence of different organic solvents to produce monodispersed iron oxide nanoparticles.

The ratios of the reactants (precursor, solvent and surfactant), reaction time, temperature and ageing period are the main parameters controlling the morphology and particle sizes of the synthesized nanoparticles. Note that a high temperature is generally required to induce nucleation and stimulate particle growth.

### 1.5.4 Microemulsion or Nanoemulsion-Mediated Synthesis

Synthesis of nano-sized magnetic particles from a micellular solution of ferrous dodecyl sulphate $Fe(DS)_2$ gives particle sizes that can be controlled by controlling the temperature and the surfactant concentration. Ultrafine magnetite nanoparticles with particle size 4 nm have been synthesized within the reverse micelle nanocavities by controlled hydrolysis of the following mixture: (i) an aqueous solution containing $FeCl_3$ and $FeCl_2$, (ii) ammonium hydroxide with aerosol optical thickness (AOT) as surfactant and (iii) heptane as the continuous oil medium (Arturo et al. 1993).

Spinel ferrites with general formula $MFe_2O_4$ (where M = Co, Ni, Mn, Zn, Cu, Mg, etc.) have been synthesized by microemulsion techniques. Although different types of magnetic nanoparticles can be synthesized using microemulsions, this method has several disadvantages: a narrow working window; low yield compared to the other techniques; and the need for a large amount of solvent, which makes the method difficult to scale-up.

### 1.5.5 Sol–gel Method

The sol–gel method is both easy and the most widely used route for the synthesis of nanomaterials. In this method, metal nitrates or acetates are dissolved completely in an appropriate organic solvent at an elevated temperature. Then the components of the solution are allowed to react with each other at a lower temperature until a gel is formed. Subsequent heating of the gel at high temperature gives the final product. This method is often used to synthesize metal oxides but can also be used for the synthesis of metallic and bimetallic nanoparticles.

### 1.5.6 Reduction of Metal-Containing Compounds

This is a general method for the synthesis of metallic nanoparticles by reduction of metal salts in the presence of an aprotic solvent, i.e., a solvent incapable of acting as a proton donor. Metallic magnetic nanoparticles can also be synthesized from metal salts by using strong reducing agents like $NaBH_4$; alkali metal complexes containing organic electron acceptors like naphthalene; and alkali metal dispersions in hydrocarbons or ethers. Note, however, that $NaBH_4$ reduction of metallic compounds containing single metallic or heterometallic precursors (e.g. precursors of Fe, Ni and Co, or precursors of Fe–Co, Co–Cu and Fe–Cu) results in nanoparticles containing high percentages of boron (20 mass% B or more).

### 1.5.7 *Ceramic Method*

The ceramic method is one of the earliest methods for synthesizing ferrite materials. Highly crystalline materials are prepared by mixing precursor materials in a particular molar ratio, followed by calcination and sintering at a very high temperature. Calcination compacts the precursor materials by removing volatiles. Sometimes calcination is used to convert metal nitrates into metal oxides. This is followed by sintering to obtain the required crystalline phase.

### 1.5.8 *Electrospinning and Co-Electrospinning*

#### 1.5.8.1 Electrospinning

The electrospinning method is used to make magnetic nanofibres. This technique is important in the light of experimental results showing the benefits of using nanofibres instead of nanoparticles, see Sect. 4.2. Figure 1.4 is a schematic of a typical electrospinning set-up. The electrospinning takes place in several steps, as summarized in Fig. 1.4.

***Electrospinning Procedure***

(1) Homogeneous polymer solutions containing precursor materials of magnetic fibres are prepared by stirring the mixtures for a desired time period. For example, to make $CoFe_2O_4$ nanofibres the precursors are inorganic salts like Fe $(NO_3)_3$ and $CoCl_2$ or gels containing Fe $(NO_3)_3$ and $CoCl_2$ and citric acid as a chelating agent.
(2) The viscosity of the solution is adjusted to a certain limit by adding solvent, and then the polymer solution is loaded into a plastic syringe that fits within the cavity of a syringe pump. The syringe needle is then connected to the anode of a high voltage power supply.

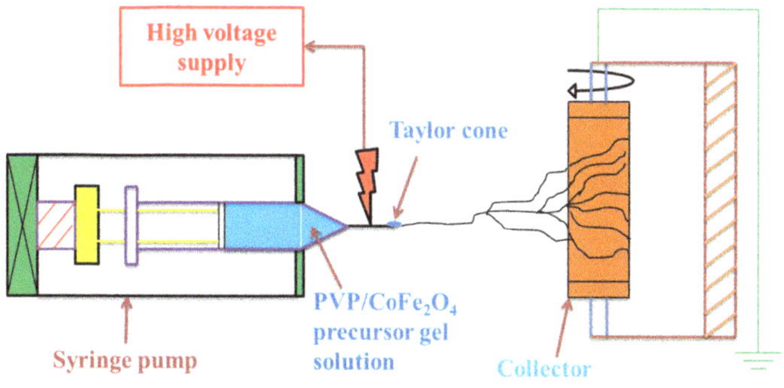


**Fig. 1.4** Schematic of an electrospinning set-up

(3) An earthed rotating drum wrapped with aluminium foil acts as collector. The distance between the needle and drum is adjusted to between 10 and 15 cm, and the flow rate of the pump is adjusted to 5–15 μL/m.
(4) Once the polymer solution starts to come out of the needle as a droplet, a positive voltage of 10–20 kV is applied between the needle and collector. The polymer solution becomes charged, and under this condition electrostatic repulsion lessens the surface tension and the droplet is stretched.
(5) At a critical point, the solution stream breaks away from the needle surface to form a Taylor cone. The Taylor cone is then stretched by the electrostatic attraction force and forms fine nanofibres on the rotating drum.
(6) The nanofibres accumulated on the rotating drum are collected by removing the aluminium foil. They are then dried in an air oven at 60–80 °C for 1–2 h and finally calcined at high temperature.

#### 1.5.8.2 Co-electrospinning

Co-electrospinning is used to obtain hollow nanofibres. Figure 1.5 gives a schematic of a typical co-electrospinning set-up, which is used as described in the steps in the below figure.

***Co-electrospinning Procedure***

(1) Two different polymer solutions are pumped through a special nozzle consisting of a central hole surrounded by an annular channel and connected to a common needle. Magnetic precursors are mixed *only* with the polymer in the annular channel. Furthermore, the polymer in the central hole has to have lower thermal stability than the polymer (plus precursors) in the surrounding channel.
(2) After fibres are produced, in the same general way as described for electrospun nanofibres, they consist of a central polymer core with an annular shell consisting of the precursor-containing polymer.

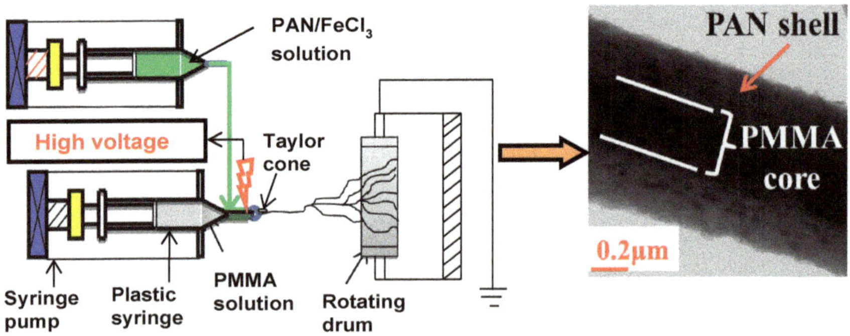


**Fig. 1.5** Schematic of a co-electrospinning set-up, in this case producing hollow carbon nanofibres

(3) The fibres are then dried and calcined at high temperature in an air/nitrogen atmosphere. During the calcination process the polymer core decomposes first, leaving a central hole. Then the surrounding polymer in the annular shell decomposes, leaving the precursor materials to form hollow magnetic nanofibres.

Hollow non-magnetic materials, e.g. carbon nanofibres, with embedded magnetic particles can also be prepared by co-electrospinning. The process is broadly similar to that of making hollow magnetic nanofibres, with three important differences: (i) the concentration of precursor materials in the polymer solution in the annular channel is less; (ii) calcination causes the polymer solution to form a carbon shell, and (iii) calcination results in discrete magnetic particles embedded in the carbon shell.

Polyacrylonitrile (PAN) (in the annular channel) and polymethylmethacrylate (PMMA) (in the central hole) polymers are often used to prepare hollow carbon fibres: see also Fig. 1.5. The PMMA decomposes completely at 550 °C in an inert atmosphere and leaves a hole, while calcination of polyacrylonitrile at 900 °C in a similar environment results in the formation of a carbon shell.

## 1.6 Fabrication of PNC MREs

There are several challenges associated with the preparation of nanoscale MREs with the desired characteristics. The main challenge is the development of large-scale and cost-effective production techniques with precise control of the dispersion of nanoparticles within the polymer matrix. Nanoparticles tend to aggregate to form large particles owing to their high surface energy, thereby cancelling all the benefits of making PNC MREs. Therefore, very specific fabrication techniques are needed. These are classified into two major categories: ex situ and in situ. These are discussed concisely in Sects. 1.6.1 and 1.6.2.

### *1.6.1 Ex Situ Fabrication*

The easiest preparation of the magnetic nanocomposites is direct incorporation of the pre-synthesized magnetic fillers into the polymer matrix when it is in either a molten or solution state. Nathani et al. (Dresco et al. 1999) have reported on the fabrication of superparamagnetic nickel ferrite/polypropylene nanocomposites by ball milling nickel ferrite in the presence of polypropylene.

Thermoplastic polymer-based magnetic nanocomposites are generally fabricated by melt blending within an extruder or in an internal mixer, since these methods are the most cost-effective. During the extrusion process, the nanoparticles are fed into the polymer melt through a hopper from a different mixing zone placed between the main polymer feeding hopper and die. For highly filled nanocomposites, a portion of the total nanofillers is sometimes fed through the main polymer feeding hopper.

Fabrication of an MRE involves three major steps: (i) mastication (breaking the polymer chains) of the rubber to reduce viscosity, (ii) mixing the magnetic particles and the other ingredients, (iii) compression moulding and controlling particle alignment as per necessity. For example, natural rubber has a high molecular weight and green strength, and so the molecular weight has to be reduced before it becomes suitable for compounding. The rubber is subjected to high shear forces in a two-roll rubber mixing mill or Banbury internal mixer, resulting in the rubber losing its elasticity and becoming a viscous gel. In this condition, the cross-linking agent(s), processing aids, plasticizers and magnetic particles are added to the rubber. Once the ingredients are mixed homogeneously, the rubber compound is transferred to a steel mould and compression-moulded in a hydraulic press at 150–160 °C for 15–20 min to complete the cross-linking process. The press is then switched off and the moulded composite was allowed to cool down to room temperature. This process produces an isotropic MRE.

To make anisotropic MREs a permanent magnet or electromagnet set-up is fitted around the upper and lower platens of the compression moulding machine, see Fig. 1.6. The mould containing the rubber compound is then placed between the platens and heated up to the softening point of the compound. The magnetic field is applied once the rubber matrix is molten or in an appropriate viscosity condition, and the temperature is held at a particular value for a certain time in order to freeze the orientation of magnetic particle chains while cross-linking of the matrix takes place. Then the magnetic field and heater are switched off and the mould is allowed to cool to room temperature. This process results in the formation of an anisotropic MRE with specifically orientated embedded magnetic particle chains.

Solution casting is another widely explored technique for fabrication of thermoplastic polymer-based magnetic nanocomposites. In this technique, a polymer is dissolved in an appropriate solvent and then magnetic nanoparticles are dispersed within the solution by mechanical stirring or ultrasonication. The solvent swells the

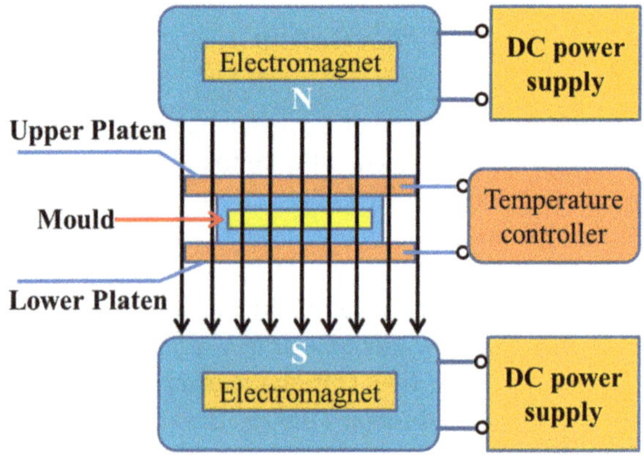


**Fig. 1.6** Fabrication of an anisotropic rubber-based MRE

polymer and helps in uncoiling or opening the gap between the polymeric chains, thereby allowing penetration of the nanoparticles inside the polymer chains. Finally, the mixture is cast either in the form of a membrane or coating on another surface, such that the magnetic nanocomposite sets by slow evaporation of the solvent.

Thermo*setting* polymer-based magnetic nanocomposites such as $Fe_3O_4$/PDMS nanocomposites are sometimes also fabricated using the solution casting technique. In this case, magnetic nanoparticles are dispersed within the liquid PDMS mixed with a volatile and non-reactive organic solvent, followed by (i) the addition of curing agents, (ii) casting and solvent evaporation and (iii) thermal curing or chemical polymerization at high temperature.

### 1.6.2 In Situ Fabrication

Although ex situ processes involve relatively easy steps, agglomeration of the nanoparticles within the polymer matrix remains the major problem associated with these methods. The alternatives are in situ methods, which denote the formation of nanostructures from an inorganic precursor in the presence of polymers dissolved in appropriate solvents. Subsequently, magnetic nanoparticles are formed in situ by chemical reactions via hydrolysis or the reduction of metallic precursors. This strategy enables modification of the interfacial interactions between the polymer matrices and magnetic nanofillers and leads to a homogeneous dispersion of the fillers within the matrix.

The most promising method for obtaining well-dispersed nanoparticles is thermal decomposition of metallic precursors to form metal or metal oxide nanoparticles, with the polymeric matrix acting as the stabilizer for the nanoparticles. This method works well for the decomposition of inorganic metal carbonyls like $Fe(CO)_5$, $Co_2(CO)_8$, $Ni(CO)_4$ and metal oleates.

An in situ polymerization technique becomes more efficient by direct dispersion of the magnetic nanoparticles in the monomer, followed by dispersion polymerization of the monomer, suspension cross-linking, or emulsion/inverse emulsion polymerization, thereby encapsulating the magnetic particles in the polymer. For example, single-step synthesis of a methacrylic acid and hydroxyethyl methacrylate copolymer containing magnetite nanoparticles has been carried out by the inverse emulsion technique (Dresco et al. 1999).

The emulsion or inverse emulsion techniques, described above, produce composite particles with very wide size distributions and always with adsorbed surfactant on the particles. However, magnetic nanoparticles encapsulated by a polymer matrix can also be prepared via a mini-emulsion process. This process firstly uses high-speed shear devices (e.g. high-pressure homogenizers or ultrasound) to produce stable and small droplets of oil (colloids) with a homogeneous size distribution in the range of 30–100 nm. These mini-droplets are then loaded with monomers and magnetic nanoparticles and subsequently polymerized to form magnetic nanocomposites. An example is given by Luo et al. (Luo et al. 2008), who synthesized polystyrene/$Fe_3O_4$ composite particles via a mini-emulsion polymerization technique using potassium persulphate as initiator, sodium dodecyl sulphate (SDS) as the surfactant, and hexadecane or sorbitan monolaurate as the co-stabilizer.

Grafting techniques enhance the scope of the mini-emulsion technique, which can be used to adsorb monomers directly onto particle surfaces. In '*grafting to*' techniques, polymers can be grafted onto the particle surfaces by the reaction between anchoring groups attached to the polymer chains, and functional groups on the particle surfaces. '*Grafting from*' follows adsorption of a monomer onto the particle surfaces followed by polymerization of the monomer using an initiator initially anchored to the particle surfaces. This surface-initiated polymerization readily enables synthesizing end-tethered polymer coronas on the nanoparticle surfaces with proper control over the molecular weight and polydispersity. The basic requirement of this technique is the use of suitable surface-active initiators while suppressing unwanted flocculation of the colloid.

### *1.6.3 Some Recent Developments*

Recently, some modified methods have been adopted for the preparation of magnetic nanocomposites, e.g. synthesis of hybrid core–shell nanoparticles. In this technique, plasma is generated by inserting a quartz glass reaction tube into the microwave cavity. Water-free volatile organic precursors like chlorides, metal alkoxides, carbonyls or metal-alkyls are evaporated outside the tube and delivered into the plasma zone of the reaction tube by mixing with a carrier gas. Core–shell nanoparticles are formed via a gas phase nucleation and growth process.

A seed-precipitation polymerization technique has been used for the preparation of core–shell composite particles such as (i) methacrylic acid and hydroxyethyl methacrylate copolymer-coated magnetic nanoparticles and (ii) vinyl-terminated polystyrene-encapsulated magnetic nanoparticles. Also, layer-by-layer deposition of nanoparticles and polymers is used for preparation of highly homogeneous thin composite films.

## 1.7 Characterization Techniques for Magnetic Fillers and MREs

There are many characterization techniques for magnetic fillers and MREs:

(1) Structural characterizations of magnetic particles/fibres are carried out by X-ray diffraction, X-ray photoelectron spectroscopy, Raman spectroscopy and energy dispersive X-ray analysis.

(2) Morphological properties of magnetic particles/fibres and MREs are investigated by optical microscopy, Raman confocal microscopy, field emission scanning electron microscopy (FE-SEM) and transmission electron microscopy (TEM).
(3) Magnetic properties of the magnetic particles/fibres and MREs are determined using vibrating sample magnetometers (VSM). Anisotropic characteristics of MREs are investigated by measuring the saturation magnetization at 0 and 90° sample orientation, i.e., by measuring the saturation magnetization parallel and perpendicular to the particle chains of anisotropic MREs, or along the length and width of isotropic MREs. Temperature dependent magnetic properties of particles/fibres are determined by recording the zero-field-cooled and field-cooled curves in superconducting quantum interference devices (SQUID). Actuation properties can be investigated by measuring the deflection and blocking force with the increase in magnetic flux density.
(4) MR properties of MREs are determined using a parallel plate rheometer with a magnetic field generating set-up. Both static and dynamic MR properties are determined by subjecting the MREs to either steady state rotary shear or an oscillatory shear mode while in a magnetic field. A dynamic mechanical analyser combined with an applied magnetic field application is sometimes used to evaluate the dynamic rheological properties.
(5) Vibration damping characteristics of MREs can be studied by measuring the vibration transmissivity.
(6) The mechanical properties of MREs (yield and tensile strengths, tensile and compression moduli, and elongation to fracture) are determined in universal testing machines (UTM).
(7) Thermal transition and thermal stabilities of MREs are determined by differential scanning calorimetry (DSC) and thermogravimetric analysis (TGA).
(8) The cross-linking reaction of MREs can be studied by functional group analysis using Fourier transform infrared spectroscopy (FTIR). The cross-linking density can be determined by thermomechanical analysis (TMA) using a novel method suggested by Setua et al. (Setua et al. 2002): firstly, a constant load is applied to the sample using a calibrated hardness indenter, and the deformation is recorded as a function of time. Then Young's modulus is calculated using Eq. (5), and finally the cross-linking density is calculated using Eq. (6):

$$E_{\mathrm{m}} = \left(F/P^{3/2}\right)\left(9/16r^{1/2}\right) \tag{1.5}$$

$$\upsilon_{\mathrm{e}} = E_{\mathrm{m}}/3RT \tag{1.6}$$

where $E_{\mathrm{m}}$ is the elastic modulus of the sample in dyne/cm$^2$, F is the load applied (dynes), $P$ is the penetration of the indenter in cm, r is the probe radius in cm, $\upsilon_{\mathrm{e}}$ is the cross-link density of the sample in mol/cm$^3$, T is the temperature in K and R is the universal gas constant in force units, i.e., $8.314 \times 10^7$ erg/(mol K).

## 1.8 Specific Scientific Studies

In our research work, which is prompted by the functional requirements of MREs, different magnetic nanofillers with high saturation magnetization and permeability were synthesized via novel synthesis techniques, and then characterized using sophisticated instrumental techniques. The following studies were made:

***Study #1***

(1) PDMS (silicone-based) PNCs were fabricated using bimetallic $FeCo_3$ nanoparticles and then evaluated for their MR properties.
(2) $FeCo_3$ nanoparticles were synthesized by hydrazine reduction of $FeSO_4{\cdot}7H_2O$ and $CoCl_2{\cdot}6H_2O$ metal salts in an aqueous medium and characterized for structural, morphological and magnetic properties.
(3) Isotropic and anisotropic PDMS–$FeCo_3$ PNC samples containing 5–20 wt% nanoparticles were fabricated by the solution casting technique and investigated for MR properties by applying a magnetic field perpendicular to the plane of a sample in a parallel plate rheometer.

The effects of particle loading (filler amounts) and particle alignment on the magneto-rheological, magnetic and anisotropic properties of the PNCs showed that the isotropic PNCs possessed higher absolute and relative MR effects compared to the anisotropic PNCs. The highest absolute and relative MR effects were obtained for isotropic composites containing 5 and 20 wt% filler materials, respectively (Mordina et al. 2014). Comparing the MR effect with the literature data, it was observed that for the same matrix and filler materials, the PNCs with 10–20 wt% nanoparticles showed equivalent performance to that of conventional MREs containing 60–80 wt% micron-sized magnetic particles.

Figure 1.7 gives a schematic representation of the fabrication of isotropic and anisotropic PNCs consisting of a PDMS matrix and $FeCo_3$ nanoparticles, and also the property evaluations.

***Study #2***

(1) $CoFe_2O_4$ nanoparticles and nanofibres were prepared by ultrasonic assisted co-precipitation as well as by the electrospinning technique (Mordina et al. 2015).
(2) PDMS-based PNCs using either particulate or fibrous $CoFe_2O_4$ were fabricated.

The effects of particle and fibre loading, the shapes and morphologies of the fillers, and their degree of orientation, on the magnetic and MR properties were investigated. The study showed that nanofibres were more effective in achieving higher MR effects than nanoparticles, owing to the fibres having a larger volume coverage ability, high aspect ratio and surface area. Around 100 to 400% enhancement in the MR properties was observed for PNCs filled with $CoFe_2O_4$ nanofibres in comparison to PNCs containing $CoFe_2O_4$ nanoparticles.

Comparing the MR effects with literature data showed that PDMS PNCs containing 5–10 wt% $CoFe_2O_4$ nanofibres exhibited MR properties similar to those

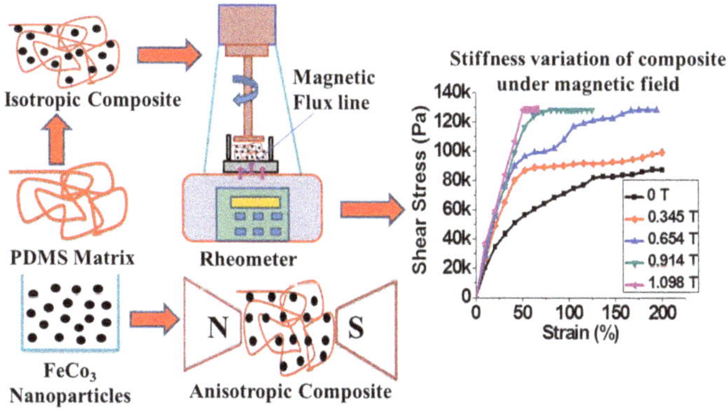


**Fig. 1.7** Schematic representation of the fabrication of isotropic and anisotropic PDMS/$FeCo_3$ nanocomposites and the magneto-rheological study

achieved by conventional PDMS composites containing 60–80 wt% micron-sized particles.

***Other Studies***

- Barium ferrite nanoparticles and graphene oxide (GO) nanosheets were synthesized by co-precipitation following a modification of Hummer's method (Mordina et al. 2016).
- The effect of GO nanosheets on the MR properties of polyacrylamide hydrogels was explored by fabricating hydrogels consisting of either barium ferrite nanoparticles or a combination of barium ferrite nanoparticles and GO nanosheets.
- Recently, studies have been made of the magnetic and actuation properties of polyacrylamide hydrogels containing different filler loadings of barium ferrite nanoparticles or a combination of barium ferrite nanoparticles and GO. The hydrogels were semi-solid and fabricated into rod shapes, either uncovered or sleeved with hard plastic covers. The rods were subsequently subjected to a bending deflection under an NdFeB permanent block magnet (Mordina et al. xxxx).

## 1.9 Applications of MREs

MREs are smart materials and possess magnetic field-dependent moduli, damping and mechanical properties. Based on these properties, MREs find widespread applications in automobiles, vibration controls, controlled vibration dampers, automotive bushings, sensors, engine mounts, stiffness-tailorable mounts, actuators (Figs. 1.9 and 1.10) and strategic (Defence) applications. Some of these applications are described in Sects. 1.9.1–1.9.7.

### 1.9.1 Tunable Vibration Absorber (Liao et al. 2011b; Kim et al. 2011; Hoang et al. 2011)

Figure 1.8 illustrates a test set-up for an MRE-based Tunable Vibration Absorber (TVA) for a cryogenic cooler. A voice coil motor is embedded inside the compressor (mass $\approx$ 2.1 kg) to replicate an actual dynamic compressor system. The voice coil actuator generates an axial direction disturbance similar to that of an actual cryogenic cooler where the constant-frequency linear compressor creates a harmonic disturbance.

The vibration absorber consists of two MRE supports that function in shear mode, and a single absorber mass made of brass. Two rollers hold the compressor on both sides to ensure that the movement of the compressor takes place only in the axial direction. Additionally, the rollers restrict the compressor from being tilted to either side owing to the magnetic force. Spring and damping elements are employed between the compressor and base, see Fig. 1.8a and b, in order to isolate the compressor from primary vibration. The MRE-based TVA functions by reducing the transmission of disturbance to the base.

The experimental study revealed that the MRE-based TVA was capable of tuning the frequencies from 32 to 60 Hz with a magnetic flux density up to 240 mT and was effective in suppressing the vibration of the compressor in a cryogenic cooler.

### 1.9.2 Actuators (Koo et al. 2012; Kashima et al. 2012)

Figure 1.9 gives a cross-sectional view of the structure of an actuator. An electromagnet is embedded within the inner soft silicone elastomer, and a slightly harder MRE encases them. Figure 1.10 demonstrates the operating principle of the actuator. When the coils are excited by passing an electric current, the produced magnetic flux passes through the MRE, which has a smaller reluctance (opposition to the passage of magnetic flux lines) compared to the silicone elastomer. The segment of the magnetic elastomer (MRE) facing the core is magnetized, leading to contraction of the part of the MRE where the magnetic flux density is maximum. Once the electromagnet is switched off, the MRE relaxes and reverts back to its initial position.

Bending-type actuation can also be obtained by fabricating an MRE actuator based on hard ferromagnetic particles such as barium ferrite ($BaFe_{12}O_{19}$), strontium ferrite ($SrFe_{12}O_{19}$) and rare earth magnetic materials such as samarium cobalt ($SmCo_5$) and neodymium iron boron magnets ($Nd_2Fe_{14}B$). Typical bending actuation behaviour of an MRE is shown in Fig. 1.11. Investigation revealed that the ($Nd_2Fe_{14}B$)-based actuator showed maximum response as compared to the ($BaFe_{12}O_{19}$, $SrFe_{12}O_{19}$ and $SmCo_5$)-based actuators (Koo et al. 2012).

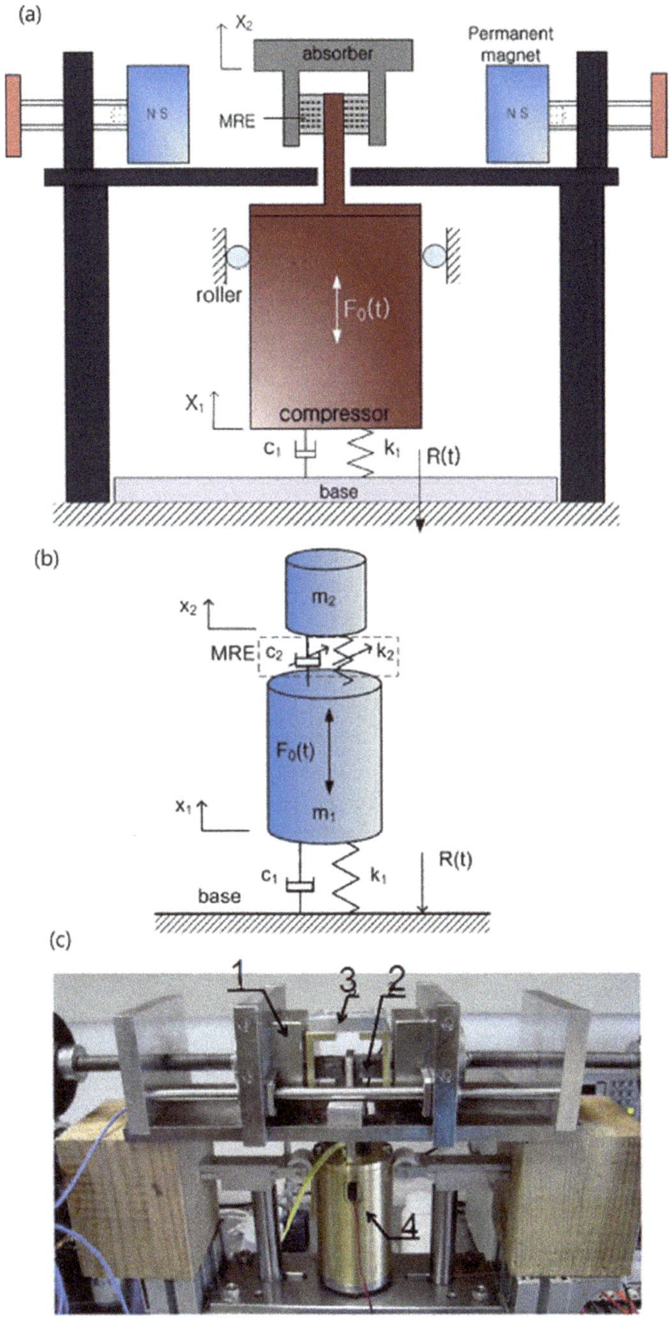


**Fig. 1.8** Cryogenic cooler system with an MRE TVA diagram: **a** and **b** model system, and **c** test setup showing 1: permanent magnet; 2: MRE; 3: absorber mass; 4: compressor

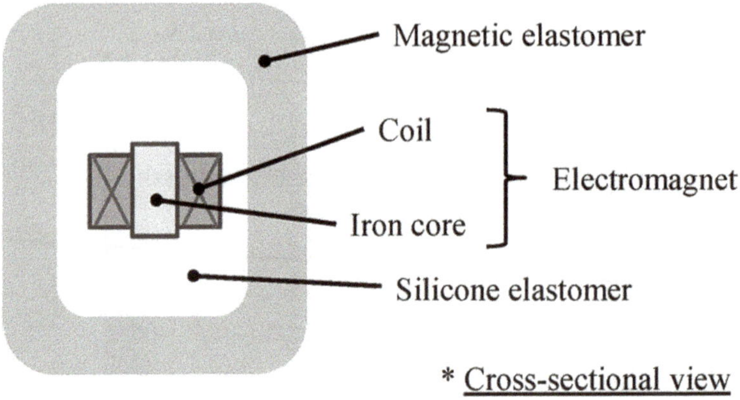


**Fig. 1.9** Structure of the actuator

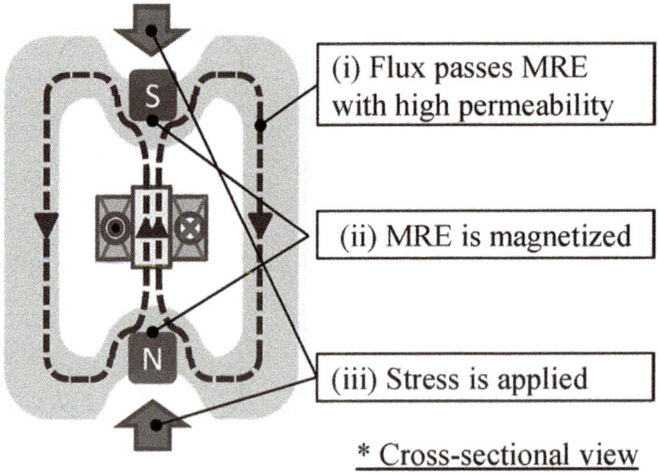


**Fig. 1.10** Operating principle of the actuator

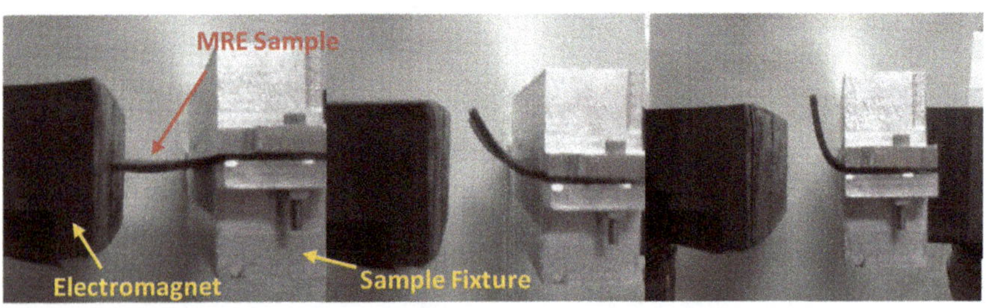


**Fig. 1.11** Bending type actuator made of a hard magnetic particle MRE

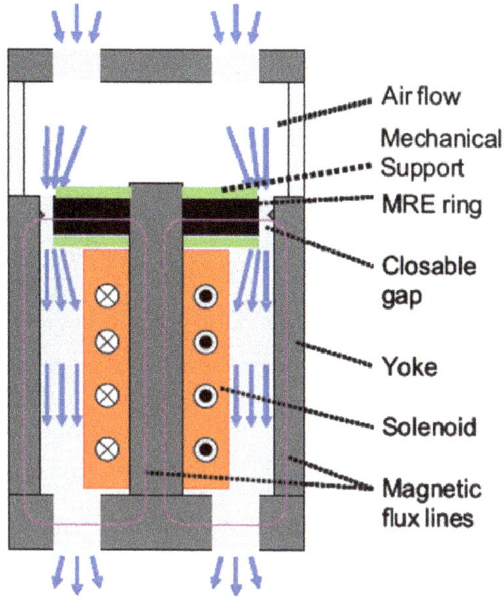


**Fig. 1.12** Schematic of magneto active polymer valve

### *1.9.3 Magneto-Active Polymer Valve for Airflow Control (Bose et al. 2011)*

A schematic of a magneto-active polymer valve and the test setup are shown in Figs. 1.12 and 1.13, respectively. The valve consists of a magnetic circuit and an MRE ring. Under an external magnetic field, the MRE ring expands radially and closes the opening around the ring. This expansion restricts the airflow through the gap. Moreover, by controlling the expansion of the MRE ring to an intermediate position, the valve can also be used proportionately. Investigation revealed that the designed valve could withstand a maximum holding pressure of 0.8 bar in the closed state.

### *1.9.4 Isolators in Vehicle Seat Vibration Control (Du et al. 2011; Li et al. 2012)*

Figure 1.14 gives a schematic of an MRE-based seat isolator. It contains different components such as the base, core, coil, MRE and two nonmagnetic rings. The core coils and base form the electromagnet assembly, which generates variable magnetic fields depending on the magnitude of the current passing through the coil. The nonmagnetic rings have dual roles: (i) they partially support the weight of the coil and prevent the coil pressing on the MRE; (ii) they restrict the magnetic flux lines from directly entering the MRE without passing through the core. In the absence of

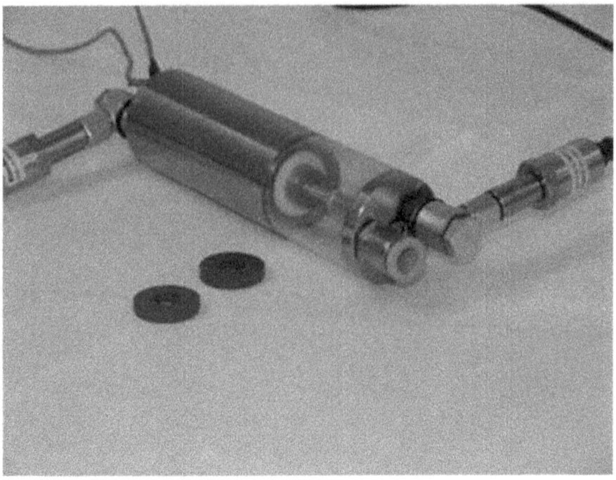

Fig. 1.13 Experimental setup for the valve performance

the rings, the magnetic flux lines would pass through the MRE surface only, therefore not generating an MR effect.

MREs can act as variable springs whose moduli can be tuned by adjusting the current intensity in the coils via variation of the electrical power as well as the magnetic field intensity. This means that the MRE modulus could be tuned in real time, and hence the stiffness of the spring can be adjusted to fit the specific requirement. In a simulation study, the root mean square (RMS) of the driver's body acceleration was used to evaluate the performance index of different isolators by taking into account the random nature of the road input.

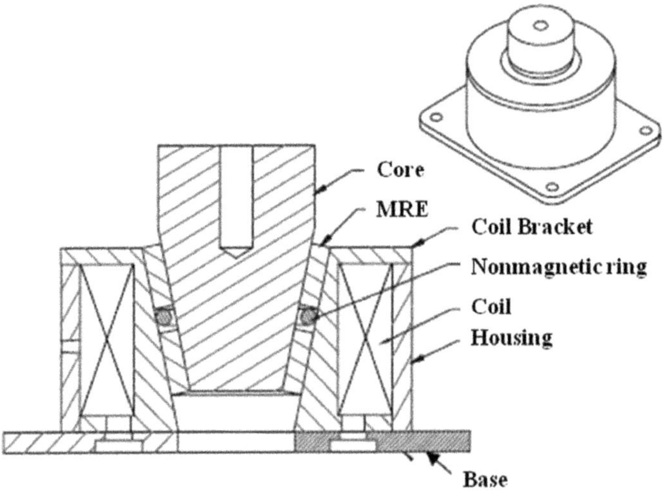


Fig. 1.14 Schematic diagram of an MRE seat isolator

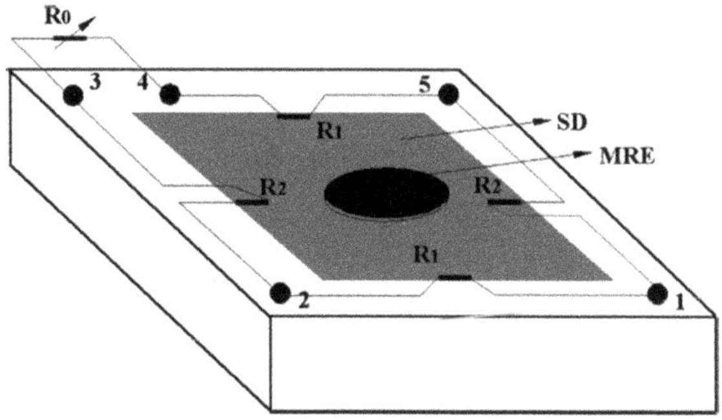


**Fig. 1.15** Schematic diagram of the magnetic field sensor. SD: Sensitive Diaphragm

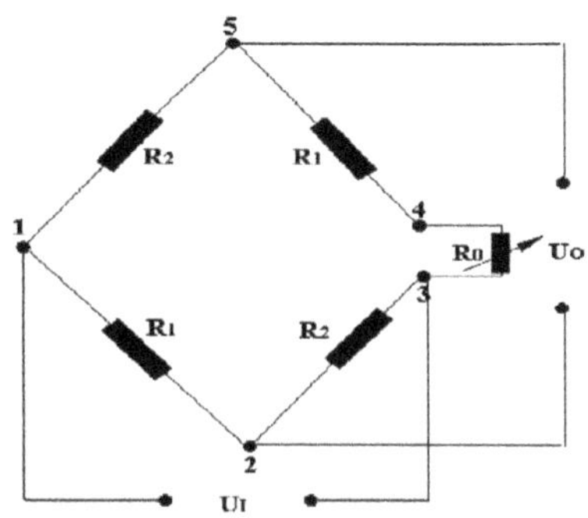


**Fig. 1.16** Circuit diagram of the Wheatstone bridge

## 1.9.5 *Magnetic Field Sensor (Du and Chen 2012)*

A model sensor is described in Fig. 1.15. It consists of a sensitive diaphragm embedded with a piezo-resistive Wheatstone bridge (shown separately in Fig. 1.16). The MRE is fixed at the centre of the sensitive diaphragm and is magnetized under a magnetic field. This leads to deflection of the diaphragm, which changes the piezo-resistance and finally unbalances the Wheatstone bridge.

Piezo-resistive sensors have a very simple design, they are easy to integrate into a structure and are very amenable to electronic signal processing as compared to optical, piezoelectric and capacitive-based sensors. Moreover, piezo-resistive sensors possess high mechanical stability and are low cost. The experimental findings revealed that the sensor possessed good linearity in the magnetic field range of 0–120 kA/m and the saturation magnetic field was about 150 kA/m.

## 1.9.6 *Rubber Bushing (Blom and Kari 2012)*

Figure 1.17 is a schematic of a magneto-sensitive rubber bushing with length L and inner and outer radii of a and b. When a torsion loading with a resulting twist angle $\varphi$ is applied to the bushing, an equivalent isolator shear strain is generated, and this gives an equivalent stress response. Then by taking the Fourier transforms of stress and strain, a relation for the strain averaged frequency as well as amplitude-dependent shear modulus can be obtained. For a suitably designed MRE rubber bushing, an almost 100% increase in stiffness is obtained even for very small twist angles. This makes possible a rapid change and adaptive increase in the resonance frequency of a simple mass–spring system by more than 40%. Such magneto-sensitive rubber bushings offer promising possibilities for developing a new generation of bushings that are capable of combating undesirable noise and vibrations very efficiently, leading to comfortable rides and less harsh dynamic loads on vehicles.

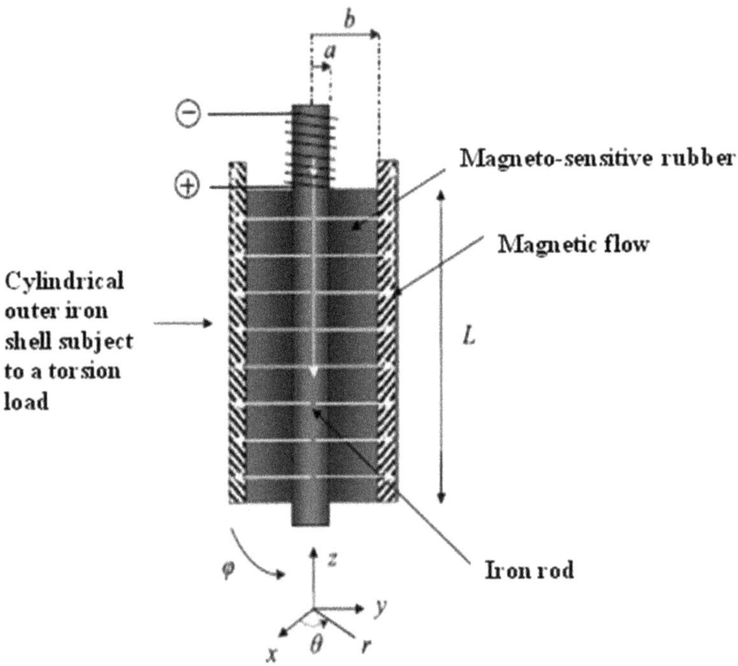


**Fig. 1.17** Axially symmetric magneto-sensitive rubber bushing

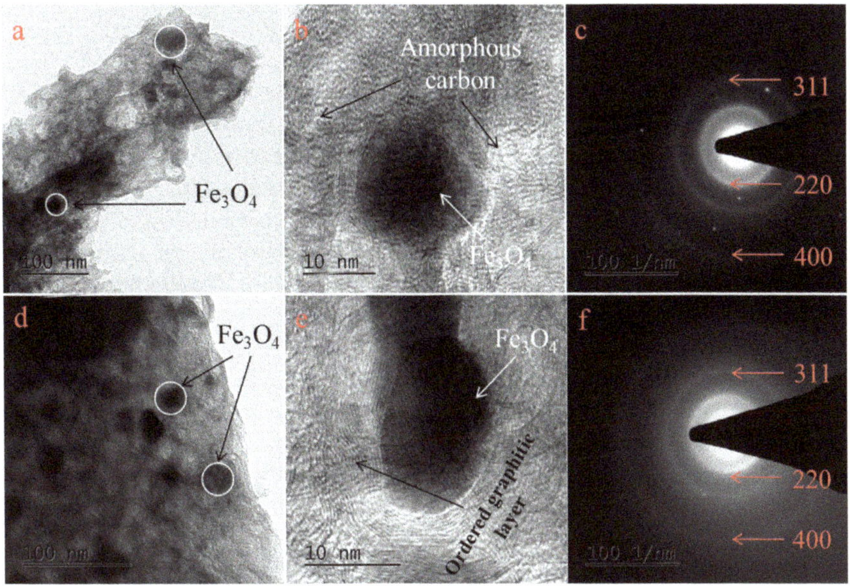


**Fig. 1.18** **a** and **b** HRTEM images, **c** SAED pattern of CNF@5 $Fe_3O_4$; **d** and **e** HRTEM images, **f** SAED pattern of CNF@7.5 $Fe_3O_4$

### *1.9.7 Microwave Absorption Properties of PDMS Nanocomposites Containing Electrospun Hollow Mesoporous Carbon Nanofibres (CNF) Embedded with $Fe_3O_4$ Nanoparticles (Mordina et al. 2017)*

The nanofibres were prepared via co-electrospinning of polyacrylonitrile/$FeCl_3$ and polymethylmethacrylate solutions, followed by stabilization and carbonization. Figure 18 shows the high-resolution TEM (HRTEM) images and selected area electron diffraction (SAED) patterns of two hollow CNFs embedded with $Fe_3O_4$ nanoparticles.

The nanofibres were incorporated in a PDMS matrix and the resulting composites functioned well in the temperature range -50 to 200 °C, owing to the PDMS matrix having a very low glass transition temperature. A maximum reflection loss of -44 dB with associated bandwidth of 2.26 GHz under -10 dB has been obtained for the nanocomposite containing only 25 wt% CNFs. These excellent properties are due to the combined effect of dielectric and magnetic loss mechanisms operating in tandem. Figure 1.19 shows the microwave absorption properties of the PDMS nanocomposites fabricated using these hollow CNFs.

## 1.10 General Summary and Remarks

This chapter provides a concise overview of the synthesis of different nanosized magnetic materials (particles, fibres and rods) and fabrication of polymer nanocomposite (PNC) magnetic-rheological elastomers MREs) based on these magnetic materials. Characterization of the structural and magnetic properties of these materials and evaluation of their magneto-rheological properties have been discussed. It is worth noting that the literature on MREs mostly concerns the use of micron-sized magnetic materials. However, PNC MREs are superior both in respect of performance as well as costs, owing to their better volume coverage and 3D chain formation capabilities at low volume percentages within the polymer matrix.

Silicone elastomers filled with iron or carbonyl iron are the most extensively used systems for development of MREs owing to their ease of fabrication and curing and the high saturation magnetization of iron. Anisotropic MREs generally exhibit higher MR properties as compared to isotropic MREs when the direction of the applied magnetic field is parallel to the direction of particle chains. The opposite is true when the magnetic field direction is perpendicular to the particle chains. Both plasticizer addition and surface modifications of the magnetic materials result in significant changes in MR effects.

MREs are promising for active vibration damping, isolators, actuator, sensors, airflow control valves, rubber bushings, etc. and are futuristic materials for automobile, structural engineering, micro-electromechanical devices and strategic applications in Defence sectors. However, a major limitation to their widespread use is the lack of efficient mass production methods. Most of the specimens cited in the literature were made using simple casting or mechanical mixing methods. These need to be scaled up. One possibility is the use of vacuum-assisted resin transfer molding (VARTM). This method is usable with a range of iron particle sizes and silicone elastomer systems and has been found to be effective and efficient within certain limits of applicability.

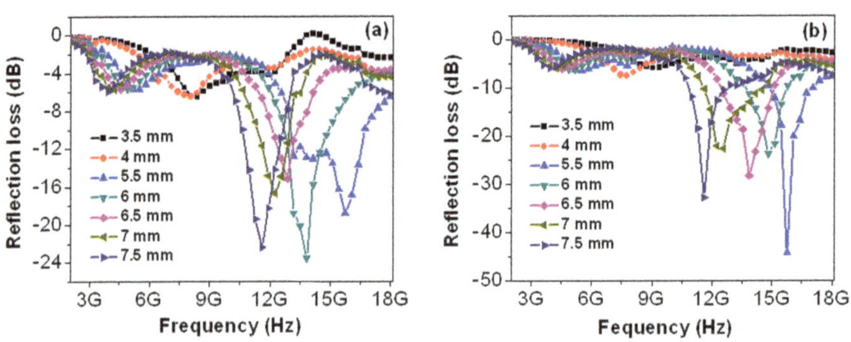


**Fig. 1.19** Reflection loss versus frequency plot **a** PDMS–25wt% CNF@5 $Fe_3O_4$ and **b** PDMS–25wt% CNF@7.5 $Fe_3O_4$

## 1.11 Bibliography

1. Boczkowska A, Awietjan S (2012) "Microstructure and properties of magnetorheological elastomers" In: Advanced Elastomers-Technology, Properties and Applications, IN TECH Open Access, ISBN 978-953-51-0739-2, Ch 6.
2. Li WH, Zhang XZ, Du H (2013) "Magnetorheological elastomers and their applications" In: Advances in elastomers 1: blends and interpenetrating networks,
3. Visakh PM, Thomas S, Chandra AK, Mathew AP(Eds), Springer, Berlin, Germany, pp 357–374.
4. Ruddy C, Ahearne E, Byrne G (2005) "A Review on Magnetorheological elastomers: properties and applications" www.ucd.ie/mecheng/ams/news-items/cillian%20Ruddy.pdf. n 565, pp 146–152.
5. Bajpai OP, Setua DK, Chattopadhyay S (2015) "Process-structure–property relationships in nanocomposites based on piezoelectric-polymer matrix and magnetic nanoparticles" In: Manufacturing of nanocomposites with engineering plastics, Mittal V(Ed), Woodhead Publishing Co, UK, pp 255–278.
6. Hu G, Liu Q, Ding R, Li G (2017) "Adv in Mechanical Eng 9(5), pp 1–15. https://doi.org/10.1177/1687814017694581, journals.sagepub.com/home/ade.

## References

Abramchuk, S.S., Grishin, D.A., Kramarenko, E.Y., Stepanov, G.V., Khokhlov, A.R.: Effect of a homogeneous magnetic field on the mechanical behavior of soft magnetic elastomers under compression. Polym Sci Ser A **48**(2), 138–145 (2006)

Alberdi-Muniain, A., Gil-Negrete, N., Kari, L.: Direct energy flow measurement in magneto-sensitive vibration isolator systems. J. Sound Vib. **331**, 1994–2006 (2012)

An, Y., Shaw, M.T.: Actuating properties of soft gels with ordered iron particles: basis for a shear actuator. Smart Mater. Struct. **12**, 157–163 (2003)

An, H.N., Sun, B., Picken, S.J., Mendes, E.: Long time response of soft magnetorheological gels. J. Phys. Chem. B **116**, 4702–4711 (2012)

Antonel, P.S., Jorge, G., Perez, E., Butera, A., Leyva, A.G., Negri, R.M.: Magnetic and elastic properties of $CoFe_2O_4$ - polydimethylsiloxane magnetically oriented elastomer nanocomposites. J. Appl. Phys. **110**, 043920 (2011)

Arturo, M., Quintela, L., Rivas, J.: Chemical reactions in microemulsions: a powerful method to obtain ultrafine particles. J. Colloid Interface Sci. **158**, 446–451 (1993)

Blom, P., Kari, L.: The frequency, amplitude and magnetic field dependent torsional stiffness of a magneto-sensitive rubber bussing. Int. J. Mech. Sci. **60**, 54–58 (2012)

Boczkowska, A., Awietjan, S.F., Pietrzko, S., Kurzydłowski, K.J.: Mechanical properties of magnetorheological elastomers under shear deformation. Compos. B **43**, 636–640 (2012)

Bose, H., Rabindranath, R., Ehrlich, J.: Soft magnetorheological elastomers as new actuators for valves. J. Int. Mater. Syst. Struct. **23**(9), 989–994 (2011)

Chen, L., Gong, X.L., Li, W.H.: Effect of carbon black on the mechanical performances of magnetorheological elastomers. Polym. Test. **27**, 340–345 (2008b)

Chen, L., Gong, X.l., Jiang, W.Q., Yao, J.J., Deng, H.X., Li, W.H.: Investigation on magnetorheological elastomers based on natural rubber. J. Mater. Sci. **42**, 5483–5489 (2007)

Chen, L., Gong,X.l., Chin Li, W.H.: Damping of magnetorheological elastomers. J. Chem. Phys. **21**, 581–585 (2008)

Deng, H., Li, X., Peng, Q., Wang, X., Chen, J., Li, Y.: Monodisperse magnetic single crystal ferrite microspheres. Angew. Chem. Int. Ed. **44**, 2782–2785 (2005)

Denver, H., Heiman, T., Martin, E., Gupta, A., Borca-Tasciuc, D.A.: Fabrication of polydimethylsiloxane composites with nickel nanoparticle and nanowire fillers and study of their mechanical and magnetic properties. J. Appl. Phys. **106**, 064909 (2009)

Dishovsky, N., Ruskova, K., Radulov, I.: In situ magnetic modification of polar elastomers. Mater. Res. Bull. **36**, 35–45 (2001)

Dong X, Ma N, Qi M, Li J, Chen R, Ou J (2012) The pressure-dependent MR effect of magnetorheologicalelastomers. Smart Mater Struct 21: 075014 (1–5).

Dresco, P.A., Zaitsev, V.S., Gambino, R.J., Chu, B.: Preparation and properties of magnetite and polymer magnetite nanoparticles. Langmuir **15**, 1945–1951 (1999)

Du, G., Chen, X.: MEMS magnetometer based on magnetorheological elastomer. Meas **45**, 54–58 (2012)

Du H, Li W, Zhang N (2011) Semi-active variable stiffness vibration control of vehicle seat suspension using an MR elastomer isolator. Smart Mater. Struct. **20**, 105003 (1–10)

Eem, S.H., Jung, H.J., Koo, J.H.: Application of MR elastomers for improving seismic protection of base-isolated structures. IEEE Trans. Magn. **47**(10), 2901–2904 (2011)

Evans, B.A., Fiser, B.L., Prins, W.J., Rapp, D.J., Shields, A.R., Glass, D.R., Superfine, R.A.: Highly tunable silicone-based magnetic elastomer with nanoscale homogeneity. J. Magn. Magn. Mater. **324**, 501–507 (2012)

Fan, Y.C., Gong, X.L., Jiang, W.Q., Zhang, W., Wei, B., Li, W.H.: Effect of maleic anhydride on the damping property of magnetorheological elastomers. Smart Mater. Struct. **19**, 055015 (1–8) (2010)

Farrell, D., Majetich, S.A., Wilcoxon, J.P.: Preparation and characterization of monodisperse Fe nanoparticles. J. Phys. Chem. B **107**, 11022–11030 (2003)

Farshad, M., Benine, A.: Magnetoactive elastomer composites. Polym. Test. **23**, 347–353 (2004)

Fuchs, A., Zhang, Q., Elkins, J., Gordaninejad, F., Evrensel, C.: Development and characterization of magnetorheological elastomers. J. Appl. Polym. Sci. **105**, 2497–2508 (2007)

Ginder, J.M., Nichols, M.E., Elie, L.D., Tardiff, J.L.: Magnetorheological elastomers: properties and applications In: Proceedings of SPIE, Smart Structures and Materials 1999: Smart Materials Technologies, Wuttig M R Edition, Newport Beach, CA, USA 3675, pp. 131–138 (1999)

Gong, X., Fan, Y., Xuan, S., Xu, Y., Peng, C.: Control of the damping properties of magnetorheological elastomers by using polycaprolactone as a temperature-controlling component. Ind. Eng. Chem. Res. **51**, 6395–6403 (2012)

Gong, X.L., Chen, L., Li, J.f.: Study of utilizable magnetorheological elastomers. Int. J. Mod. Phys. B **21** (28 & 29), 4875–4882 (2007)

Hoang, N., Zhang, N., Du, H.: An adaptive tunable vibration absorber using a new magnetorheological elastomer for vehicular power train transient vibration reduction. Smart Mater. Struct. **20**, 015019 (1–10) (2011)

Hu, Y., Wang, Y.L., Gong, X.L., Gong, X.Q., Zhang, X.Z., Jiang, W.Q., Zhang, P.Q., Chen, Z.Y.: New magnetorheological elastomers based on polyurethane/Si-rubber hybrid. Polym. Test. **24**, 324–329 (2005b)

Hu, F.Q., Wei, L., Zhou, Z., Ran, Y.L., Li, Z., Gao, M.Y.: Preparation of biocompatible magnetite nanocrystals for in-vivo magnetic resonance detection of cancer. Adv. Mater. **18**, 2553–2556 (2006)

Hu, Y., Wang, Y.L., Gong, X.Q., Gong, X.L., Zhang, X.Z., Jiang, W.Q., Zhang, P.Q., Chen, Z.Y.: Magnetorheological elastomers based on polyurethane/si-rubber hybrid. Int. J. Mod. Phys B **19** (7, 8 & 9), 1114–1120 (2005)

Hyeon, T., Lee, S., Park, J., Chung, Y., Na, H.B.: Synthesis of highly crystalline and monodisperse maghemite nanocrystallites without a size-selection process. J. Am. Chem. Soc. **123**, 12798–12801 (2001)

Jiang, W.Q., Yao, J.J., Gong, X.L., Chen, L.: Enhancement in magnetorheological effect of magnetorheological elastomers by surface modification of iron particles. Chin. J. Chem. Phys. **21**(1), 87–92 (2008)

Jolly, M.R., Carlson, J.D., Muñoz, B.C., Bullions, T.A.: The magnetoviscoelastic response of elastomer composites consisting of ferrous particles embedded in a polymer matrix. J Intel Mater Syst Struct **7**(6), 613–622 (1996)

Jolly, M.R., Carlson, J.D., Munoz, B.C.: A model of the behaviour of magnetorheological materials. Smart Mater. Struct. **5**, 607–614 (1996)

Kaleta, J., Krolewicz, M., Lewandowski, D.: Magnetomechanical properties of anisotropic and isotropic magnetorheological composites with thermoplastic elastomer matrices. Smart Mater. Struct. **20**, 085006(1–12) (2011)

Kallio, M., Aalto, S., Lindroos, T.: Preliminary test on a MRE device. In: Conference Proceeding, AMAS Workshop on smart mater struct, SMART'03, pp. 353–360. Jadwisin, Poland (2003)

Kallio, M., Lindroos, T., Aalto, S., Jarvinen, E., Karna, T., Meinander, T.: Dynamic compression testing of a tunable spring element consisting of a magnetorheological elastomer. Smart Mater. Struct. **16**, 506–514 (2007)

Kashima, S., Miyasaka, F., Hirata, K.: Novel soft actuator using magnetorheological Elastomer. IEEE Trans. Magn. **48**(4), 1649–1652 (2012)

Kim, Y.K., Koo, J.H., Kim, K.S., Kim, S.H.: Suppressing harmonic vibrations of a miniature cryogenic cooler using an adaptive tunable vibration absorber based on magneto-rheological elastomers. Rev. Sci. Instrum. **82**, 035103 (2011)

Koo, J.H., Dawson, A., Jung, H.J.: Characterization of actuation properties of magnetorheological elastomers with embedded hard magnetic particles. J Intl Mater Syst Struct **23**, 1049–1054 (2012)

Landa, R.A., Antonel, P.S., Ruiz, M.M., Perez, O.E., Butera, A., Jorge, G., Oliveira, C.L.P., Negri, R.M.: Magnetic and elastic anisotropy in magnetorheological elastomers using nickel-based nanoparticles and nanochains. J. Appl. Phys. **114**, 213912 (2013)

Lejon, J., Kari, L.: Preload, frequency, vibrational amplitude and magnetic field strength dependence of magnetosensitive rubber. Plast. Rubber Compos. **38**(8), 321–326 (2009)

Li, J., Gong, X., Xu, Z.B., Jiang, W.: The effect of pre-structure process on magnetorheological elastomer performance. Int. J. Mater. Res. **99**, 1358–1364 (2008)

Li, W., Zhang, X., Du, H.: Development and simulation evaluation of a magnetorheological elastomer isolator for seat vibration control. J. Intl. Mater. Syst. Struct. **23**, 1041–1048 (2012)

Liao, G.J., Gong, X.L., Xuan, S.H., Kang, C.J., Zong, L.H.J.: Development of a real-time tunable stiffness and damping vibration isolator based on magnetorheological elastomer. Int. Mater Syst Struct **23**, 25–33 (2011a)

Liao, G.J., Gong, X.L., Kang, C.J., Xuan, S.H.: The design of an active–adaptive tuned vibration absorber based on magnetorheological elastomer and its vibration attenuation performance. Smart Mater. Struct. **20**, 075015 (1–10) (2011)

Lokander, M., Stenberg, B.: Improving the magnetorheological effect in isotropic magnetorheological rubber materials. Polym. Test. **22**, 677–680 (2003a)

Lokander, M., Stenberg, B.: Performance of isotropic magnetorheological rubber materials. Polym. Test. **22**, 245–251 (2003b)

Luo, Y.D., Dai, C.A., Chiu, W.Y.: Polystyrene/$Fe_3O_4$ composite latex via miniemulsion polymerization-nucleation mechanism and morphology. J. Polym. Sci. A Polym. Chem. **46**, 1014–1024 (2008)

Mitsumata, T., Ohori, S.: Magnetic polyurethane elastomers with wide range modulation of elasticity. Polym. Chem. **2**, 1063–1067 (2011)

Moeen, S.J., Vaezi, M.R., Yousefi, A.A., Ghasemi, E.: Synthesis and rheological properties of nickel-zinc ferrite polymer nanocomposites. J. Appl. Polym. Sci. **123**, 2534–2539 (2012)

Mordina, B., Tiwari, R.K., Setua, D.K., Sharma, A.: Magnetorheology of polydimethylsiloxane elastomer/$FeCo_3$ nanocomposite. J. Phys. Chem. C **118**, 25684–25703 (2014)

Mordina, B., Tiwari, R.K., Setua, D.K., Sharma, A.: Superior elastomeric nanocomposites with electrospun nanofibers and nanoparticles of $CoFe_2O_4$ for magnetorheological applications. RSC Adv. **5**, 19091–19105 (2015)

Mordina, B., Tiwari, R.K., Setua, D.K., Sharma, A.: Impact of graphene oxide on the magnetorheological behaviour of $BaFe_{12}O_{19}$ nanoparticles filled polyacrylamide hydrogel. Polymer **97**, 258–272 (2016)
Mordina, B., Tiwari, R.K., Setua, D.K., Sharma, A.: $Fe_3O_4$ nanoparticles embedded hollow mesoporous carbon nanofibers and polydimethylsiloxane-based nanocomposites as efficient microwave absorber. J. Phys. Chem. C **121**, 7810–8782 (2017)
Mordina, B., Setua, D.K.: Sharma AActuation, magnetic and microstructural characteristics of polyacrylamide/$BaFe_{12}O_{19}$/ graphene oxide based nanocomposite hydrogels. Compos. Part A (under review)
Mousavian, S., Ebadi-Dehaghani, H., Ashouri, D., Sadeghipour, H., Jabbari, F.: Effect of polymer matrix on the magnetic properties of polymer bonded magnets filled $Fe_3O_4$ nanoparticles. J. Polym. Res. **19**, 9991 (2012)
Nathani, H., Gubbala, S., Misra, R.D.K.: Magnetic behavior of nickel ferrite–polyethylene nanocomposites synthesized by mechanical milling process. Mater. Sci. Eng. B **111**, 95–100 (2004)
Padalka, O., Song, H.J., Wereley, N.M., Filer, J.A., II., Bell, R.C.: Stiffness and damping in Fe Co, and Ni nanowire-based magnetorheological elastomeric composite. IEEE Trans. Magn. **46**(6), 2275–2277 (2010)
Park, J., An, K., Hwang, Y., Park, J.G., Noh, H.J., Kim, J.Y., Park, J.H., Hwang, N.M., Hyeon, T.: Ultra-large-scale syntheses of monodisperse nanocrystals. Nat. Mater. **3**, 891–895 (2004)
Park, B.J., Fang, F.F., Choi, H.J.: Magnetorheology: Materials and application. Soft Matter **6**, 5246–5253 (2010)
Popp KM, Zhang XZ, Li WH, Kosasih PB (2009) MRE properties under shear and squeeze modes and applications. J. Phys., Conf. Ser. **149**, 012095 (1–4)
Rockenberger, J., Scher, E.C., Alivisatos, A.P.: A new nonhydrolytic single-precursor approach to surfactant-capped nanocrystals of transition metal oxides. J. Am. Chem. Soc. **121**, 11595–11596 (1999)
Rwei, S.P., Wang, L.Y., Yang, P.W.: Synthesis and magnetorheology study of iron oxide and iron cobalt oxide suspensions. J. Nanomater. **2013**, 1–7 (2013)
Setua, D.K., Soman, C., Bhowmick, A.K., Mathur, G.N.: Oil resistant thermoplastic elastomers of nitrile rubber and high density polyethylene blends. Polym. Eng. Sci. **42**, 10–18 (2002)
Setua, D.K.: "Smart elastomers" In: Bhowmick, A.K. (ed.) Current Topics in Elastomer Research, pp 276–287. Taylor & Francis Gr LLC, CRC Press USA (2008)
Shiga, T., Okada, A., Kurauchi, T.: Magnetroviscoelastic behavior of composite gels. J. Appl. Polym. Sci. **58**, 787–792 (1995)
Song, K.H., Park, B.J., Choi, H.J.: Effect of magnetic nanoparticle additive on characteristics of magnetorheological fluid. IEEE Trans. Magn. **45**, 4045–4048 (2009a)
Song, H.J., Wereley, N.M., Bell, R.C., Planinsek, J.L., Filer II, J.A.: Field dependent response of magnetorheological elastomers utilizing spherical Fe particles versus Fe nanowires. J Phys: Conf. Ser. **149**, 012097 (1–4) (2009)
Stepanov, G.V., Chertovich, A.V., Kramarenko, A.Y.: Magnetorheological and deformation properties of magnetically controlled elastomers with hard magnetic filler. J. Magn. Magn. Mater. **324**, 3448–3451 (2012)
Sun, S., Murray, C.B., Weller, D., Folks, L., Moser, A.: Monodisperse FePt nanoparticles and ferromagnetic FePt nanocrystal superlattices. Science **287**, 1989–1992 (2000)
Szabo, D., Szeghy, G., Zrı´nyi M,: Shape transition of magnetic field sensitive polymer gels. Macromol **31**, 6541–6548 (1998)
Tian, T.F., Li, W.H., Alici, G., Du, H., Deng, Y.M.: Microstructure and magnetorheology of graphite-based MR elastomers. Rheol. Acta **50**, 825–836 (2011)
Tian, Y., Liu, Y., He, M., Zhao, G.: Sun Y (2013) High damping properties of magnetic particles doped rubber composites at wide frequency. Mater. Res. Bull. **48**, 2002–2005 (2013)
Varga, Z., Filipcsei, G., Zrınyi, M.: Magnetic field sensitive functional elastomers with tuneable elastic modulus. Polymer **47**, 227–233 (2006)

Vicente, J., Bossis, G., Lacis, S., Guyot, M.: Permeability measurements in cobalt ferrite and carbonyl iron powders and suspensions. J. Magn. Magn. Mater. **251**, 100–108 (2002)
Von, L.P., Lofland, S.E., Koo, J.H., Kadlowec, J., Dermond, M.: Dynamic characterization of bimodal particle mixtures in silicone rubber magnetorheological materials. Polym. Test. **27**, 931–935 (2008)
Von, L.P., Lofland, S.E., Biggs, J., Roche, J., Mineroff, J., Babcock, M.: Investigating new symmetry classes in magnetorheological elastomers: cantilever bending behaviour. Smart Mater. Struct. **20**, 105022 (1–10) (2011)
Wang, D., Chen, J.S., Sun, L.: Homogenization of magnetostrictive particle-filled elastomers using an interface-enriched reproducing kernel particle method. Finite Elem. Anal. Des. **39**, 765–782 (2003)
Wang, X., Zhuang, J., Peng, Q., Li, Y.: A general strategy for nanocrystal synthesis. Nature **437**, 121–124 (2005)
Wang, Y., Hu, Y., Chen, L., Gong, X., Jiang, W., Zhang, P., Chen, Z.: Effects of rubber/magnetic particle interactions on the performance of magnetorheological elastomers. Polym. Test. **25**, 262–267 (2006)
Wang, Y., Hu, Y., Gong, X., Jiang, W., Zhang, P., Chen, Z.: Preparation and properties of magnetorheological elastomers based on silicon rubber/polystyrene blend matrix. J. Appl. Polym. Sci. **103**, 3143–3149 (2007)
Wu, J., Gong, X., Fan, Y., Xia, H.: Improving the magnetorheological properties of polyurethane magnetorheological elastomer through plasticization. J. Appl. Polym. Sci. **123**, 2476–2484 (2012)
Wu, J., Gong, X., Fan, Y., Xia, H.: Anisotropic polyurethane magnetorheological elastomer prepared through in situ polycondensation under a magnetic field. Smart Mater. Struct. **19**, 105007 (1–8) (2010)
Xu, Y., Gong, X., Xuan, S., Zhang, W., Fan, Y.: A high-performance magnetorheological material: preparation, characterization and magnetic-mechanic coupling properties. Soft Matter **7**, 5246–5254 (2011)
Yanez-Flores, I.G., Betancourt-Galindo, R., Matutes Aquino, J.A., Rodriguez-Fernandez, O.: Preparation and characterization of magnetic PVC nanocomposites. J. Non-Cryst. Solids **353**, 799–801 (2007)
Yin, H.M., Sun, L.Z., Chen, J.S.: Magneto-elastic modeling of composites containing chain-structured magnetostrictive particles. J. Mech. Phys. Solids **54**, 975–1003 (2006)
Zhu, L., Xie, D., Ma, J., Shao, J., Shen, X.: Fabrication of polydimethylsiloxane composites with nickel particles and nickel fibers and study of their magnetic properties. Smart Mater. Struct. **22**, 045015 (1–7) (2013)
Zrínyi, M., Barsi, L., Büki, A.: Deformation of ferrogels induced by nonuniform magnetic fields. J. Chem. Phys. **104**(21), 8750–8756 (1996)

# Chapter 2 High Temperature Resistant Thermosetting Resin Materials

**Ajit S. Singh, Vishal Das, Swati Chopra, Arvind Kr. Pandey, and Eswara Prasad Namburi**

**Abstract** This chapter reviews worldwide research and technological developments on thermally resistant thermosetting resins for various high-end uses in the aerospace, automobile and marine industries. It provides a cursory look on commercially available high-temperature resins like bismaleimide, polyimides and cyanate esters resins, including their preparation, properties, criticalities, specific applications in the Defence sector, and their drawbacks. Of the high-temperature resins, phthalonitrile (PN) resins have recently been shown as the most thermally stable thermosetting resins, capable of sustaining temperatures > 400 °C. They meet most of the stringent requirements of Defence, with many additional attributes like flame retardance, high char yield, excellent adhesion properties and ambient temperature shelf life. Further, this chapter covers the work done by DMSRDE in this area of high-temperature resins (phthalonitrile resins) and describes the efforts for synthesizing these resins by molecular tailoring to obtain an optimal combination of flame retardance, and high-temperature resistance for replacing currently used epoxy and phenolic-based systems in aerospace and marine defence systems.

**Keywords** High temperature · Flame retardance · Phthalonitrile · Cyanate ester · Polyimide · Char yield

---

A. S. Singh · V. Das · S. Chopra · A. Kr. Pandey
Defence Materials and Stores Research and Development Establishment (DMSRDE), Defence Research and Development Organisation (DRDO), Kanpur, India

E. P. Namburi (✉)
Former Outstanding Scientist and Ex-Director, Defence Materials and Stores Research and Development Establishment (DMSRDE), Defence Research and Development Organisation (DRDO), Kanpur, Uttar Pradesh, India
e-mail: neswarap@gmail.com

 
E. P. Namburi et al. (eds.), *Novel Defence Functional and Engineering Materials (NDFEM) Volume 1*, Indian Institute of Metals Series,
https://doi.org/10.1007/978-981-99-9791-6_2

## 2.1 Introduction

Enormous growth in polymer science and technology has led to the development of heat-resistant polymers via precise organic molecular tailoring. With rapid technological advancement and paradigm shifts in materials requirements, high-temperature polymers (resins) have established themselves in the growing industrial demand for stringent quantitative as well as qualitative requirements. These resins are required for many key applications, e.g. sealants and adhesives for aircraft, binder materials for brake systems, structural composites for advanced aircraft, space vehicles, weapon systems and electronics. Many different classes of heat-resistant polymer systems have been reported in open domain as well as many proprietary items are also commercially available.

What constitutes a heat-resistant resin is arguable, but the most acceptable definition is a material that can retain mechanical properties for thousands of hours at 230 °C, hundreds of hours at 300 °C, minutes at 540 °C and seconds up to 760 °C.

Also, it is important that these materials retain useful properties under particular conditions like pressure or vacuum; mechanical loading, radiation, chemical or electrical exposure, all at temperatures ranging from cryogenic to above 500 °C.

There are chemical and physical factors that cumulatively make a particular polymer heat resistant. The chemical factors include primary bond strength, resonance stabilization, mechanism of bond cleavage, molecular symmetry/structural regularity, rigid inter-chain structure, branching and cross-linking. Physical factors include secondary Van der Waals forces, molecular dipolar interaction, hydrogen bonding, molecular weight, molecular weight distribution, crystallinity, molecular dipolar interaction and purity.

For tailoring a heat-resistant polymer, the prime consideration is the mainframe skeleton, which dictates the thermal stability. Table 2.1 lists the dissociation energies for several types of bonds: it is evident that aromatic and heterocyclic bonds are more thermally stable compared to the others. In more detail, the bond dissociation energy of a resonance stabilized system is much more than that of other systems owing to the cumulative effect of resonance stabilization. The N–N bond dissociation energy is relatively low, but inclusion of N–N bonds in heterocyclic rings results in reasonably high thermal stability owing to resonance stabilization. Other potentially interesting bonds are inorganic bonds like B–N, Si–N and Ti–O, but these bonds are vulnerable to attack by water and oxygen. Therefore, only a few useful high-temperature inorganic polymers have been reported.

Considering the secondary physical forces, dipole–dipole interactions and hydrogen bonding add a further 6–9.8 kcal/mol towards molecular stability and substantially affect the cohesive energy density, which directly influences the stiffness, glass transition temperature (Tg), melting point and solubility. The strong electron withdrawing groups in polymer chains impart more stability compared to electron-donating groups.

Sometimes we encounter issues of molecular symmetry or regularity wherein moieties are joined at the same position in each repeat unit. The presence of isomers

**Table 2.1** Bond dissociation energies

| Type of bond | Bond dissociation energy in (kcal/mol) | Polymer/material representative |
|---|---|---|
| C–C | 83.6 | Polyethylene, |
| C = C | 145.8 | PPV |
| C–F | 102.8 | Teflon |
| 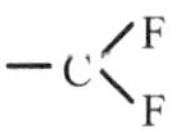 | 120.4 | PVDF |
| 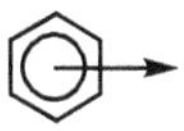 Resonance stabilization of particular molecule additions | 39.2–68.6 | Polystyrene, PMR-15 |
| N–N | 38.2 | Azopolyimides, NLO azopolymers |
| B–N | 92.3 kcal/mol | Polyborazine |
| Si–N | 104.4 kcal/mol | Polysilazanes |
| Ti–O | 160.6 kcal/mol | Silico-titanium polymers |

lowers the $T_g$. From a physical perspective, crystallinity means ordered structure in the thermoplasts, while branching causes disorder, leading to poor thermal stability. In polymers, the oriented chain macromolecules possess the highest thermal stability, $T_g$ coupled with low solubility and difficult processability. Van der Waals forces, dipole–dipole interactions and hydrogen bonding in the linear chains impart close packing leading to crystalline domains in the polymer microstructure.

Polymers have manifested themselves in a variety of forms ranging from thermoplastics, thermosetting resins, elastomers, fibres and so on. The combination of advanced polymeric fibres and fabrics in conjunction with thermoset/thermoplastic polymers has opened up new avenues for product designers and engineers in the form of composite materials and structures. The unique macromolecular and micromolecular combinations result in property matrices, which can also replace traditional ceramic and metallic materials in load-bearing engineering applications Considering the stringent quality requirements and performance criteria for critical uses, the suitability of polymeric materials is adjudged by their overall performance and with limited or no trade-off in properties. At times it becomes mandatory to demonstrate high strength (tensile strength, flexural strength and impact resistance) along with other attributes like high-temperature stability, high continuous use temperature, corrosion resistance, chemical resistance, and ease of processing.

High temperature-resistant resins are the key requirements of present defence systems. Most of our systems like aerospace, missile and naval sector demand high-temperature resistant materials for a variety of sub-systems and applications. Currently, the research community and industry are focusing on the development of high-temperature resins based on a multi-path approach: the materials are selected on the grounds of their suitability in terms of performance, useful extended life and ease of fabrication. These general aspects can be further specified as follows:

(1) Properties: thermal, mechanical and thermo-mechanical properties; ambient temperature storage life.
(2) Processing: hand-layup, filament winding, pultrusion, autoclave, resin transfer moulding and resin infusion.

## 2.2 Types of High-Temperature Resistant Resins

### *2.2.1 Thermosetting Polyimides*

These are a special class of polymers prepared by thermal condensation of non-stoichiometric proportions of diamines and di-anhydrides, followed by thermo-curable end-group capping. Various polyimides have been synthesized in the last 60 years, e.g. PMR-15, LaRC-160, PMR-II-50, AFR-700, RP-46, PETI-330, AFR-PE-4, DMBZ-15 and are being used in the aerospace industry. PMR-15 is still today the resin of choice for major jet engine manufacturers. However, major disadvantages of these systems include the use of carcinogenic diamines like MDA and ODA during the manufacturing process; waste management; poor shelf-life of the resin in solution form; presence of toxic solvents in the resin system (e.g. methanol); evolution of volatiles during the curing process; lack of inherent tack in dry resin samples; and processing difficulties like high-temperature curing followed by staggered post-curing cycles at various temperatures.

### *2.2.2 Bismaleimides*

These are currently being used in the airframes of the U.S Air Force Lockheed Martin F-22 *Raptor* and F-35 *Lightning II*; the *F-17* flap hinge fairing structure; the thrust reversing inner fixed cowl structure for Pratt and Whitney 4168 engines; and the tail boom structures for helicopters.

The main drawbacks of bismaleimide resins are their brittleness, toxic and mutagenic precursors, poor solubility in organic solvents, narrow processing windows and high temperature curing. The high cost of resin manufacturing limits the use of these resins to high-end aerospace and defence sectors.

### *2.2.3 Other Types*

Besides polyimides and bismaleimides only cyanate esters and phthalonitrile resins fall in the category of high-temperature resins (except for some phenolic and benzoxazine resins).

### 2.2.4 General Property Comparisons

Table 2.2 compares many of the important properties of current thermally resistant resins. It may be seen that the phthalonitrile (PN) resins are presently the most suitable candidates. This is reinforced by the data in Fig. 2.1.

**Table 2.2** General properties of common thermosetting polymer matrices (Nair et al. 2001; Kessler 2012; Wilson 1988; Broughton 2012; Sastri et al. 1996)

| Property | Epoxy | Phenolic | Toughened BMI | Cyanate esters | PMR-15 | Phthalonitriles |
|---|---|---|---|---|---|---|
| Continuous use temperature (°C) | Rt-180 | Rt-250 | Rt ~ 350 | Rt ~ 250 | Rt-315 | Rt- 400 |
| Cure temperature (°C) | Rt-220 | 150–190 | 220–300 | 175–250 | 280–320 | 250–350 |
| Tensile modulus (GPa) | 3.1–3.8 | 3–5 | 3.4–4.1 | 3.1–3.4 | 4.4 | 4–5 |
| Onset degradation temperature (°C) | 250–350 | 300–400 | 360–400 | 400–420 | ~400 | >450 |
| $T_g$ (°C) | ~50–300 | ~250–260 | ~150–350 | ~190–400 | ~327 | ~350–>450 |
| Toxic gas emission (combustion) | Yes | Yes | Yes | Yes | Minimum | Negligible |
| Solubility in common solvents | Excellent in both non-polar and polar aprotic solvents | Excellent in polar protic and aprotic solvents | Excellent only in polar aprotic solvents | Excellent in both non-polar and polar aprotic solvents | Excellent in polar protic and aprotic solvents | Excellent only in polar aprotic solvents |
| Storage life and conditions | −18 °C to sub-zero temperature for 6 months for liquid resins | −18°C for 6 months in solution form | 4–5 years in solid form | 1–2 years under strict moisture-free conditions | ~1 year in solution form at ambient conditions | >5 years Stage B resin at ambient conditions |
| Natural tackiness | Yes | Yes | No | Yes in case of commercially available Bisphenol A-based resin | Yes in solution form | No in monomeric resins and yes in oligomeric resins in solution form |

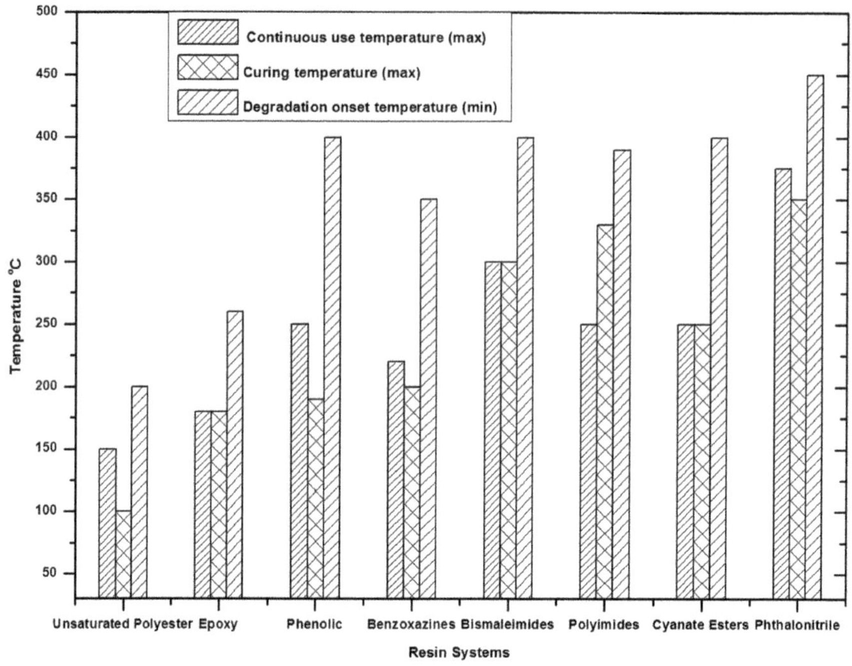


**Fig. 2.1** Bar graph showing several properties of various resin systems (Nair et al. 2001; Kessler 2012; Wilson 1988; Broughton 2012; Sastri et al. 1996; Yagci et al. 2009)

Figure 2.1 provides a bar graph comparison of several material properties of commercially available and reported resins. It is evident that the phenolics, cyanate ester and phthalonitrile resin family provide the maximum usage temperature. However, as indicated in Table 2.2, phenolic resins have shelf life and toxic gas issues. Also, cyanate ester preparation involves toxic cyanogen bromide and hence phthalonitrile dominates as the highest end use available high temperature resin systems.

Thus, in the light of the many practical considerations (*see the lists at the end of* Sect. 2.1, and Table 2.2 and Fig. 2.1), the present chapter discusses high-temperature resistant phthalonitrile (PN) resins. This is done with emphasis on the present status and challenges.

**Table 2.3** Mechanical properties of dihydroxy biphenyl-based PN resin (Sastri et al. 1996)

| | |
|---|---|
| Tensile strength (MPa) | 80 (±7%) |
| Tensile modulus (GPa) | 44 (±5%) |
| Elongation to fracture (%) | 1.2 (±5%) |
| Flexural strength (MPa) | 80 (±9%) |
| Flexural modulus (GPa) | 42 (±3%) |
| Flexural toughness, $G_{1C}$ (J/m$^2$) | 120–130 |

## 2.3 Phthalonitrile (PN) Resins

### *2.3.1 Historical Perspective*

Phthalonitrile functionality was not studied as a reactive chain forming group for the polymeric domain until the 1970s. Identification of the phthalonitrile moiety as a thermocurable group goes back to this decade: its use as a polymeric matrix resin was actually identified by Keller (Laskoski et al. 2006) from the US Naval Research Laboratory (NRL). Owing to certain restrictions, phthalonitrile resins are presently commercially available only from US-based firms like Maverick Corporation, Renegade Materials, JFC Technologies and Cardolite Corporation.

These polymers show excellent thermal, mechanical and thermo-oxidative stability. They also have certain unique properties like low dielectric constants, radar transparency, flame retardancy, low water intake and superior adhesive properties, making them suitable for canister, radar and electronic packaging applications.

In the past two decades, Keller and his NRL group have remained the nodal agency for developing (i) various PN-based polymers and (ii) composites using carbon and glass fibres. These composites have shown excellent properties for aerospace and marine applications. For example, the NRL has evaluated PN/glass fabric composites for flammability, using tests required for submarine application as per specification (MIL-STD-2031).

Examples of the mechanical properties of a neat PN resin (dihydroxy biphenyl-based resin) are given in Table 2.3.

### *2.3.2 Synthetic Development of PN Resins*

The synthetic advancement in PN monomer synthesis can be broadly classified into the following categories:

- Monomeric resins
- Self-curing resins
- Oligomeric/polymeric resins
- Mixed curing functionality-based resins.

### 2.3.3 Monomeric PN Resins

These resins have a very simple structure and are synthesized by the nucleophilic substitution reaction of a typical aromatic dihydroxy compound with 4-nitrophthalonitrile, in the presence of a strong base like $K_2CO_3$ and in polar aprotic solvents like DMSO, DMF, DMAC, NMP etc. (see Fig. 2.2). The resin so obtained is called a Stage A resin, which can be converted to a more stable pre-polymeric Stage B form that is generally dark green in colour and can be stored under ambient conditions for several years.

The actual basic skeleton of the polymeric structure formed upon curing is still not clear, but FT-IR spectra of the partially cured resin showed that there are different possible cross-linked structures (skeletons) formed during polymerization (see Fig. 2.3).

In this context, the synthesis of 4-nitrophthalonitrile can also be included, since it is the primary component for introduction of a phthalonitrile group. The 4-nitrophthalonitrile can be synthesized by the multistep reaction pathway, Fig. 2.4 (Young and Onyebuagu 1990).

Numerous monomeric PN resins involving various types of bisphenols have been synthesized. An extensive literature survey has been made by the present authors, and the results are given in Table 2.4. These include dihydroxybiphenyl, resorcinol, naphthalene, Bisphenol A, Bisphenol F, fluorene, benzophenone, phenolphthalein, sulphone and phloroglucinol.

Researchers have synthesized and assessed the thermo-mechanical and rheological properties of (i) highly fluorinated resins, phosphate esters, phosphazene and siloxanes; (ii) heteroaromatics like oxadiazole, benzimidazole and pyridine and (iii) flexible hinge precursor PN resins.

Keller and others have also synthesized conducting PN resins, which upon curing + thermal pyrolysis result in conducting carbon-based materials. Owing to the increase in cost of precursor materials (because of their dependency on petroleum products and the issue of waste products management), research is now focusing on the use of natural or semi-synthetic biologically available raw materials; or bisphenols like resveratrol and dihydroresveratrol to synthesize bisphthalonitrile resins.

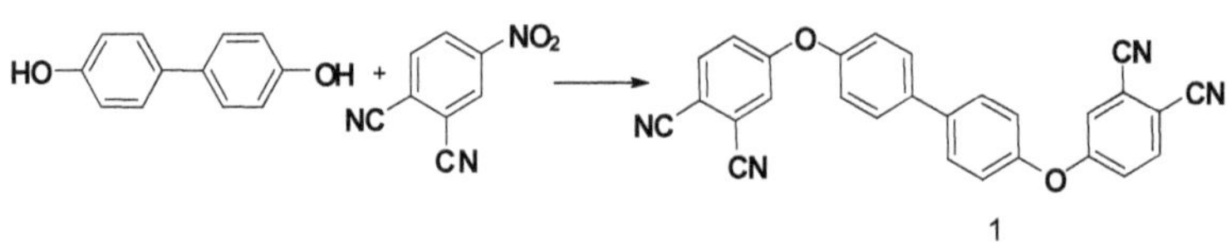


1

Bisphthalonitrile Monomer

**Fig. 2.2** General method of synthesis of PN resins

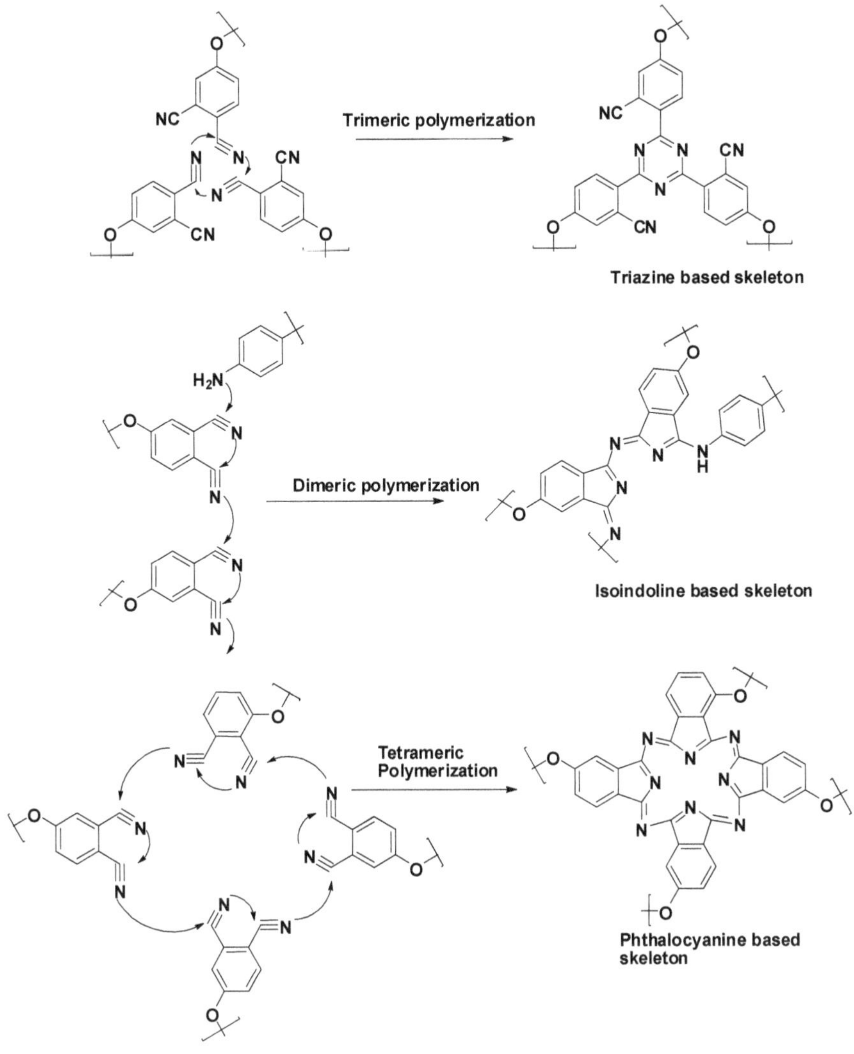


**Fig. 2.3** Formation of triazine, isoindoline and phthalocyanine skeletons

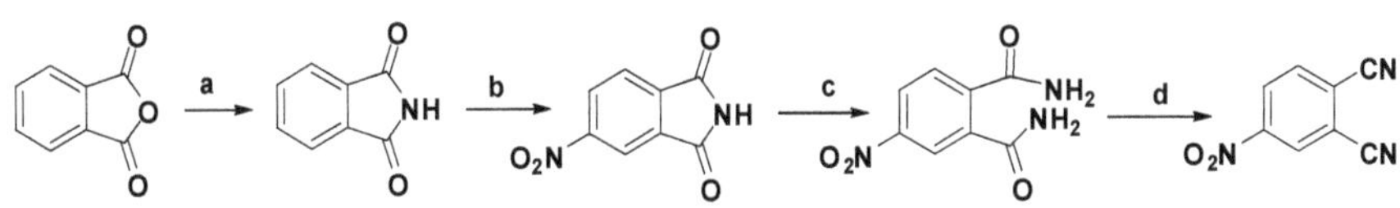

Reagents and conditions: a) $(NH_4)_2CO_3$, 300 °C, sublimation; b) fuming $HNO_3$, conc $H_2SO_4$, 15-35 °C; c) 30% $NH_3$ solution, rt, 24 hrs; d) $SOCl_2$, DMF,reflux

**Fig. 2.4** Preparation of 4-nitrophthalonitrile

**Table 2.4** Different types of monomeric PN resins

| Name/Type | Structure and number | References |
|---|---|---|
| Dihydroxybiphenyl | 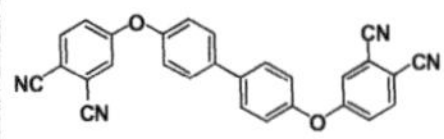 1 | Laskoski et al. (2006) |
| Resorcinol | 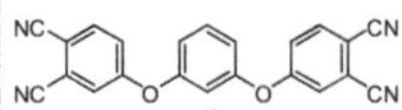 2 | Keller and Dominguez (2005) |
| Naphthalene | 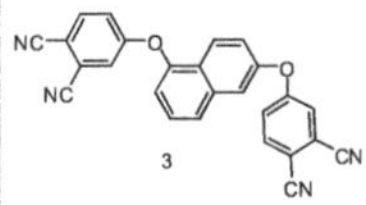 3 | Zhao et al. (2014) |
| Bisphenol A | 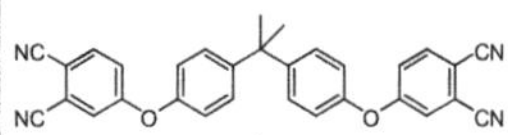 4 | Ting et al. (1982) |
| Bisphenol F | 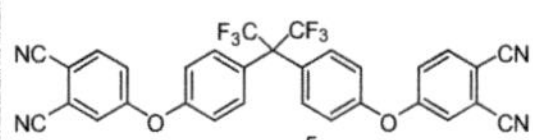 5 | Ting et al. (1982) |
| Fluorene | 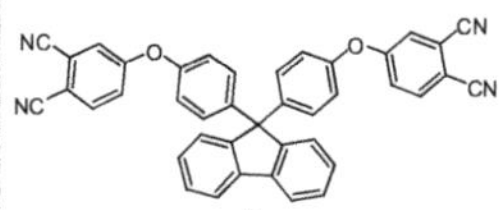 6 | Achar et al. (1986) |
| Benzophenone | 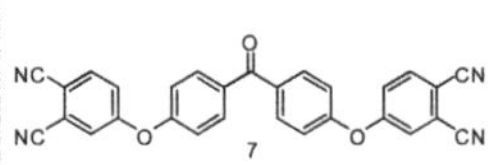 7 | Ting et al. (1982) |
| Phenolphthalein | 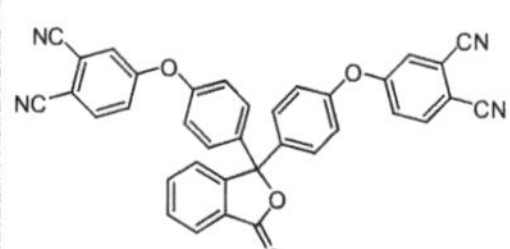 8 | Achar et al. (1986) |
| Sulphone | 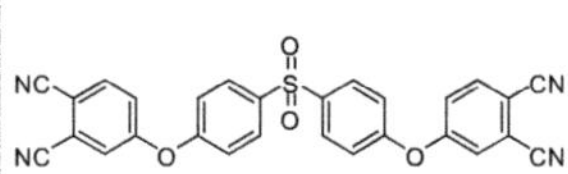 9 | Ting et al. (1982) |
| Phloroglucinol | 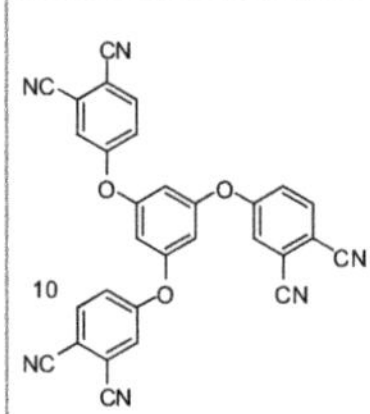 10 | Köç et al. (2016) |

(continued)

**Table 2.4** (continued)

| Name/Type | Structure and number | References |
|---|---|---|
| Highly fluorinated | 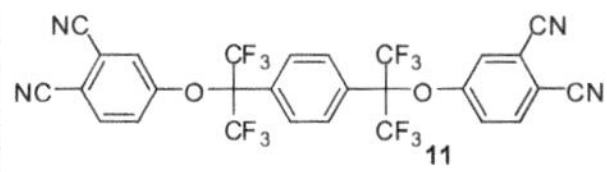 11 | Keller and Griffith (1979) |
| Phosphate ester | 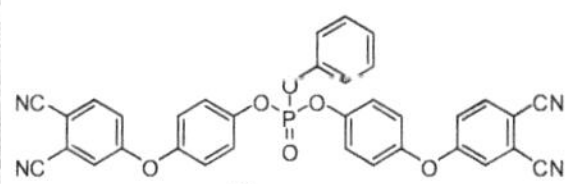 12 | Bulgakov et al. (2016a) |
| Phosphate ester | 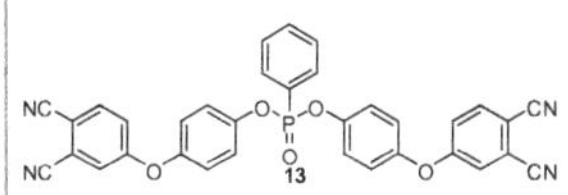 13 | Bulgakov et al. (2016a) |
| Phosphazene | 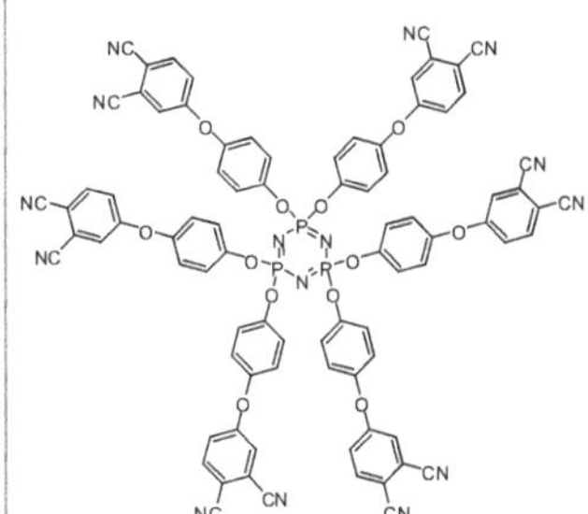 14 | Zhao et al. (2015a) |
| Siloxane | 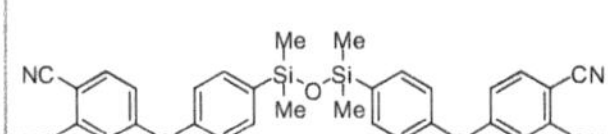 15 | Dzhevakov et al. (2016) |
| Siloxane | 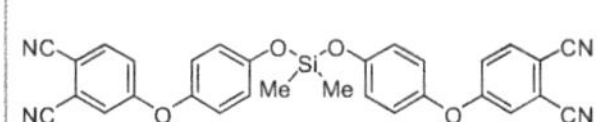 16 | Babkin et al. (2015) |
| Siloxane | 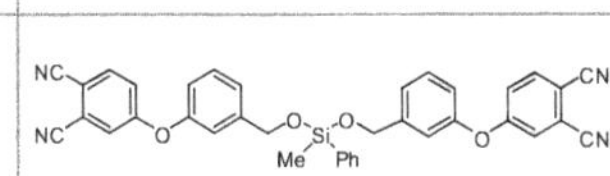 17 | Babkin et al. (2016) |
| Oxadiazole | 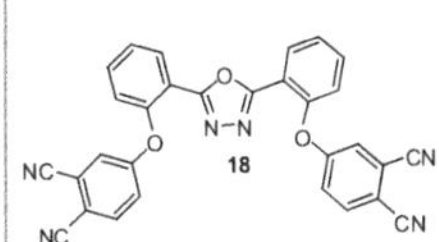 18 | Badshah et al. (2014) |
| Bismaleimide | 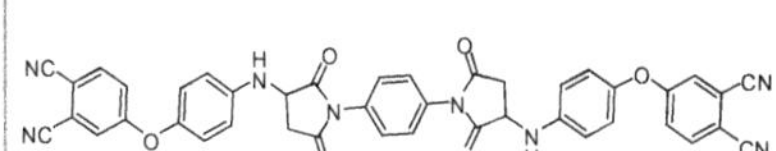 22a | Kaliavaradhan and Muthusamy (2016) |

(continued)

**Table 2.4** (continued)

| Name/Type | Structure and number | References |
|---|---|---|
| Benzimidazole | 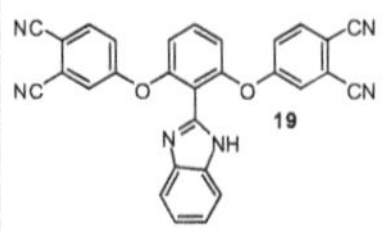 19 | Wu et al. (2012) |
| Pyridine | 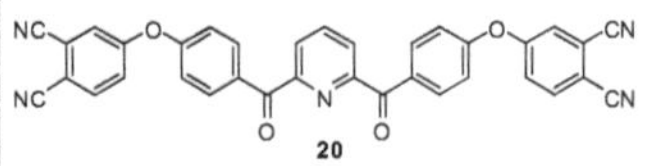 20 | Xi et al. (2016) |
| Flexible hinge precursor phthalonitrile | 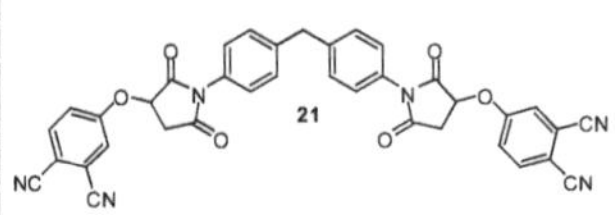 21 | Zhang et al. (1997) |
| Flexible hinge precursor phthalonitrile | 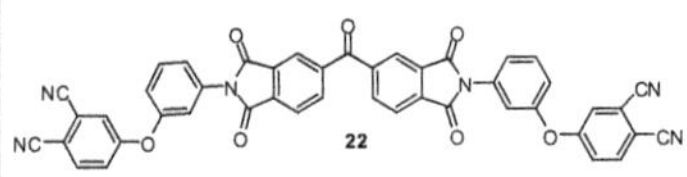 22 | Keller (1993) |
| Flexible hinge silazane precursor phthalonitrile | 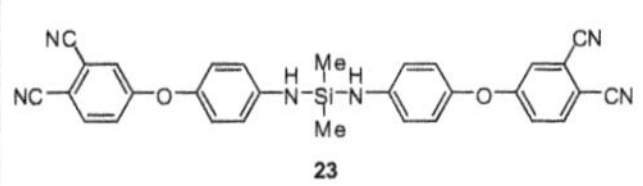 23 | Zhang et al. (2014) |
| Conducting phthalonitrile | 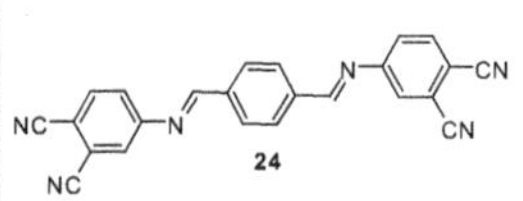 24 | Walton et al. (1985) |
| Conducting phthalonitrile | 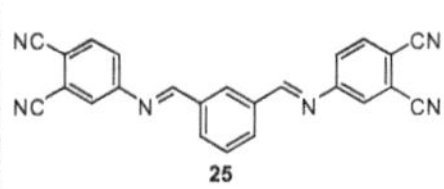 25 | Walton et al. (1985) |
| Resveratrol | 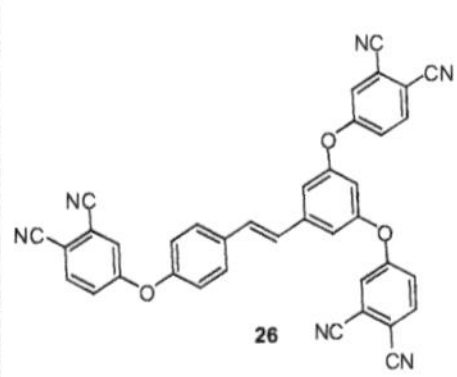 26 | Laskoski et al. (2016a) |
| Dihydroresveratrol | 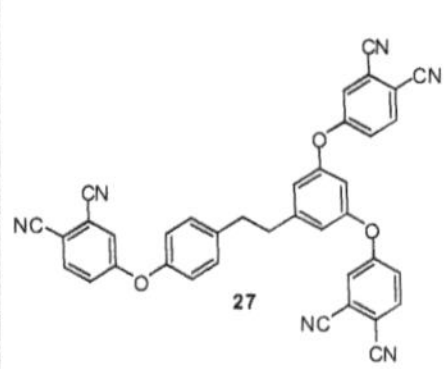 27 | Laskoski et al. (2016a) |

### 2.3.4 *Self-Curing PN Resins*

Table 2.5 lists these types of resins. Polymerization occurs via cyano groups by an additional cure mechanism. The polymerization of the neat resin is generally sluggish and requires several days at elevated temperatures, hence curing additives having free amino or hydroxyl groups are required to be added, which is one of the drawbacks of these resins. Scientists have therefore looked for synthesis of self-curing PN resins without use of any external curing initiators. They have incorporated free amino, phenolic hydroxyl and active hydrogen groups into monomers, which can in-situ initiate the curing reaction without the addition of an external curing agent.

### 2.3.5 *Oligomeric/Polymeric PN Resins*

Table 2.6 lists these types of resins. Monomeric PN resins face some processing difficulties like high melting points and curing temperatures, and lack of natural tack. To overcome these problems, oligomeric PN resins have been synthesized and assessed for their properties. PN functionality has been introduced in polyethers, polyether sulphone, polyether sulphone ketone, polyether ether ketone, polyether ketones, novolac, polyamides, polyether nitrile, polyimides, aromatic ether phosphine oxides, polyether triazine, poly (phthalazinone ether nitrile) and phenolic resin, in order to improve the processing parameters and enhance the thermal stability.

These resins are not generally isolated from solution form and are used *as such* for composite fabrication. This has an additional advantage because in solution form they have natural tack, which is essential for stacking layers of fabric.

### 2.3.6 *Mixed Curing Functionality-Based PN Resins*

Table 2.7 lists these types of resins. Although introduction of oligomeric PN resins improves some processing parameters like melting temperatures, the high curing temperatures remain an issue. To overcome this problem, researchers have introduced certain low-curing-temperature functionalities like phenolic, cyanate ester, propargyl ether, alkyne, vinyl, allyl and benoxazine groups into PN-based monomers, followed by evaluation of their properties. Of these possibilities, benzoxazine-based monomers have been most studied because there is zero volume shrinkage upon curing.

**Table 2.5** Self-curing types of PN resins

| Name/Type | Structure | References |
|---|---|---|
| Free amino group | 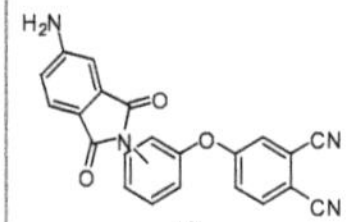 28 | Amir et al. (2010) |
| Free amino group | 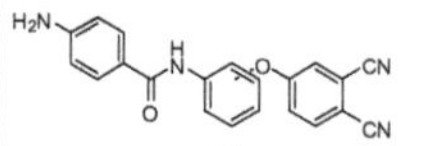 29 | Badshah et al. (2013) |
| Free amino group | 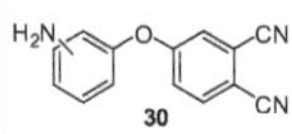 30 | Boyle et al. (1995); Sheng et al. (2016); Guo et al. (2012a) |
| Free amino group | 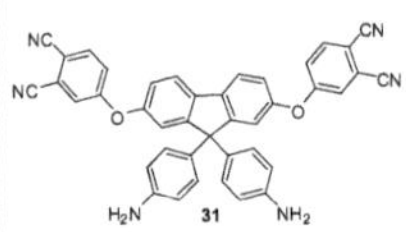 31 | Zhang et al. (2015) |
| Free amino group | 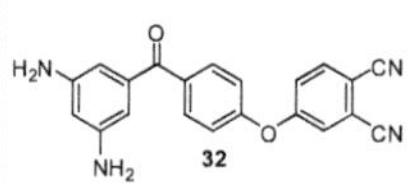 32 | Zhou et al. (2009) |
| Free amino group | 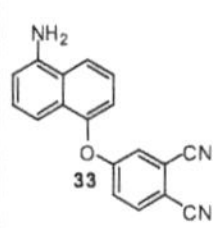 33 | Zhao et al. (2015b) |
| Phenolic hydroxyl group | 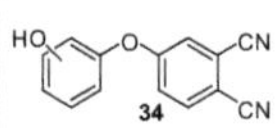 34 | Wang et al. (2015) |
| Phenolic hydroxyl group | 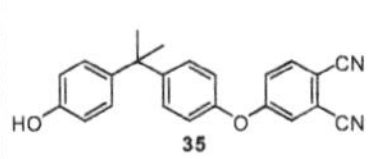 35 | Zeng et al. (2009a) |
| Phenolic hydroxyl group | 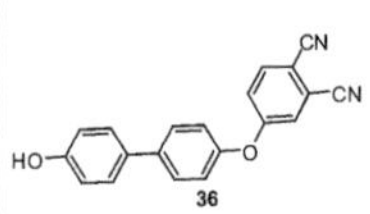 36 | Zeng et al. (2009a) |
| Active hydrogen group | 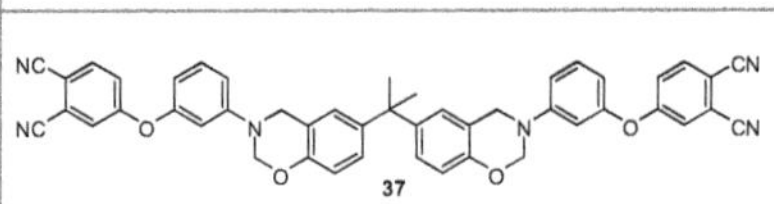 37 | Zuo and Liu (2010) |
| Active hydrogen group | 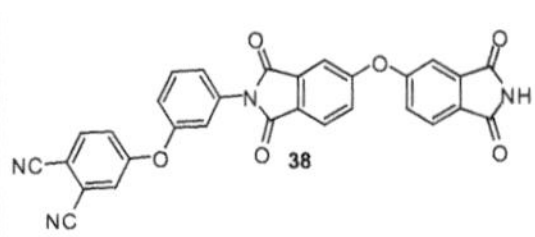 38 | Hu et al. (2015) |

**Table 2.6** Oligomeric/polymeric types of PN resins

| Name/Type | Structure | References |
|---|---|---|
| Polyether | 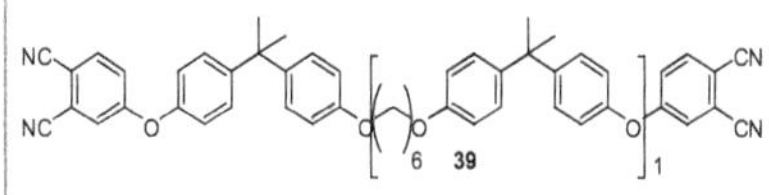 39 | Laskoski et al. (2015a) |
| Polyether | 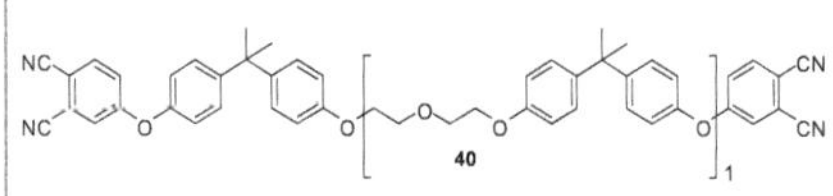 40 | Laskoski et al. (2015a) |
| Polyether | 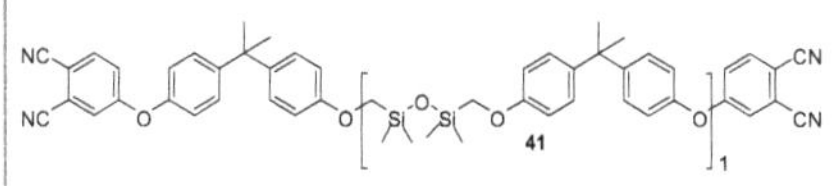 41 | Laskoski et al. (2015a) |
| Polyether | 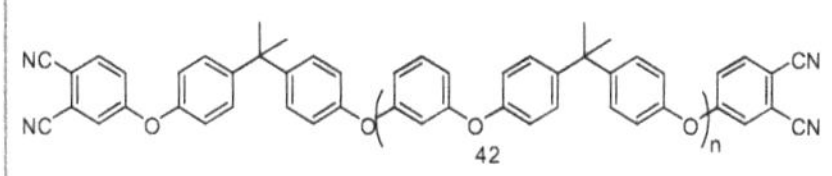 42 | Laskoski et al. (2015b) |
| Polyether | 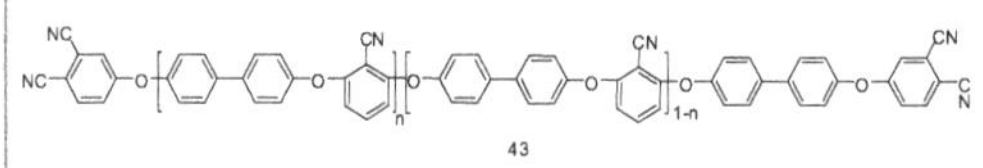 43 | Zou et al. (2012) |
| Polyether sulphone | 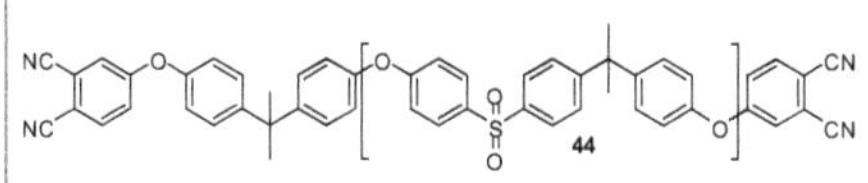 44 | Laskoski et al. (2015c) |
| Polyether sulphone | 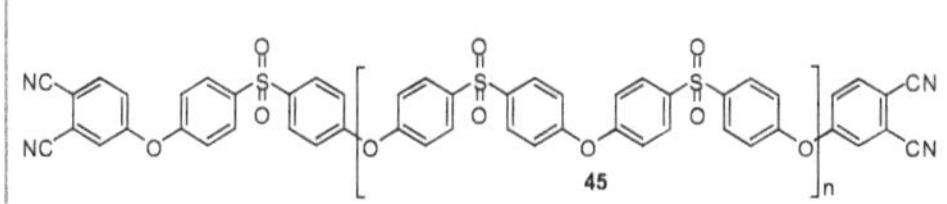 45 | Laskoski et al. (2016b) |
| Polyether sulphone ketone | 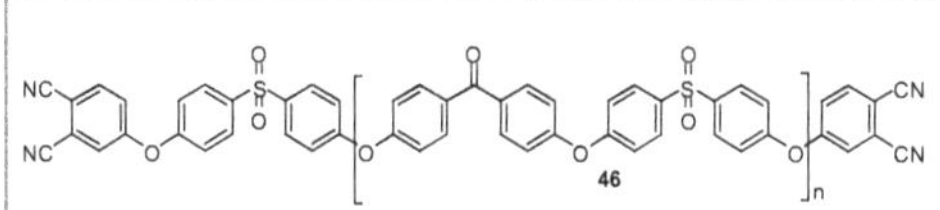 46 | Laskoski et al. (2016b) |
| Polyether ether ketone | 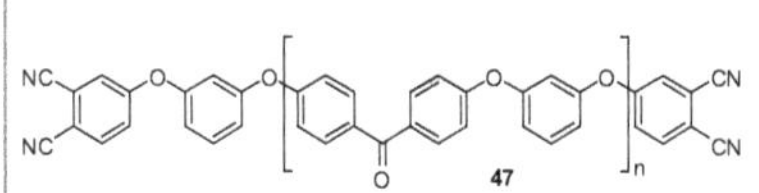 47 | Liu et al. (2015) |
| Polyether ketone | 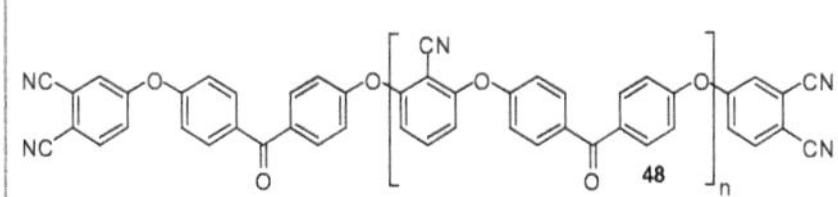 48 | Liu et al. (2014) |
| Polyether ketone | 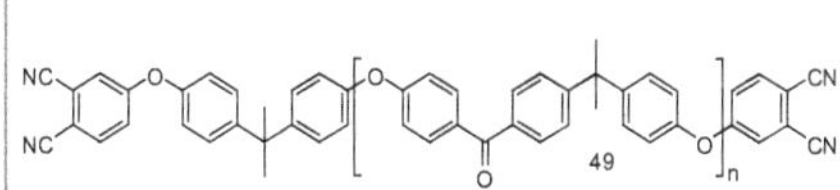 49 | Laskoski et al. (2005) |

(continued)

**Table 2.6** (continued)

| Name/Type | Structure | References |
|---|---|---|
| Novolac | 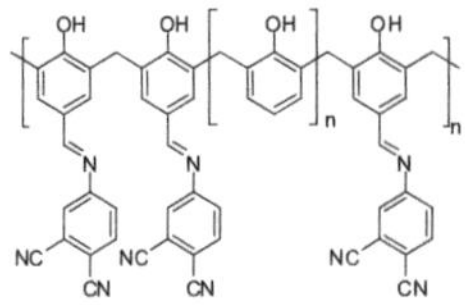 50 | Yang et al. (2007) |
| Polyamide | 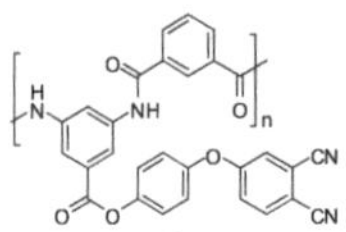 51 | Zeng et al. (2009b) |
| Polyamide | 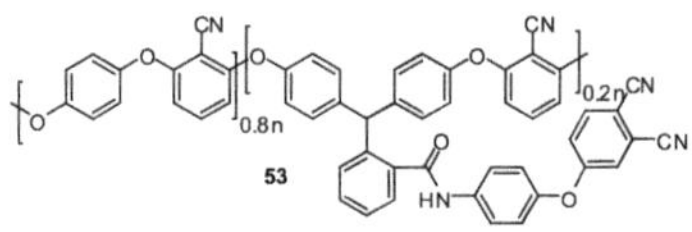 53 | Yang et al. (2013) |
| Polyether nitrile | 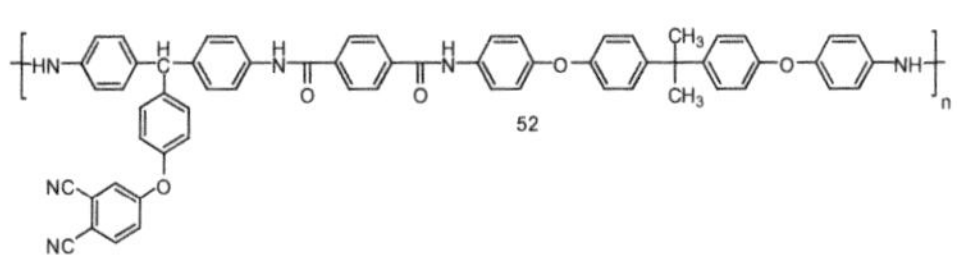 52 | Carja et al. (2012) |
| Polyimide | 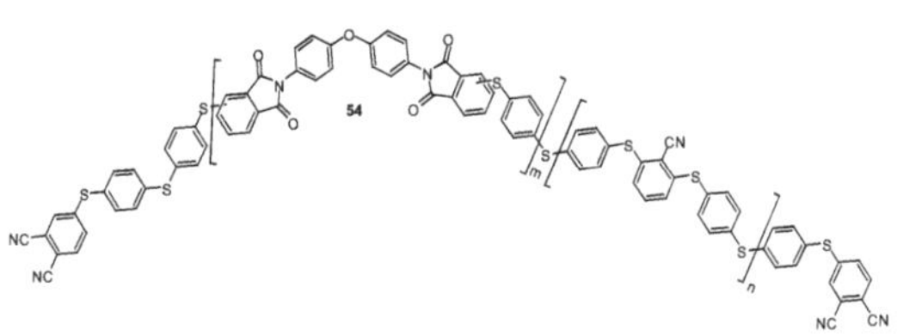 54 | Mushtaq et al. (2016) |
| Polyimide | 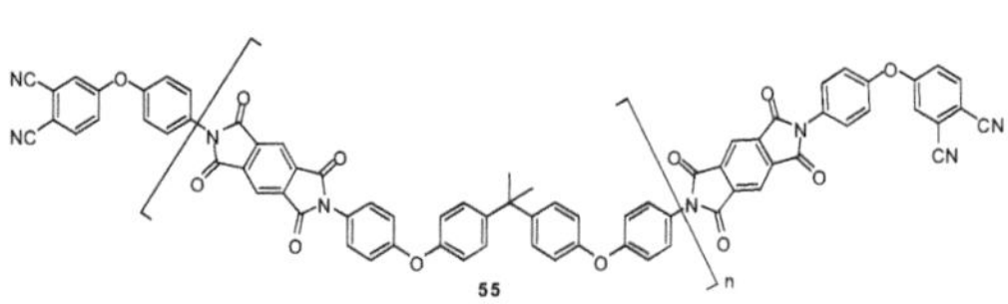 55 | Selvakumar and Sarojadevi (2009) |
| Polyimide | 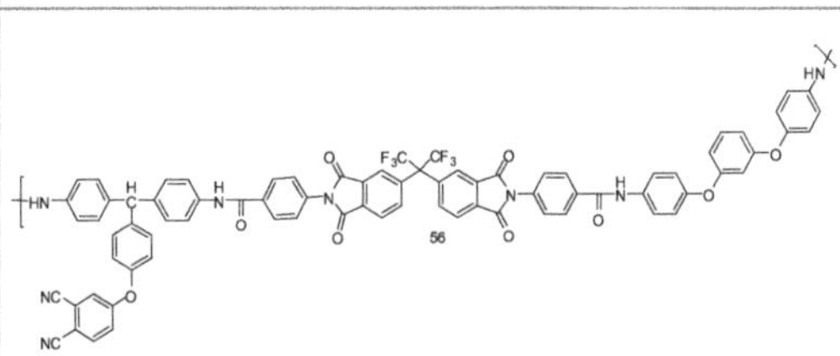 56 | Hamciuc et al. (2013) |

(continued)

**Table 2.6** (continued)

| Name/Type | Structure | References |
|---|---|---|
| Polyimide | 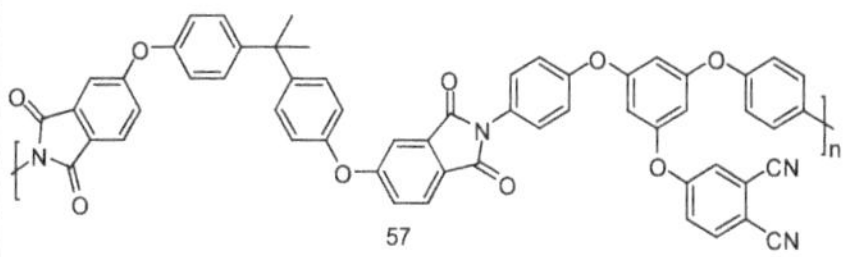 57 | Hu et al. (2014a) |
| Aromatic ether phosphine oxide | 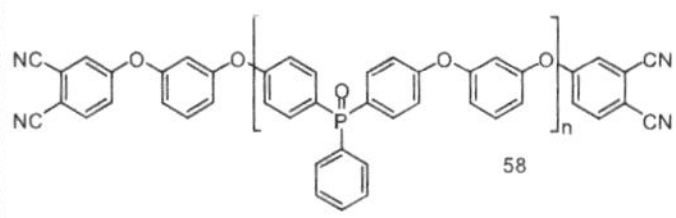 58 | Laskoski et al. (2007a) |
| Aromatic ether phosphine oxide | 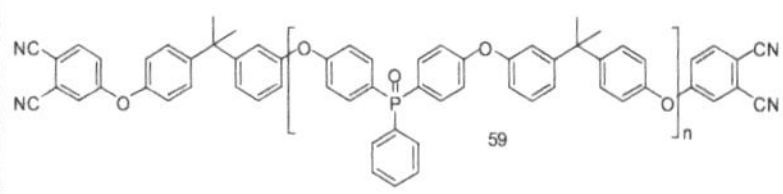 59 | Laskoski et al. (2007a) |
| Polyether triazine | 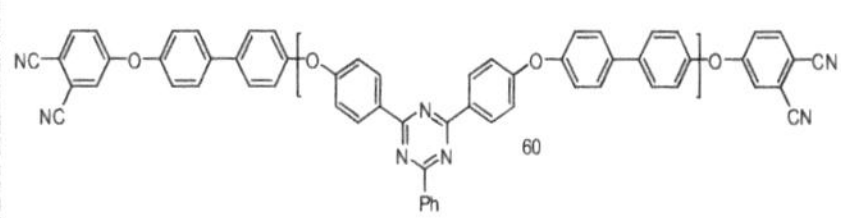 60 | Zong et al. (2015) |
| Poly (phthalazinone ether nitrile) | 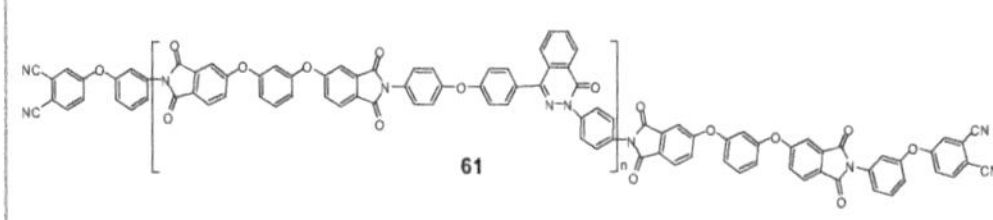 61 | Weng et al. (2017) |
| Phenolic resin | 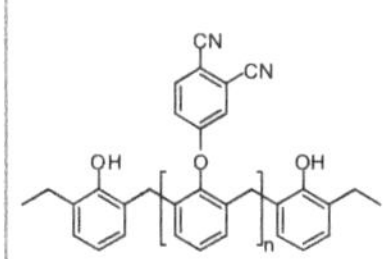 62 | Augustine et al. (2013) |

## 2.4 Semi-Interpenetrating Networks (S-IPNs) and Interpenetrating Networks (IPNs)

Besides basic research on synthesis, scientists have also focused on using commercially available polymers to modify PN properties by blend formation or preparation of s-IPNs and IPNs. The chief objective of these formulations was to bring down the curing temperatures of pre-polymers or to improve the processing parameters as well as the thermal and mechanical properties. Table 2.8 summarizes some significant investigations.

**Table 2.7** Mixed curing functionality-based types of PN resins

| Name/Type | Structure | References |
|---|---|---|
| Phenolic resin group | 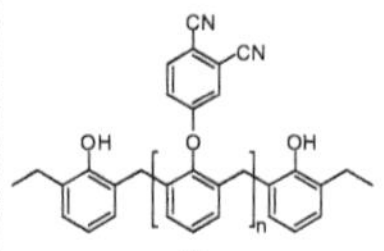 62 | Augustine et al. (2013) |
| Cyanate ester group | 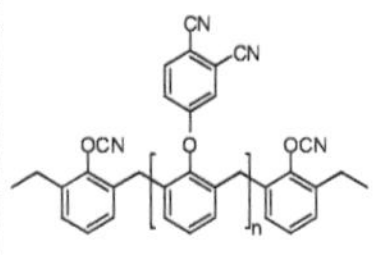 63 | Augustine et al. (2015a) |
| Propargyl ether group | 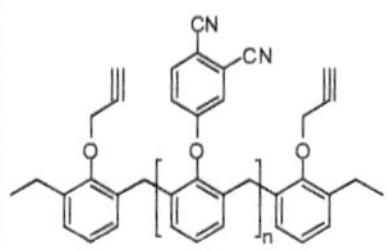 64 | Augustine et al. (2015b) |
| Alkyne group | 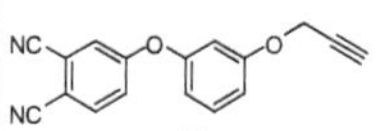 65 | Bulgakov et al. (2017) |
| Alkyne group | 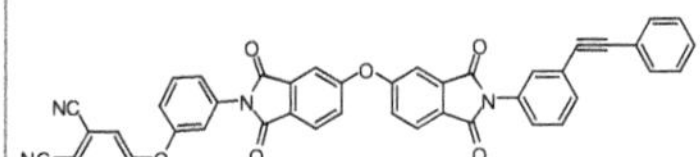 66 | Yuan et al. (2015) |
| Vinyl group | 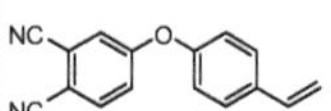 68 | Sumner et al. (2004) |
| Allyl goup | 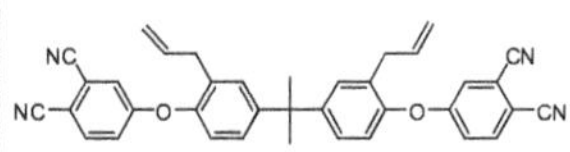 69 | Zou et al. (2014) |
| Alkyne group | 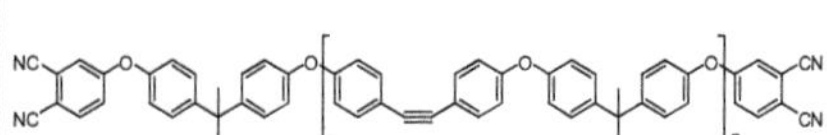 67 | Laskoski et al. (2013) |
| Benzoxazine group | 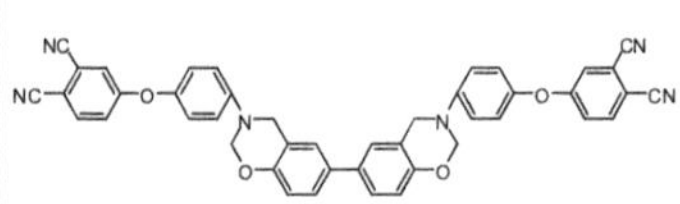 70 | Zuo and Liu (2010) |
| Benzoxazine group | 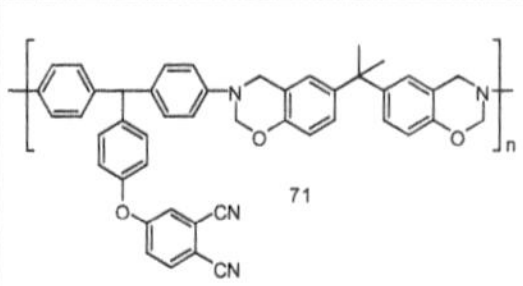 71 | Zheng et al. (2016) |

**Table 2.8** Investigations of modifying PN properties

| Name/Type | Properties investigated | References |
|---|---|---|
| PN resin + amino PN resin | Co-curing behaviour | Sheng et al. (2014) |
| Novolac-based PN + epoxy + PEEK | IPN and s-IPN preparation | Augustine et al. (2014) |
| PN + curable propargyl group monomers | IPN evaluation | Augustine et al. (2015c) |
| PN + hydroxy-terminated PEEK | S-IPN evaluated vs hydroxy-terminated PEEK | Augustine et al. (2015d) |
| Alicyclic imide-containing monomer | Comparison of properties and curing vs PN | Yuan et al. (2016) |
| Benzimidazole-based PN monomer | Co-curing behaviour with PN resin | Hu et al. (2014b) |
| PN resins + phenyl nitrile functionalized benzoxazines | Co-curing behaviour | Brunovska and Ishida (1999) |
| PN resins + nitrile functionalized benzoxazines | Co-curing behaviour | Singh et al. (2018) |
| PN resins + oligomeric PN resins | Co-curing | Dominguez and Keller (2007) |
| PN resin–epoxy resin blends | Co-curing | Dominguez and Keller (2008) |
| Novolac + PN blends | Co-curing; thermomechanical and visco-elastic properties | Guo et al. (2012b) |
| Dihydroxy biphenyl-based PN + hydroxyl-terminated poly aryl ether nitrile | Co-curing | Huang et al. (2016a) |
| Bismaleimides + amino-substituted PN resins | Curing profiles | Huang et al. (2016b) |
| Novolac resins; PN resins | Flame retardance | Sumner et al. (2002) |
| PN-terminated poly aryl ether nitrile + phloroglucinol-based tris PN resins | Cross-linking behaviour | Tong et al. (2013) |
| PN-terminated oligomeric biphenylethernitrile + bisphthalonitrile blends | Preparation of blends and thermal properties | Yang et al. (2011a) |
| PN-terminated poly aryl ether nitrile + trifunctional PN | Thermal and processing properties | Yang et al. (2015) |
| PN resin + epoxy resin blends | Halogen-free flame-retardant properties | Zhao et al. (2012) |
| Co-polymers of allyl functional PNs and bismaleimide | Processing ability | Zou et al. (2017) |

## 2.5 Curing Catalysts

Because the curing reaction of cyano groups is very slow and often results in decomposition, various curing additives have been developed, such as organic amines, by Keller (Sastri and Keller 1998) and other groups, Fig. 2.5. These have electron donating functionalities. Similarly, phenols on account of having electron donating groups also act as excellent curing agent (Wang et al. 2015).

Organic acids (Burchill and Keller 1993) and their amine salts (Burchill 1994), transition metal salts (Li et al. 2009), phosphazenes (Keller 1994), active methylene compounds (Ji et al. 2016), Lewis acids (Wu et al. 2017), reactive functional groups of graphene oxide (Robert et al. 2015), benzoxazines (Singh et al. 2018) and ionic liquids (Cheng et al. 2017) have been added to catalyze the curing reaction via a nucleophilic substitution reaction. Scientists have also carried out co-curing of amino- and hydroxyl-functionalized PN monomer resins with bis-phthalonitrile resins (Yang and Liu 2010).

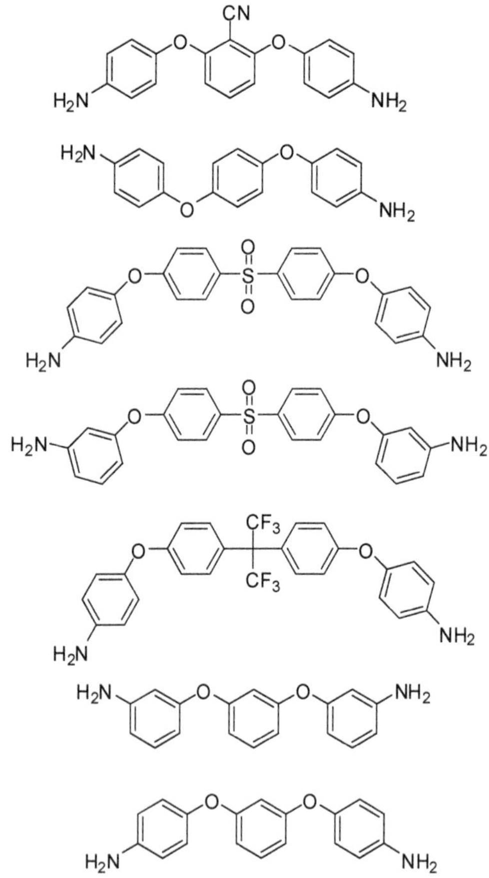


**Fig. 2.5** Some organic diamines used for curing phthalonitrile resins. (Yang et al. 2011b)

## 2.6 Composites and Nanocomposites

In order to improve the functional properties of PN resin systems and their use for structural applications, various macro to nanocomposites have been fabricated using micro to nanofillers and fibres to fabrics. The formulations with respect to resin system and filler along with their corresponding properties have been summarized in Table 2.9.

## 2.7 DMSRDE Contributions to PN Resin Developments

Based on a thorough literature study on high-temperature resins, the DMSRDE has contributed towards the synthesis of various PN resin systems by changing the functionality/pendant groups namely biphenyl, adamantane and phosphate. These PN resin systems have been combined with carbon fibre fabric to produce composites, and their properties have been evaluated and compared with those of some currently used commercial composite systems, i.e. carbon fibre/epoxy and glass fibre/epoxy.

### *2.7.1 Dihydroxy Biphenyl-Based Phthalonitrile Resin (BPN)*

Singh et al. at DMSRDE (Singh et al. 2018) have also synthesized and evaluated Limiting Oxygen Index (LOI) values and isothermal ageing of a dihydroxy biphenyl-based phthalonitrile resin (BPN). The following properties were determined:

- LOI values from 45.1 to 46.5 when cured using different catalysts.
- Isothermal ageing of cured samples showed only a 1.5% mass loss after 300 h at 300 °C.
- Water absorption of 1.5–2% in 24 h.

They have also evaluated the comparative effect of indigenously synthesized curing additives like mono amino substituted phthalonitrile (APN) and benzoxazine (NBZ) with that of a commercial curing additive (m-DDS), see Fig. 2.6. The results showed that the benzoxazine-based curing additive provided a better processing and curing window. Other results are shown in Figs. 2.7, 2.8, 2.9, 2.10, 2.11 and 2.12:

(1) The cured resins did not show a Tg up to 400 °C, see Fig. 2.7.
(2) The DMTA data showed that there was no change in modulus up to 400 °C when samples were tested from 50 to 400 °C at a frequency of 10 Hz, see Figs. 2.8 and 2.9.
(3) The dielectric constants and loss factor of the cured resins have been reported for the first time (Figs. 2.10 and 2.11).
(4) The curing mechanism using a benzoxazine-based co-curing initiator has been established, see Fig. 2.12. This shows that free hydroxyl groups formed during

**Table 2.9** Investigations of PN resin composites

| Name/Type | Properties investigated | References |
|---|---|---|
| Bisphenol A-based PN resin + silicon nitride nanoparticles | Thermal and thermo-mechanical | Derradji et al. (2018) |
| Self-curing + Bisphenol A-based PN resin + silicon carbide microparticles | Mechanical and microhardness | Derradji et al. (2018) |
| Bisphenol A-based PN resin + silane surface modified $Al_2O_3$ nanoparticles | Mechanical and thermal property enhancement effect | Derradji et al. (2018) |
| Bisphenol A-based PN resin + silane surface modified titania nanoparticles | Mechanical and corrosion resistance behaviour | Derradji et al. (2018) |
| Oligomeric PN resin + silane modified $ZrO_2$ nanoparticles | Thermal and mechanical | Derradji et al. (2018) |
| PN + boron nitride nanoparticles | Thermal conductivity | Derradji et al. (2018) |
| PN resins + silane modified ZnO nanoparticles | UV shielding applications | Derradji et al. (2018) |
| PN resin + titanium aluminium carbide | Thermo-mechanical, conductance and toughness | Derradji et al. (2018) |
| PN resins + chemically bonded iron carbonyl | Magnetic | Jia et al. (2011) |
| PN resin-epoxy resin blend + graphene oxide | Co-curing | Kaliavaradhan et al. (2016) |
| IPN of PN resins + graphene nanoplatelets | Thermal and mechanical | Lei et al. (2013) |
| Polyaryl ether ketone PN resin + modified and unmodified montmorillonite nanoclay composites | Viscoelastic | Liu et al. (2016) |
| PEN-Ph based PN resin + $Fe_3O_4$ microspheres prepared via autoclave method | X and Ku microwave band absorption | Tong et al. (2014) |
| 4-nonyl phenol-based PN resin + $Fe_3O_4$ microspheres | Micro-wave absorption | Xu et al. (2014) |
| PN resin + unidirectional carbon fibre | Curing cycle, mechanical and rheological | Sastri et al. (1996) |
| PN resin + glass fibre | metallographic and flammability | Sastri et al. (1997) |
| PN resin + ortho diallyl bisphenol A + E glass fabric | Thermal and flexural | Luo et al. (2016) |
| Benzoxazine functionalized PN resin + glass fibre | Morphological, thermal and mechanical | Xu et al. (2013a; Xu et al. (2013b) |
| PEN + epoxy + PN resin copolymer + glass fibre | Processing and mechanical | Zhao et al. (2013) |

(continued)

**Table 2.9** (continued)

| Name/Type | Properties investigated | References |
|---|---|---|
| Bisphenol A-based PN resin + glass fibre | Tensile and flexural | Chen et al. (2013) |
| Silicon containing PN resin + fibres | Mechanical and visco-elastic | Bulgakov et al. (2016b) |

the initial curing of benzoxazine act as initiators for the cross-linking reaction of the PN resin.

### 2.7.2 *Other Investigations*

The authors have also synthesized various soluble and low melting temperature phosphorus derivatives of the PN resins for fabrication of composites having excellent processing and curing windows and have also assessed their properties. An Indian patent has been filed and will soon be published.

Studies with respect to the self-curing behaviour of amino-, hydroxyl- and carboxylic acid-functionalized PN resins have also been made with regard to using them as curing agents. The curing kinetics of certain PN resins have been investigated for understanding the curing profiles and to be able to modify the actual curing conditions.

We have also prepared semi-interpenetrating networks of PN resins with engineering thermoplastics in various ratios. These have been evaluated for optimum compositions, and the results will be published soon.

## 2.8 Applications

Research results at the NRL have shown that glass- and carbon fabric-based composites using PN resins can be employed for submarines, ship hulls and even for fighter aircraft and helicopter bodies as well as their interiors: see Fig. 2.12 for some examples.

PN resins have many potential defence applications, and scientists are trying to replace multiple resin formulations used in missiles with single resin systems to avoid several processing parameters required for structural and ablative needs. There are numerous other related fields in which these resins can be used, such as radomes used under high temperature conditions, since these resins have low dielectric constants.

On account of their low water absorption property owing to their highly aromatic skeletons, PN resin composites are candidates for marine applications, especially propellers which require low moisture uptake and acoustic properties. These resins can also replace phenolics and PMR resins in the fabrication of engine cages where

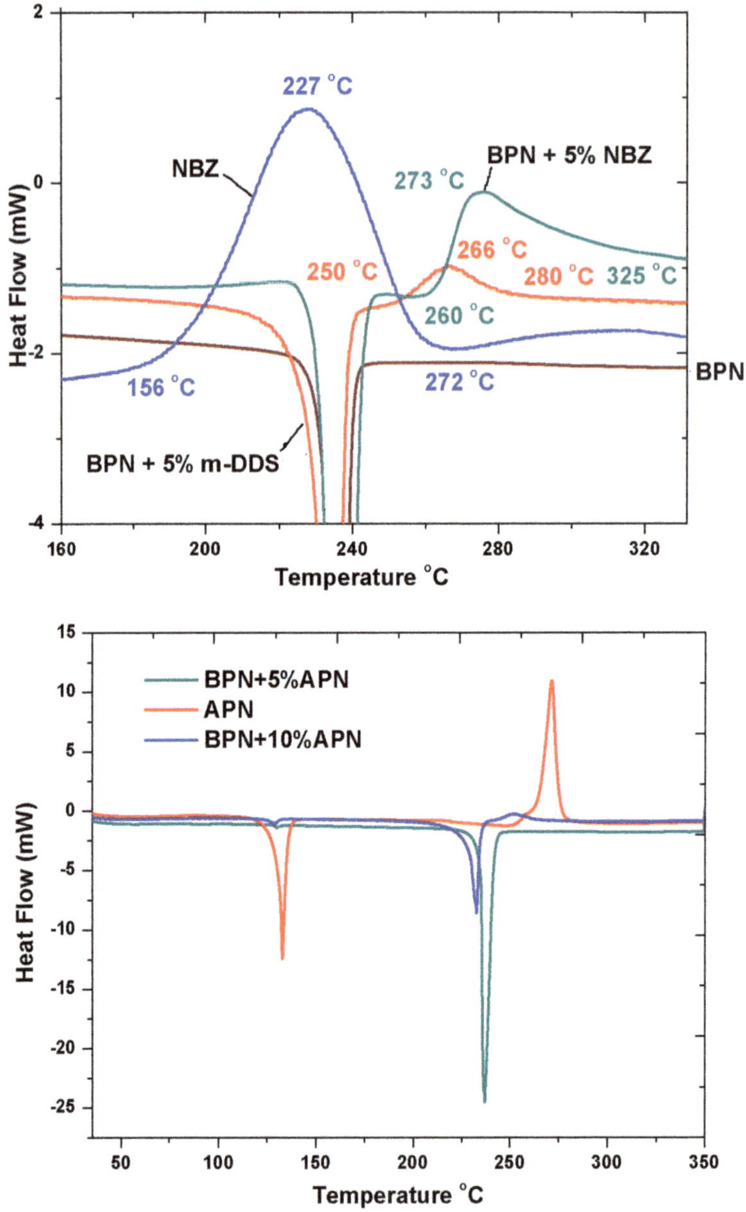


**Fig. 2.6** DSC analysis graphs for uncured samples of BPN with m-DDS, NBZ and APN

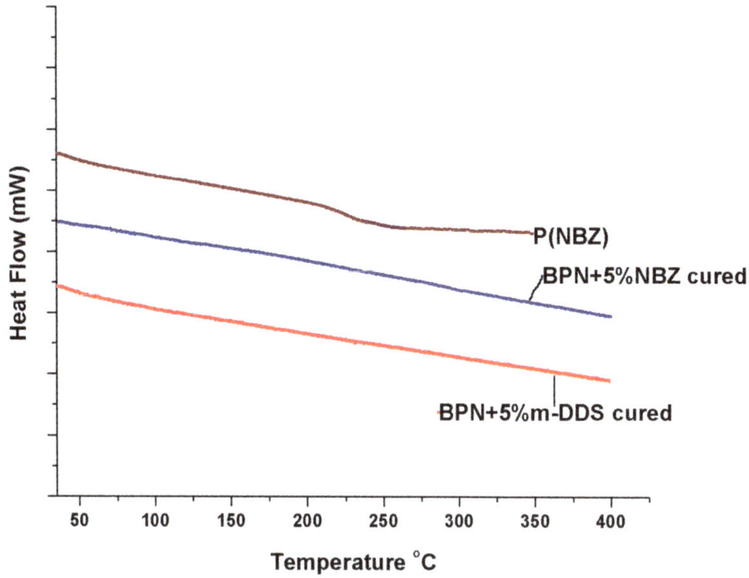


**Fig. 2.7** DSC analysis graph of cured samples of BPN with m-DDS and NBZ

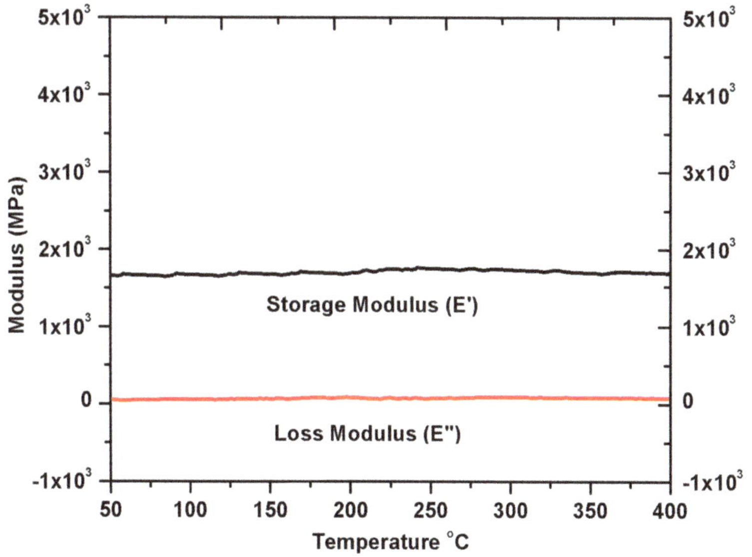


**Fig. 2.8** DMTA Analysis of cured sample of BPN + 5% m-DDS

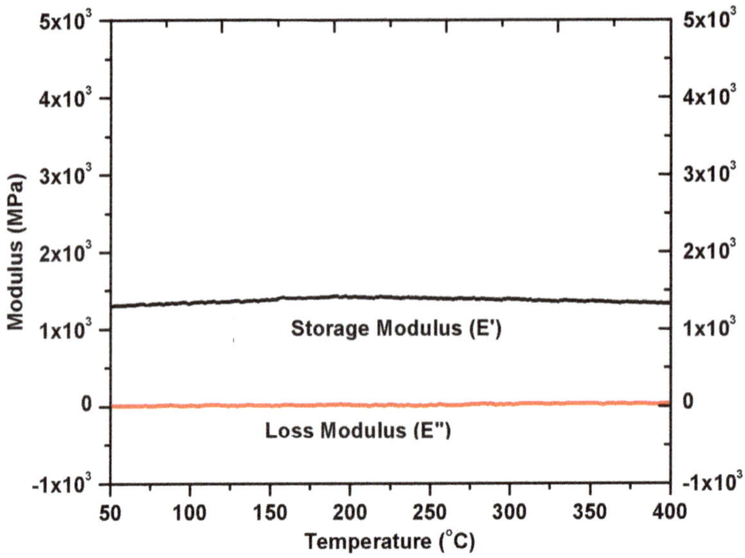


**Fig. 2.9** DMTA analysis of cured sample of BPN + 5% NBZ

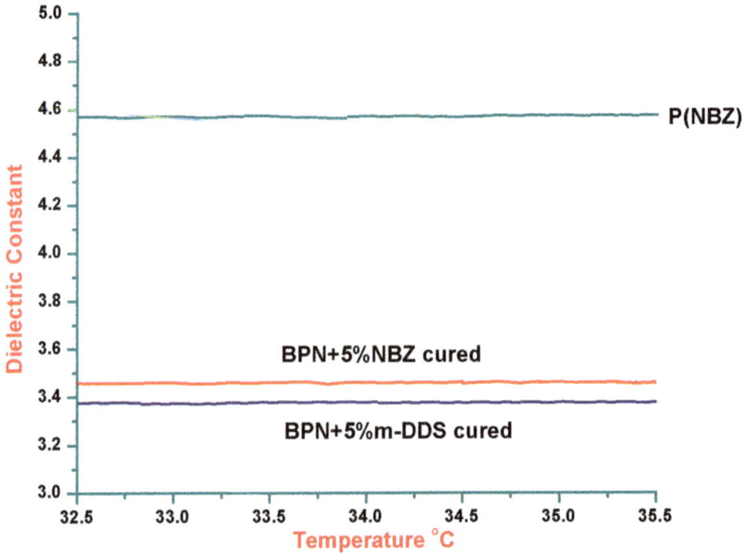


**Fig. 2.10** Dielectric analysis of cured samples

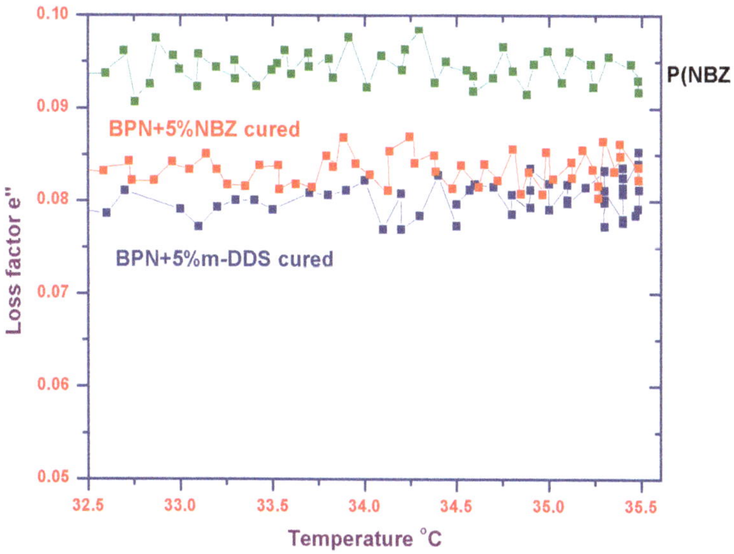


**Fig. 2.11** Loss factor analysis of cured samples

temperatures are very high. The US Navy is also testing PN resins for replacing vinyl esters in the development of fire-resistant composite helicopter hangers.

Other applications for PN resins are:

- Synthesis of CNTs and conducting carbonaceous material under specific conditions (Laskoski et al. 2007b).
- Conducting polymers have also been synthesized using PN matrix systems (Walton et al. 1985).
- Magnetic composites prepared by chemically bonding iron carbonyl with PN resins (Jia et al. 2011).
- Thermally conducting composites using boron nitride and corrosion-resistant coatings using silane-modified titanium oxide particles (Derradji et al. 2018).
- Matrix resins for various other formulations (Fig. 2.13).

## 2.9 Additional Remarks

The main challenges associated with PN resins include high costs; high melting points; poor processing parameters like narrow process and curing windows; high temperature curing with staggered post-curing cycles; brittleness and low elongation to fracture.

Another disadvantage is the use of non-environmentally friendly solvents like DMSO, DMF, NMP and DMAC during manufacturing. Some researchers have

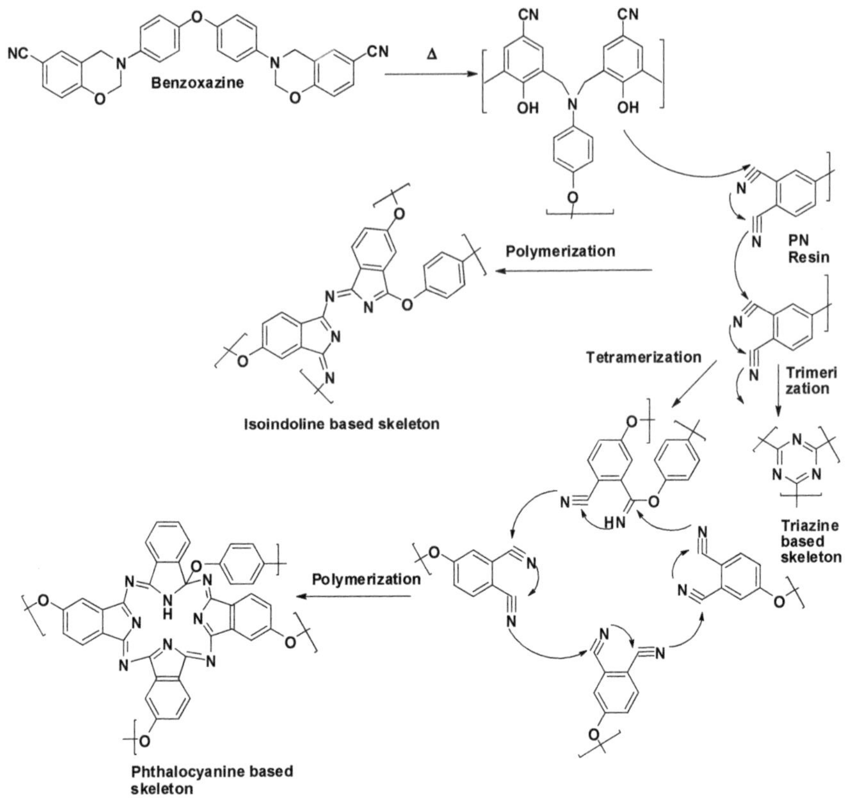


**Fig. 2.12** Most probable curing mechanism for PN resin with benzoxazine (NBZ)

started studying waste management of byproducts of PN resin manufacturing, since large amounts of water are required during production.

The DMSRDE has also tried to fabricate dihydroxy biphenyl functionalized PN resin and carbon fibre-based composites using solvent hand layups and filament winding. We found that such fabrication is extremely difficult on account of (i) poor solubility and (ii) the wettability of the fabric with resin because of its solid nature and lack of natural tack. The high post-curing temperatures are also a major issue that needs to be addressed. The DMSRDE research efforts are continuing for modification of the resin formulation and process, and carbon fabric-based composites are undergoing testing for optimization of the processing parameters.

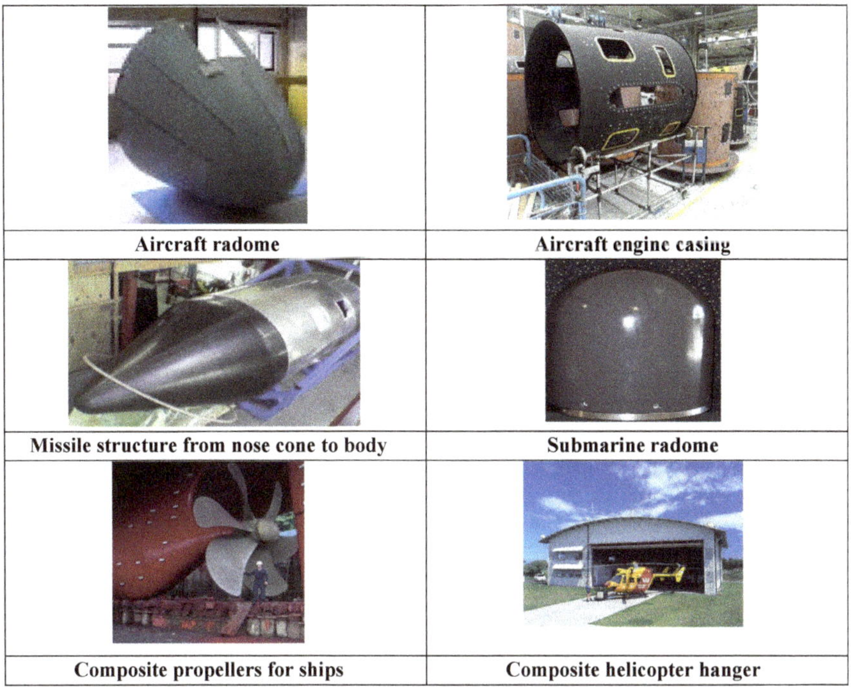


**Fig. 2.13** Some possible applications using PN resins

## 2.10 Conclusions

The present chapter includes and considers the following aspects of high-temperature resins:

- Definition and qualifying criteria for high-temperature resins: materials that can retain mechanical properties for thousands of hours at 230 °C, hundreds of hours at 300 °C, minutes at 540 °C and seconds up to 760 °C. Also, these materials must retain useful properties under particular conditions like pressure or vacuum; mechanical loading, radiation, chemical or electrical exposure, all at temperatures ranging from cryogenic to above 500 °C.

On this basis, only polyimides, bismaleimides, cyanate esters and phthalonitrile (PN) resins qualify.

- Historical background, synthetic development, modifications, applications with respect to the defence and aerospace industries along with current manufacturing challenges.
- Emphasis is given to the relatively new PN resin systems. This includes a summary description of the work done on these resin systems at the DMSRDE in the last

few decades, with the aim of developing high-temperature resins for defence requirements.

**Acknowledgements** The authors are thankful to Dr. D.K. Setua and Dr. SV Kamat (DG, NS&M of DRDO) for their kind support and encouragement, respectively. They feel indebted to Dr. R.J.H. Wanhill for a large number of technical corrections as well as for an almost redo of the manuscript to improve the readability and presentation. They also thank DRDO for funding and facilities.

## References

Achar, B., Fohlen, G., Parker, J.: Synthesis and characterization of the novel type of polymerizable bisphthalonitrile monomers. J. Polym. Sci., Part A: Polym. Chem. **24**(8), 1997–2010 (1986)

Amir, B., Zhou, H., Liu, F., Aurangzeb, H.: Synthesis and characterization of self-catalyzed imide-containing pthalonitrile resins. J. Polym. Sci., Part A: Polym. Chem. **48**(24), 5916–5920 (2010)

Augustine, D., Mathew, D., Nair, C.R.: Phenol-containing phthalonitrile polymers–synthesis, cure characteristics and laminate properties. Polym. Int. **62**(7), 1068–1076 (2013)

Augustine, D., Vijayalakshmi, K.P., Sadhana, R., Mathew, D., Nair, C.P.R.: Hydroxyl terminated PEEK-toughened epoxy–amino novolac phthalonitrile blends–Synthesis, cure studies and adhesive properties. Polymer **55**(23), 6006–6016 (2014)

Augustine, D., Mathew, D., Nair, C.P.R.: Phthalonitrile resin bearing cyanate ester groups: synthesis and characterization. RSC Adv. **5**(111), 91254–91261 (2015a)

Augustine, D., Mathew, D., Nair, C.P.R.: One component propargyl phthalonitrile novolac: synthesis and characterization. Eur. Polymer J. **71**, 389–400 (2015b)

Augustine, D., Mathew, D., Nair, C.P.R.: Mechanistic and kinetic aspects of the curing of phthalonitrile monomers in the presence of propargyl groups. Polymer **60**, 308–317 (2015c)

Augustine, D., Mathew, D., Nair, C.P.R.: End functionalized thermoplastic-toughened phthalonitrile composites: influence on cure reaction and mechanical and thermal properties. Polym. Int. **64**(1), 146–153 (2015d)

Babkin, A.V., Zodbinov, E.B., Bulgakov, B.A., Kepman, A.V., Avdeev, V.V.: Low-melting siloxane-bridged phthalonitriles for heat-resistant matrices. Eur. Polymer J. **66**, 452–457 (2015)

Babkin, A.V., Zodbinov, E.B., Bulgakov, B.A., Kepman, A.V., Avdeev, V.V.: Thermally stable phthalonitrile matrixes containing siloxane fragments. Polym. Sci., Ser. B **58**(3), 298–306 (2016)

Badshah, A., Kessler, M.R., Heng, Z., Hasan, A.: Synthesis and characterization of phthalonitrile resins from ortho-linked aromatic and heterocyclic monomers. Polym. Int. **63**(3), 465–469 (2014)

Badshah, A., Kessler, M. R., Heng, Z., Zaidi, J.H., Hameed, S., Hasan, A.: An efficient approach to prepare ether and amide-based self-catalyzed phthalonitrile resins. Polym. Chem. **4**(12), 3617–3622 (2013)

Boyle, M.E., Adkins, J.D., Snow, A.W., Cozzens, R.F., Brady, R.F., Jr.: Synthesis and characterization of melt-polymerizable aminophthalocyanine monomers. J. Appl. Polym. Sci. **57**(1), 77–85 (1995)

Broughton, W.: Assessing the moisture resistance of adhesives for marine environments. In: Adhesives in Marine Engineering, pp. 155–186 (2012)

Brunovska, Z., Ishida, H.: Thermal study on the copolymers of phthalonitrile and phenylnitrile-functional benzoxazines. J. Appl. Polym. Sci. **73**(14), 2937–2949 (1999)

Bulgakov, B.A., Babkin, A.V., Dzhevakov, P.B., Bogolyubov, A.A., Sulimov, A.V., Kepman, A.V., Chertovich, A.V.: Low-melting phthalonitrile thermosetting monomers with siloxane-and phosphate bridges. Eur. Polym. J. **84**, 205–217 (2016a)

Bulgakov, B.A., Sulimov, A.V., Babkin, A.V., Kepman, A.V., Malakho, A.P., Avdeev, V.V.: Dual-curing thermosetting monomer containing both propargyl ether and phthalonitrile groups. J. Appl. Polym. Sci. **134**(18), 44786–44794 (2017)

Bulgakov, B.A., Babkin, A.V., Bogolyubov, A.A., Afanas'eva, E.S., Kepman, A.V.: Mechanical and physicochemical properties of matrices for fiber reinforced plastics based on low-melting phthalonitrile monomers. Russ. Chem. Bull. **65**(1), 287–290 (2016)

Burchill, P.J.: On the formation and properties of a high-temperature resin from a bisphthalonitrile. J. Polym. Sci., Part A: Polym. Chem. **32**(1), 1–8 (1994)

Burchill, P.J., Keller, T.M.: Curing phthalonitriles with acid and amine. US Patent 5237045 (1993)

Carja, I.D., Hamciuc, C., Hamciuc, E., Vlad-Bubulac, T., Lisa, G.: New highly thermostable aromatic polyamides with pendant phthalonitrile groups. Macromol. Res. **20**(10), 1011–1020 (2012)

Chen, Z., Guo, H., Tang, H., Yang, X., Xu, M., Liu, X.: Preparation and properties of bisphenol A-based bis-phthalonitrile composite laminates. J. Appl. Polym. Sci. **129**(5), 2621–2628 (2013)

Cheng, K., Lv, J.B., Ma, J.Z., Hu, J.H., Chen, C., Zeng, K., Yang, G.: The curing behavior and properties of phthalonitrile resins using ionic liquids as a new class of curing agents. Express Polym. Lett. **11**(11), 924–934 (2017)

Derradji, M., Jun, W. Wenbin, L.: Phthalonitrile Resins and Composites: Properties and Applications. William Andrew (2018)

Dominguez, D.D., Keller, T.M.: Properties of phthalonitrile monomer blends and thermosetting phthalonitrile copolymers. Polymer **48**(1), 91–97 (2007)

Dominguez, D.D., Keller, T.M.: Phthalonitrile-epoxy blends: Cure behavior and copolymer properties. J. Appl. Polym. Sci. **110**(4), 2504–2515 (2008)

Dzhevakov, P.B., Korotkov, R.F., Bulgakov, B.A., Babkin, A.V., Kepman, A.V., Avdeev, V.V.: Synthesis and polymerization of disiloxane Si-O-Si-linked phthalonitrile monomer. Mendeleev Commun. **6**(26), 527–529 (2016)

Guo, H., Chen, Z., Zhang, J., Yang, X., Zhao, R., Liu, X.: Self-promoted curing phthalonitrile with high glass transition temperature for advanced composites. J. Polym. Res. **19**(7), 9918–9926 (2012a)

Guo, H., Lei, Y., Zhao, X., Yang, X., Zhao, R., Liu, X.: Curing behaviors and properties of novolac/bisphthalonitrile blends. J. Appl. Polym. Sci. **125**(1), 649–656 (2012b)

Hamciuc, E., Hamciuc, C., Carja, I.D., Vlad-Bubulac, T., Ignat, M.: Phthalonitrile-containing poly (amide imide)s with nanoactuation properties. Polym. Eng. Sci. **53**(2), 334–342 (2013)

Hu, J., Zhang, J., Zou, Y., Zeng, K., Yang, G.: Significant approaches to promote post-cure reaction of bulky polyimides with pendant phthalonitrile unit. Macromol. Res. **22**(10), 1074–1083 (2014a)

Hu, J., Yuan, P., Zeng, K., Yang, G.: Study of the curing kinetics of a benzimidazole/phthalonitrile resin system. Thermochim. Acta **590**, 30–39 (2014b)

Hu, J., Liu, Y., Jiao, Y., Ji, S., Sun, R., Yuan, P., Yang, G.: Self-promoted phthalimide-containing phthalonitrile resins with sluggish curing process and excellent thermal stability. RSC Adv. **5**(21), 16199–16206 (2015)

Huang, Y., Luo, Y., Xu, M., Lei, Y., Liu, X.: Studied on mechanical, thermal and dielectric properties of BPh/PEN-OH copolymer. Compos. B Eng. **106**, 294–299 (2016a)

Huang, Y., Jiang, M., Xu, M., Zou, X., Luo, Y., Lei, Y., Liu, X.: Curing behavior and processability of BMI/3-APN system for advanced glass fiber composite laminates. J. Appl. Polym. Sci. **133**(27), 43640–43647 (2016b)

Ji, S., Yuan, P., Hu, J., Sun, R., Zeng, K., Yang, G.: A novel curing agent for phthalonitrile monomers: curing behaviors and properties of the polymer network. Polymer **84**, 365–370 (2016)

Jia, K., Xu, M., Zhao, R., Liu, X.: Chemically bonded iron carbonyl for magnetic composites based on phthalonitrile polymers. Polym. Int. **60**(3), 414–421 (2011)

Kaliavaradhan, K., Muthusamy. S.: Synthesis and characterization of various phenylene diamine-based bismaleimide-containing phthalonitrile resins. Polym. Bull. **73**(7), 1921–1938 (2016)

Kaliavaradhan, K., Rukmanikrishnan, B. Muthusamy, S.: Investigation of the effect of graphene oxide on the properties of epoxy-phthalonitrile nanocomposites. Adv. Polym. Technol. 1–12 (2016)

Keller, T.M.: Imide containing phthalonitrile resins. Polymer **34**(5), 952–955 (1993)

Keller, T.M., Dominguez, D.D.: High temperature resorcinol-based phthalonitrile polymer. Polymer **46**(13), 4614–4618 (2005)

Keller, T.M., Griffith, J.R.: The synthesis of highly fluorinated phthalonitrile resins and cure studies. J. Fluorine Chem. **13**(4), 315–324 (1979)

Keller, T.M.: Phosphazene-containing amine as curing agent for phthalonitrile-based polymer. U.S. Patent No. 5,208,318. Washington (1994)

Kessler, M.R.: Cyanate Ester Resins, 2nd edn, pp. 1–15. Wiley Encyclopedia of Composites (2012)

Köç, M., Zorlu, Y., İşci, Ü., Berber, S., Ahsen, V., Dumoulin, F.: A library of dimeric and trimeric phthalonitriles linked by a single aromatic ring: comparative structural and DFT investigations. Cryst. Eng. Comm. **18**(8), 1416–1426 (2016)

Laskoski, M., Dominguez, D.D., Keller, T.M.: Synthesis and properties of a bisphenol A based phthalonitrile resin. J. Polym. Sci., Part A: Polym. Chem. **43**(18), 4136–4143 (2005)

Laskoski, M., Dominguez, D.D., Keller, T.M.: Synthesis and properties of aromatic ether phosphine oxide containing oligomeric phthalonitrile resins with improved oxidative stability. Polymer **48**(21), 6234–6240 (2007a)

Laskoski, M., Keller, T.M., Qadri, S.B.: Direct conversion of highly aromatic phthalonitrile thermosetting resins into carbon nanotube containing solids. Polymer **48**, 7484–7489 (2007b)

Laskoski, M., Dominguez, D.D., Keller, T.M.: Alkyne-containing phthalonitrile resins: Controlling mechanical properties by selective curing. J. Polym. Sci., Part A: Polym. Chem. **51**(22), 4774–4778 (2013)

Laskoski, M., Neal, A., Schear, M.B., Keller, T.M., Ricks-Laskoski, H.L., Saab, A.P.: Oligomeric aliphatic-aromatic ether containing phthalonitrile resins. J. Polym. Sci., Part A: Polym. Chem. **53**(18), 2186–2191 (2015a)

Laskoski, M., Schear, M.B., Neal, A., Dominguez, D.D., Ricks-Laskoski, H.L., Hervey, J., Keller, T.M.: Improved synthesis and properties of aryl ether-based oligomeric phthalonitrile resins and polymers. Polymer **67**, 185–191 (2015b)

Laskoski, M., Keller, T.M., Ricks-Laskoski, H.L., Giller, C.B., Hervey, J.: Improved synthesis of oligomeric sulfone-based phthalonitriles. Macromol. Chem. Phys. **216**(17), 1808–1815 (2015c)

Laskoski, M., Clarke, J.S., Neal, A., Harvey, B.G., Ricks-Laskoski, H.L., Hervey, W.J., Keller, T.M.: Sustainable high-temperature phthalonitrile resins derived from resveratrol and dihydroresveratrol. Chem. Select **1**(13), 3423–3427 (2016a)

Laskoski, M., Clarke, J.S., Neal, A., Ricks-Laskoski, H.L., Hervey, W.J., Keller, T.M.: Synthesis of bisphenol A-free oligomeric phthalonitrile resins with sulfone and sulfone-ketone containing backbones. J. Polym. Sci., Part A: Polym. Chem. **54**(11), 1639–1646 (2016b)

Laskoski, M., Dominguez, D.D., Keller, T.M.: Processable phthalonitrile resins with high-thermal and oxidative stability. In: ACS Symposium Series. Oxford University Press (2006)

Lei, Y., Zhao, R., Yang, X., Liu, X.: Mechanical and thermal properties of graphite nanoplatelets reinforced polyarylene ether nitriles/bisphthalonitrile IPN system. J. Appl. Polym. Sci. **127**(5), 3595–3600 (2013)

Li, W.T., Zuo, F., Jia, K., Liu, X.B.: Study of catalytic effect of ammonium molybdate on the bisphthalonitrile resins curing reaction with aromatic amine. Chin. Chem. Lett. **20**(3), 348–351 (2009)

Liu, T., Yang, Y., Wang, T., Wang, H., Zhang, H., Su, Y., Jiang, Z.: Synthesis and properties of poly (aryl ether ketone)-based phthalonitrile resins. Polym. Eng. Sci. **54**(7), 1695–1703 (2014)

Liu, T., Wang, Y., Su, Y., Yu, H., Zhao, N., Yang, Y., Jiang, Z.: Preparation and properties of film materials of poly (aryl ether ketone)-based phthalonitrile resins. Polym. Eng. Sci. **55**(10), 2313–2321 (2015)

Liu, T., Li, Y., Su, Y., Yu, H., Zhao, N., Yang, Y., Jiang, Z.: Preparation and properties of poly (aryl ether ketone) based phthalonitrile resins/clay nanocomposites. Polym. Compos. **37**(10), 3003–3014 (2016)

Luo, Y., Xu, M., Pan, H., Jia, K., Liu, X.: Effect of ortho-diallyl bisphenol A on the processability of phthalonitrile-based resin and their fiber-reinforced laminates. Polym. Eng. Sci. **56**(2), 150–157 (2016)

Mushtaq, N., Chen, G., Sidra, L.R., Liu, Y., Fang, X.: Synthesis and crosslinking study of isomeric poly (thioether ether imide) s containing pendant nitrile and terminal phthalonitrile groups. Polym. Chem. **7**(48), 7427–7435 (2016)

Nair, C.P.R., Mathew, D., Ninan, K.N.: Cyanate ester resins, recent developments. Adv. Polym. Sci. **155**, 1–99 (2001)

Robert, T.M., Augustine, D., Mathew, D., Nair, C.P.R.: Graphene oxide induced fast curing of amino novolac phthalonitrile. RSC Adv. **5**(2), 1198–1204 (2015)

Sastri, S.B., Keller, T.M.: Phthalonitrile cure reaction with aromatic diamines. J. Polym. Sci., Part A: Polym. Chem. **36**(11), 1885–1890 (1998)

Sastri, S.B., Armistead, J.P., Keller, T.M.: Phthalonitrile-carbon fiber composites. Polym. Compos. **17**(6), 816–822 (1996)

Sastri, S.B., Armistead, J.P., Keller, T.M., Sorathia, U.: Phthalonitrile-glass fabric composites. Polym. Compos. **18**(1), 48–54 (1997)

Selvakumar, P., Sarojadevi, M.: Development of oligomeric phthalonitrile resins for advanced composite applications. Macromol. Symposia **277**(1), 190–200 (2009). Weinheim: WILEY-VCH Verlag.

Sheng, H., Peng, X., Guo, H., Yu, X., Naito, K., Qu, X., Zhang, Q.: Synthesis of high performance bisphthalonitrile resins cured with self-catalyzed 4-aminophenoxy phthalonitrile. Thermochim. Acta **577**, 17–24 (2014)

Sheng, L., Yin, C., Xiao, J.: A novel phthalonitrile monomer with low post cure temperature and short cure time. RSC Adv. **6**(27), 22204–22212 (2016)

Singh, A.S., Shukla, S.K., Pandey, A.K., Tripathi, D.N., Prasad, N.E.: Synthesis and evaluation of catalytic curing behavior of novel nitrile-functionalized benzoxazine for phthalonitrile resins. Polym. Bull. **75**(8), 3781–3800 (2018)

Sumner, M.J., Sankarapandian, M., McGrath, J.E., Riffle, J.S., Sorathia, U.: Flame retardant novolac-bisphthalonitrile structural thermosets. Polymer **43**(19), 5069–5076 (2002)

Sumner, M.J., Weyers, R.Y., Rosario, A.C., Riffle, J.S., Sorathia, U.: Synthesis and characterization of vinyl ester networks containing phthalonitrile moieties. Polymer **45**(15), 5199–5206 (2004)

Ting, R., Keller, T.M., Price, T.R. Porlanski C.F.: Characterization of the cure of diether-linked phthalonitrile resins. In: ACS Symposium Series. Chapter 26, pp. 337–350, A. Chem. Soc. (1982)

Tong, L., Pu, Z., Huang, X., Chen, Z., Yang, X., Liu, X.: Crosslinking behavior of polyarylene ether nitrile terminated with phthalonitrile (PEN-t-Ph)/1, 3, 5-Tri-(3, 4-dicyanophenoxy) benzene (TPh) system and its enhanced thermal stability. J. Appl. Polym. Sci. **130**(2), 1363–1368 (2013)

Tong, L., Pu, Z., Huang, X., Liu, X.: X, Ku-band microwave-absorption properties of polyarylene ether nitriles terminated with phthalonitrile/$Fe_3O_4$ hybrid submicron spheres. Mater. Sci. Eng. B **184**, 98–104 (2014)

Walton, T.R., Griffith, J.R., Reardon, J.P.: Electrically conducting polymers from conjugated bis-phthalonitrile monomers. J. Appl. Polym. Sci. **30**(7), 2921–2939 (1985)

Wang, J., Hu, J., Zeng, K., Yang, G.: Preparation of self-promoted hydroxy-containing phthalonitrile resins by an in situ reaction. RSC Adv. **5**(127), 105038–105046 (2015)

Weng, Z.H., Qi, Y., Zong, L.S., Liu, C., Wang, J.Y., Jian, X.G.: Multiple-$SO_3H$ functioned ionic liquid as efficient curing agent for phthalonitrile-terminated poly (phthalazinone ether nitrile). Chin. Chem. Lett. **28**(5), 1069–1073 (2017)

Wilson, D.: PMR-15 processing, properties and problems- a review. Br. Polym. J. **20**, 405–416 (1988)

Wu, D., Zhao, Y., Zeng, K., Yang, G.: A novel benzimidazole-containing phthalonitrile monomer with unique polymerization behavior. J. Polym. Sci., Part A: Polym. Chem. **50**(23), 4977–4982 (2012)

Wu, Z., Han, J., Li, N., Weng, Z., Wang, J., Jian, X.: Improving the curing process and thermal stability of phthalonitrile resin via novel mixed curing agents. Polym. Int. **66**(6), 876–881 (2017)

Xi, Z., Chen, X., Yu, X., Ma, Y., Ji, P., Naito, K., Zhang, Q.: Synthesis and properties of a novel high temperature pyridine-containing phthalonitrile polymer. J. Polym. Sci., Part A: Polym. Chem. **54**(24), 3819–3825 (2016)

Xu, M., Yang, X., Zhao, R., Liu, X.: Copolymerizing behavior and processability of benzoxazine/epoxy systems and their applications for glass fiber composite laminates. J. Appl. Polym. Sci. **128**(2), 1176–1184 (2013a)

Xu, M., Liu, M., Dong, S., Liu, X.: Design of low temperature self-cured phthalonitrile-based polymers for advanced glass fiber composite laminates. J. Mater. Sci. **48**(23), 8108–8116 (2013b)

Xu, M., Shi, X., Zou, X., Pan, H., Liu, M., Jia, K., Liu, X.: Novel $Fe_3O_4$/phthalonitrile alkyl-containing hybrid microspheres and their microwave absorption application in phthalonitrile composites. J. Magn. Magn. Mater. **371**, 20–28 (2014)

Yagci, Y., Kiskan, B., Ghosh, N.N.: Recent advancement on polybenzoxazine: a newly developed high performance thermoset. J. Polym. Sci. Part A: Polym. Chem. **47**, 5565–5576 (2009)

Yang, X.L., Liu, X.B.: Study on curing reaction of 4-aminophenoxyphthalonitrile/bisphthalonitrile. Chin. Chem. Lett. **21**(6), 743–747 (2010)

Yang, Y., Min, Z., Yi, L.: A novel addition curable novolac bearing phthalonitrile groups: synthesis, characterization and thermal properties. Polym. Bull. **59**(2), 185–194 (2007)

Yang, X., Lei, Y., Zhong, J., Zhao, R., Liu, X.: Preparation and thermal properties of novel phthalonitrile oligomer containing biphenyl ethernitrile/bisphthalonitrile blends. J. Appl. Polym. Sci. **119**(2), 882–887 (2011a)

Yang, X., Zhang, J., Lei, Y., Zhong, J., Liu, X.: Effect of different aromatic amines on the crosslinking behavior and thermal properties of phthalonitrile oligomer containing biphenyl ethernitrile. J. Appl. Polym. Sci. **121**(4), 2331–2337 (2011b)

Yang, J., Yang, X., Zhan, Y., Zou, Y., Zhao, R., Liu, X.: Synthesis and properties of crosslinked poly (arylene ether nitriles) containing pendant phthalonitrile. J. Appl. Polym. Sci. **127**(3), 1676–1682 (2013)

Yang, R., Li, K., Tong, L., Jia, K., Liu, X.: The relationship between processing and performances of polyarylene ether nitriles terminated with phthalonitrile/trifunctional phthalonitrile composites. J. Polym. Res. **22**(11), 210 (2015)

Young, J.G., Onyebuagu, W.: Synthesis and characterization of di-disubstituted phthalocyanines. J. Org. Chem. **55**(7), 2155–2159 (1990)

Yuan, P., Liu, Y., Zeng, K., Yang, G.: Synthesis and characterization of a new imide compound containing phthalonitrile and phenylethynyl end-groups. Des. Monomers Polym. **18**(4), 343–349 (2015)

Yuan, P., Ji, S., Hu, J., Hu, X., Zeng, K., Yang, G.: Systematic study on highly efficient thermal synergistic polymerization effect between alicyclic imide moiety and phthalonitrile: scope, properties and mechanism. Polymer **102**, 266–280 (2016)

Zeng, K., Zhou, K., Zhou, S., Hong, H., Zhou, H., Wang, Y., Yang, G.: Studies on self-promoted cure behaviors of hydroxy-containing phthalonitrile model compounds. Eur. Polymer J. **45**(4), 1328–1335 (2009a)

Zeng, K., Hong, H., Zhou, S., Wu, D., Miao, P., Huang, Z., Yang, G.: A new soluble aramide with pendant phthalonitrile units and polymer property enhancement by nitrile cure reactions. Polymer **50**(21), 5002–5006 (2009b)

Zhang, J., Liu, X., Wen, H., Xie, M., Cai, X.: Investigation of the properties of phthalocyanine resin containing bismaleimide groups. Polym. Int. **42**(4), 363–366 (1997)

Zhang, Z., Li, Z., Zhou, H., Lin, X., Zhao, T., Zhang, M., Xu, C.: Self-catalyzed silicon-containing phthalonitrile resins with low melting point, excellent solubility and thermal stability. J. Appl. Polym. Sci. **131**(20), 40919–40926 (2014)

Zhang, T., Wang, J., Derradji, M., Ramdani, N., Wang, H., Lin, Z.W., Liu, W.B.: Synthesis, curing kinetics and thermal properties of a novel self-promoted fluorene-based bisphthalonitrile monomer. Thermochim. Acta **602**, 22–29 (2015)

Zhao, X., Lei, Y., Zhao, R., Zhong, J., Liu, X.: Preparation and properties of halogen-free flame-retarded phthalonitrile-epoxy blends. J. Appl. Polym. Sci. **123**(6), 3580–3586 (2012)

Zhao, X., Guo, H., Lei, Y., Zhao, R., Zhong, J., Liu, X.: Effect of polyarylene ether nitriles on processing and mechanical behaviors of phthalonitrile-epoxy copolymers and glass fiber laminated composites. J. Appl. Polym. Sci. **127**(6), 4873–4878 (2013)

Zhao, F., Liu, R., Kang, C., Yu, X., Naito, K., Qu, X., Zhang, Q.: A novel high-temperature naphthyl-based phthalonitrile polymer: synthesis and properties. RSC Adv. **4**(16), 8383–8390 (2014)

Zhao, F.H., Liu, R.J., Yu, X.Y., Tang, C.C., Qu, X.W., Zhang, Q.X.: Synthesis of a novel naphthyl-based self-catalyzed phthalonitrile polymer. Chin. Chem. Lett. **26**(6), 727–729 (2015b)

Zhao, F., Liu, R., Yu, X., Naito, K., Qu, X., Zhang, Q.: A high temperature polymer of phthalonitrile-substituted phosphazene with low melting point and good thermal stability. J. Appl. Polym. Sci. **132**(39) (2015)

Zheng, J., Zhang, Y., Wang, Y., Gan, J., Shen, L., Luo, F., Lu, M.: Synthesis and characterization of a lateral phthalonitrile functionalized main-chain polybenzoxazine precursor. Macromol. Res. **24**(5), 409–414 (2016)

Zhou, S., Hong, H., Zeng, K., Miao, P., Zhou, H., Wang, Y., Yang, G.: Synthesis, characterization and self-promoted cure behaviors of a new phthalonitrile derivative 4-(4-(3, 5-diaminobenzoyl) phenoxy) phthalonitrile. Polym. Bull. **62**(5), 581–591 (2009)

Zong, L., Liu, C., Guo, Y., Wang, J., Jian, X.: Thermally stable phthalonitrile resins based on multiple oligo (aryl ether) s with phenyl-s-triazine moieties in backbones. RSC Adv. **5**(94), 77027–77036 (2015)

Zou, Y., Yang, J., Zhan, Y., Yang, X., Zhong, J., Zhao, R., Liu, X.: Effect of curing behaviors on the properties of poly (arylene ether nitrile) end-capped with phthalonitrile. J. Appl. Polym. Sci. **125**(5), 3829–3835 (2012)

Zou, X., Xu, M., Jia, K., Liu, X.: Synthesis, polymerization, and properties of the allyl-functional phthalonitrile. J. Appl. Polym. Sci. **131**(23), 41203–41210 (2014)

Zou, X., Xu, M., Jia, K., Liu, X.: Copolymerizing behavior and processability of allyl-functional bisphthalonitrile/bismaleimide system. Polym. Compos. **38**(8), 1591–1599 (2017)

Zuo, F., Liu, X.: Synthesis and curing behavior of a novel benzoxazine-based bisphthalonitrile monomer. J. Appl. Polym. Sci. **117**(3), 1469–1475 (2010)

# Chapter 3
# 0D, 1D, 2D & 3D Nano Materials: Synthesis and Applications

**Debmalya Roy, Alok Kr. Srivastava, Kingsuk Mukhopadhyay, and Eswara Prasad Namburi**

**Abstract** Grains are the building blocks of many materials. Grains are in turn comprised of many atoms. The visibility of grains to the naked eye depends on their sizes. Conventional material grain sizes fall in the sub-millimetre to centimetres range. On the other hand, average grain sizes of nanomaterials are less than 100 nm. Property improvements such as lower weight or higher strength can be achieved by using nanomaterials. They constitute a bridge between atomic/molecular and bulk systems. New or enhanced size- and shape-dependent properties can be achieved by using nanomaterials, and hence they are of interest for everyday applications, including electronics and medicine, and also for military defence. From an application viewpoint it is also necessary to develop a convenient method for selective functionalization of nanostructures. In this chapter we focus on the concept, importance, properties and synthetic processes of 0D, 1D, 2D and 3D nanostructured materials (NSMs) followed by summaries of their functionalization and applications in Defence.

D. Roy · K. Mukhopadhyay
Defence Materials and Stores Research and Development Establishment (DMSRDE), Defence Research and Development Organisation (DRDO), Kanpur, Uttar Pradesh, India

A. Kr. Srivastava
(On Deputation From DMSRDE, DRDO), National Test House, Min. Consumer Affairs, GoI, Kolkata, West Bengal, India

E. P. Namburi (✉)
Former Outstanding Scientist and Ex-Director, Defence Materials and Stores Research and Development Establishment (DMSRDE), Defence Research and Development Organisation (DRDO), Kanpur, Uttar Pradesh, India
e-mail: neswarap@gmail.com

 
E. P. Namburi et al. (eds.), *Novel Defence Functional and Engineering Materials (NDFEM) Volume 1*, Indian Institute of Metals Series,
https://doi.org/10.1007/978-981-99-9791-6_3

## 3.1 Introduction

Nanoscience and nanotechnology have witnessed huge interest, nationally and internationally, both in fundamental research and for industrial applications over the last couple of decades. Hence the foundation of nanoscience and nanotechnology and low dimensional materials has become one of the most important explored topics across multiple research fields.

Nanostructures occur naturally in nature, for example in meteorites. They have also been produced as smoke particles, from the earliest times of human civilization. However, the scientific story of nanomaterials began relatively recently, with the synthesis of colloidal gold particles by Michael Faraday in 1857. Moving on, nanostructured catalysts have been investigated for over 70 years. By the early 1940s, silica nanoparticles (precipitated and fumed) were being manufactured and sold as substitutes for ultrafine carbon black in rubber reinforcements (Adams 1988; Akerman 2002; Allred et al. 2005; Alwitt 2002; Asoh et al. 2001; Brodie and Muray 1982; Feynman 1959).

Amorphous silica nanoparticles have various applications, e.g. in automobile tyres, optical fibres and catalyst supports. Metallic nanopowders for magnetic recording tapes were developed in the 1960s and 1970s. In 1976 Granqvist and Buhrman (Alwitt 2002; Asoh et al. 2001; Brodie and Muray 1982; Feynman 1959) produced nanocrystals for the first time by the now popular inert-gas evaporation technique.

The expansion of nanophase engineering has seen a rapidly growing number of structural and functional materials, both inorganic and organic. Their development has allowed manipulation of mechanical, catalytic, electrical, magnetic, optical and electronic functions. The production of nanophase, or cluster-assembled, materials is usually based upon creating separated small clusters that are then (i) fused into a bulk-like material or (ii) embedded into compact liquid or solid matrix materials. The first example, nanophase silicon, is different from normal silicon in terms of its physical and electronic properties, and hence could be applied in macroscopic semiconductor processes for creating new devices. Another example is the doping of ordinary glass with quantized semiconductor "'colloids" to produce a high-performance optical medium with applications in optical computing (Alwitt 2002; Asoh et al. 2001; Brodie and Muray 1982; Feynman 1959; Guizard et al. 1992; Herlin-Boime et al. 2004).

## 3.2 Nanomaterials Importance

Much interest has been generated about nanomaterials because of their unusual mechanical, electrical, optical and magnetic properties. Some NSM examples are given below (Feynman 1959; Guizard et al. 1992; Herlin-Boime et al. 2004; Iijima 1991; Hofmann 2009):

(1) Ceramics: Higher elevated temperature ductility compared to coarse-grained ceramics.
(2) Metal powders: For the production of porous coatings, gas-tight materials and dense parts.
(3) Nanostructured semiconductors: These show various non-linear optical properties. Semiconductor Q-particles also show quantum confinement effects which may lead to special properties, e.g. (i) luminescence in silicon powders and (ii) silicon germanium quantum dots as infrared optoelectronic devices. Nanostructured semiconductors are also used as window layers in solar cells.
(4) Metal oxide films: These are of increasing interest for gas sensors (NOx, CO, $CO_2$, $CH_4$ and aromatic hydrocarbons) with enhanced sensitivity and selectivity. Nanostructured manganese oxide ($MnO_2$) is used in rechargeable batteries for cars or consumer goods. Nanocrystalline silicon films find application in highly transparent contacts in thin film solar cells. And titanium oxide porous films result in strong absorption in dye-sensitized solar cells; this is useful in providing high transmission and significant surface area enhancement.
(5) Magnetic nanocomposites: Used for mechanical force transfer (ferrofluids), high density information storage, and magnetic refrigeration.
(6) Catalysts: Metal clusters and colloids of mono- or pluri-metallic composition can be used as precursors for new types of heterogeneous catalysts.
(7) Polymer-based composites: These have high contents of inorganic particles, resulting in high dielectric constants. They are important for optoelectronics.

## 3.3 Fundamental Issues in Nanomaterials (NSMs) (Guizard et al. 1992; Herlin-Boime et al. 2004; Iijima 1991; Hofmann 2009; Pokropivny et al. 2007)

The fundamental issues for NSMs are the abilities to control:

- The scale (size) of the system and the modulation dimensionality.
- Interactions between nano-sized building blocks.
- Defects, concentration gradients and other anomalies to achieve the required composition

In detail, further development of NSMs requires considering the following aspects:

- Synthesis and/or fabrication methods for both the starting materials (powders) and NSMs.
- Property dependences on NSM building block sizes, microstructures and interfaces.
- Mono- and pluri-metallic nanomaterials catalytic applications.
- The transfer of developed technologies into industrial applications, including upscaling the processes.

## 3.4 Nanomaterials Properties (Brodie and Muray 1982; Feynman 1959; Guizard et al. 1992; Herlin-Boime et al. 2004; Iijima 1991; Hofmann 2009; Pokropivny et al. 2007; Tiwari et al. 2012)

The sizes and shapes of NSMs have strong influences on their physical and chemical properties. Also, although NSMs are a kind of 'halfway house' between atomic scale and meso-or-macro scale materials, their properties can be very different. Examples are mechanical strength and electrical conductivity. The differences are mainly due to nanomaterials having high surface energies, large fractions of surface atoms, reduced imperfections, and spatial confinement. Heat-treatment (thermal annealing) enables NSMs to 'self-purify', whereby the impurities and material defects are reduced. The increased material perfection enhances the chemical stability for certain nanomaterials, along with better structural, mechanical, electrical, dielectric, transport and magnetic properties than those of bulk materials.

Other factors are quantum effects, which at the NSM level influence optical/electrical/magnetic material properties; quantum dots, and quantum-well lasers for optoelectronics. The quantum effects in NSMs can be used to modify bulk material properties. For example, metallic nanoparticles and nanowires can be used as active catalysts and chemical sensors to enhance the sensitivity and sensor selectivity. Also, NSMs have different energy band structures and charge carrier densities compared with bulk materials, and hence different electronic and optical properties. For example, quantum dot- and quantum wire-based lasers and light emitting diodes (LED) are very promising innovations. Quantum dot devices also have promise for high density information storage.

Summarising, NSM electrical, optical, structural, chemical, magnetic and catalytic properties are sufficiently different to offer their use for many novel applications. (Brodie and Muray 1982; Feynman 1959; Guizard et al. 1992; Herlin-Boime et al. 2004; Iijima 1991; Hofmann 2009; Pokropivny et al. 2007).

## 3.5 Classifications (Guizard et al. 1992; Herlin-Boime et al. 2004; Hofmann 2009; Pokropivny et al. 2007; Arivalagan et al. 2011; Tiwari et al. 2012)

Materials and/or devices manufactured by means of the controlled manipulation of atomic level microstructures can be considered in three categories:

(1) Materials and/or devices with reduced dimensions and/or dimensionality based on isolated, substrate-supported, or embedded nanometre-sized particles, thin wires or thin films. Chemical vapour deposition (CVD), physical vapour deposition (PVD), various aerosol techniques, and precipitation from the vapour,

supersaturated liquids or solids are most frequently used to produce this type of microstructure.

(2) Materials and/or devices in which the nanometre-sized microstructure is based on a thin (nanometre-sized) surface region of a bulk material. The most widely applied procedures include PVD, CVD and ion implantation.

(3) Bulk solids containing nanometre-scale microstructures. These are solids having the chemical compositions, atomic arrangements, and/or building block sizes (e.g. crystallites or atomic/molecular groups) on a length scale of a few nanometres throughout the bulk.

Besides the above broad categories there is another important way of classifying nanomaterials. A *spatial dimension* reduction, or *particular crystallographic direction confinement* of particles or quasi-particles within a structure, lead to modified physical properties. The resulting NSMs can be classified as follows:

(a) 3D-systems confined in three dimensions, e.g. structures typically composed of consolidated equiaxed crystallites.
(b) 2D-systems confined in two dimensions, e.g. filamentary structures.
(c) 1D-systems confined in one dimension, e.g. layered or laminate structures.
(d) 0D-zero-dimensional structures, e.g. nano-pores and nano-particles.

The hardness difference between diamond and graphite is an outstanding example of the correlation between atomic structure and bulk material properties. Another and very different property change is the colour of cadmium sulphide crystals when their size is reduced to the nanoscale.

We note here that a distinction should be made between NSMs and nanostructures (NSs).

The term NSM refers to form, dimensionality *and* composition. The term NS refers to form and dimensionality only. This term is better used for (a) and (b) above when the compositions are more or less invariant.

## 3.6 "Nanostructure Induced Effects": Influence on Properties

There are many ways of synthesizing nanosized particles and fabricating nanostructured materials. Most products are fine particles (powders) which are amorphous or crystalline, or sometimes a metastable or unexpected phase.

Roughly two kinds of "*nano-structure induced effects*" are observed: (i) the *size effect*, involving different discrete electronic levels and physical properties; (ii) the *surface or interface induced effect*, involving the increased specific surface in particle systems. The surface effect is important for chemical processing, in particular heterogeneous catalysis.

Although these distinctions are necessary and useful, they are very general. Detailed understanding of the interrelationships between properties, production

**Table 3.1** Properties and possible applications of NSMs (Brodie and Muray 1982; Feynman 1959; Guizard et al. 1992; Herlin-Boime et al. 2004; Iijima 1991; Hofmann 2009; Pokropivny et al. 2007; Arivalagan et al. 2011; Tiwari et al. 2012)

| Property | Possible Applications |
|---|---|
| *Surface/Interface* | |
| Large specific surface area | Catalysis, sensors |
| Large surface area, small heat capacity | Heat-exchange materials; combustion catalysts |
| Lower sintering temperature | Sintering accelerators |
| Specific interface area, large boundary area | Nanostructured materials |
| Superplastic behaviour of ceramics | Ductile ceramics |
| Cluster coating and metallization | Special resistors, temperature sensors |
| Multi-shelled particles | Chemical activity of catalysts; optical elements |
| *Bulk* | |
| Single magnetic domain | Materials used for magnetic recording |
| Small mean free path of electrons in a solid | Special conductors for light or heat absorption, scattering |
| High and selective optical absorption of metal particles | Colours, filters, solar absorbers, photovoltaics, photographic material, phototropic material |
| Formation of ultrafine pores due to superfine agglomeration of particles | Molecular filters |
| Uniform mixture of different kinds of superfine particles | New generation materials |
| Grain size too small for stable dislocations | High strength and super-hardened metals |

and characterisation methods is lacking. In particular, directed tailoring of NSMs (Brodie and Muray 1982; Feynman 1959; Guizard et al. 1992; Herlin-Boime et al. 2004; Iijima 1991; Hofmann 2009; Pokropivny et al. 2007; Arivalagan et al. 2011; Tiwari et al. 2012) will be needed for future technical and industrial applications. Suggestions for NSMs uses and technology impacts are given in Tables 3.1 and 3.2.

## 3.7 Properties Survey for Nanomaterials

As stated previously, high relative surface area and quantum effects are two principal factors that distinguish NSMs from bulk materials (Hofmann 2009; Pokropivny et al. 2007; Arivalagan et al. 2011; Tiwari et al. 2012). Also, they are uniquely classified into 0D, 1D, 2D and 3D types, each with its own combination of physical and chemical properties. Broadly speaking, we may consider the properties of NSMs in three main groups:

(1) Mechanical properties**:** High strength and extreme strain resistance owing to the very low numbers of defects.

**Table 3.2** Technology and present/potential impacts of NSMs (Brodie and Muray 1982; Feynman 1959; Guizard et al. 1992; Herlin-Boime et al. 2004; Iijima 1991; Hofmann 2009; Pokropivny et al. 2007; Arivalagan et al. 2011; Tiwari et al. 2012)

| Technology | Present Impact | Potential Impact |
|---|---|---|
| Nanodevices | GMR reading heads | • Terabit memory and microprocessing<br>• Single molecule DNA sizing and sequencing<br>• Biomedical sensors<br>• Low noise, low threshold lasers<br>• High brightness displays involving nanotubes |
| High surface Area | • Molecular sieves<br>• Drug delivery<br>• Tailored catalysts<br>• Absorption/desorption | • Molecule-specific sensors<br>• Particle induced delivery<br>• Energy storage (fuel cells, batteries)<br>• Grätzel-type solar cells, gas sensors |
| Bio-medical aspects | Functionalised nanoparticles | • Cell labelling by fluorescent nanoparticles<br>• Local heating involving magnetic nanoparticles |
| Dispersions and coatings | • Thermal barriers<br>• Optical (visible and UV) barriers<br>• Imaging enhancement<br>• Ink-jet materials<br>• Coated abrasive slurries<br>• Information-recording | • Enhanced thermal barriers<br>• Multifunctional nanocoatings<br>• Fine particle structure<br>• Super-absorbent materials<br>• Higher efficiency and lower contamination<br>• Higher density information storage |
| Consolidated materials | • Low-loss soft magnetic materials<br>• High hardness, tough cutting tools<br>• Nanocomposite cements | • Superplastic forming of ceramic materials<br>• Ultrahigh-strength, tough structural materials<br>• Magnetic refrigerants, nano-filled polymer composites, ductile cements |

(2) Electrical properties: Intermediate between metals and semiconductors, and size-dependent. High conductivity is associated with less defects.

(3) Thermal conductivity: High thermal conductivities, about 10× more than those of metals. This is due to covalent bond vibrations and low numbers of defects.

## 3.8 Synthesis of NSMs (Guizard et al. 1992; Herlin-Boime et al. 2004; Iijima 1991; Hofmann 2009; Pokropivny et al. 2007; Wang 2013; Sarikaya et al. 2003a, 2003b; Ikkala and Brinke 2002; Rao and Nath 2003; Jensen 1989; Kroto et al. 1985; Lakshmi et al. 1997; Yang et al. 2010, 2015; Pang et al. 2015; Luther 2004; Madou 2002; Mao et al. 2004; Mooney and Radding 1982; Oh et al. 2005; Piner et al. 1999; Salata 2004; Sun et al. 2000; Ye et al. 2004; Ensafi et al. 2016; Ghaedi et al. 2016; Lu et al. 2013; Li et al. 2016; Liu et al. 2012; Dai et al. 2016; 27)

### *3.8.1 Top-Down, Bottom-Up and Hybrid Approaches*

There are many techniques for synthesizing and fabricating NSMs with controls on size, shape, dimensionality and structure. The most generally used methods are physical and chemical, and these are divided into *top-down* and *bottom-up* approaches (Guizard et al. 1992; Herlin-Boime et al. 2004; Iijima 1991; Hofmann 2009; Pokropivny et al. 2007; Arivalagan et al. 2011; Tiwari et al. 2012) Sometimes these are combined in a *hybrid* approach. All three are discussed briefly here:

(1) Top-down approach: Traditional workshop or microfabrication methods are often used in this approach, and externally controlled tools are used for cutting, milling and shaping to obtain the desired shape and order. Common examples of top-down processes are attrition and milling for making nanoparticles.
(2) Bottom-up approach: In this approach more complex molecular assemblies can be obtained via atomic/molecular/cluster-wise build-ups. Synthesis of nanoparticles by colloid dispersions is a typical example.

    The bottom-up approach is basically chemical synthesis. It enables the preparation of NSMs with very small sizes that cannot be achieved using the top-down approach. Also, NSMs produced by this method have fewer defects in comparison with the top-down approach, and the materials are close to their equilibrium state.
(3) Hybrid approach: Used in special lithography. For example, bottom-up thin film formation can be followed by top-down etching.

### *3.8.2 Nanoparticle Synthesis*

Commercial or industrial manufacture of nanoparticles consists of:

- Solid phase mechanical processes, e.g. grinding, milling and alloying.
- Sol–gel technique.

**Table 3.3** Typical synthesis methods for nanoparticles, following the top-down and bottom-up approaches (Guizard et al. 1992; Herlin-Boime et al. 2004; Iijima 1991; Hofmann 2009; Pokropivny et al. 2007; Arivalagan et al. 2011; Tiwari et al. 2012; Wang 2013; Sarikaya et al. 2003a, 2003b; Ikkala and Brinke 2002; Rao and Nath 2003; Jensen 1989; Kroto et al. 1985; Lakshmi et al. 1997; Yang et al. 2010, 2015; Pang et al. 2015; Luther 2004; Madou 2002; Mao et al. 2004; Mooney and Radding 1982; Oh et al. 2005; Piner et al. 1999; Salata 2004; Sun et al. 2000; Ye et al. 2004; Ensafi et al. 2016; Ghaedi et al. 2016; Lu et al. 2013; Li et al. 2016; Liu et al. 2012; Dai et al. 2016; 27).

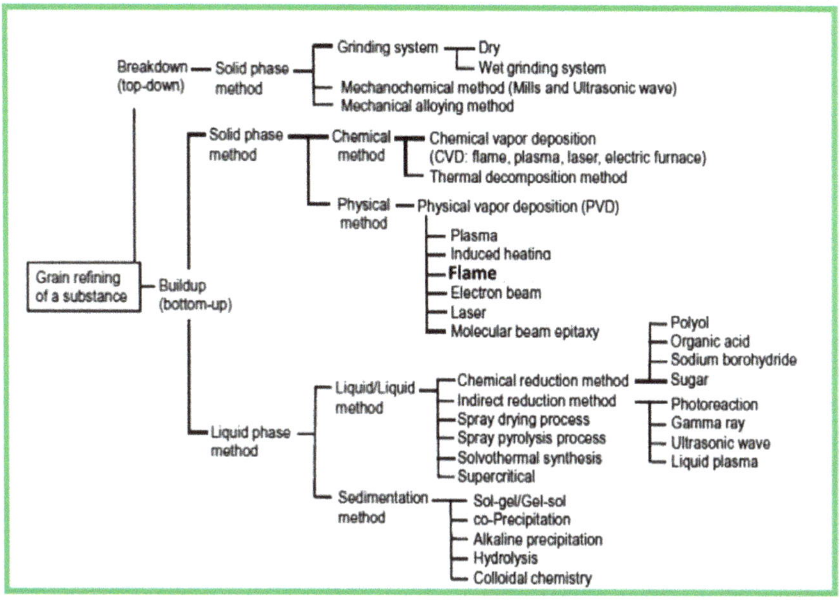


- Liquid phase methods involving chemical reactions in solvents, resulting in precipitates and colloids.
- Gas phase processes, including vapour deposition, plasma synthesis, flame pyrolysis and high temperature evaporation.

However, there is no fixed or standard method to get particular kinds of nanoparticles. Even different methods could be used for the same materials, keeping in mind the required properties (Guizard et al. 1992; Herlin-Boime et al. 2004; Iijima 1991; Hofmann 2009; Pokropivny et al. 2007; Arivalagan et al. 2011; Tiwari et al. 2012). A summary of nanoparticles top-down and bottom-up synthesis methods is given in Table 3.3. It is obvious that there are very many methods. These indirectly give an impression of the interest in, and importance of, nanoparticle R&D.

#### 3.8.2.1 Mechanical Grinding

Grinding is the most widely used example of the 'top-down' approach because of its simplicity, the straightforward equipment and processing, and known capability of being upscaled. However, contamination from milling media and/or the atmosphere, together with difficulties in consolidating the powder product without coarsening the nanocrystalline microstructure, are major disadvantages.

#### 3.8.2.2 Wet Chemical Synthesis of Nanomaterials

This is based either on top-down or bottom-up methods. Single crystal etching in an aqueous solution for producing nano-pored materials is the best-known top-down example. Sol–gel and precipitation techniques are the best examples of bottom-up methods.

#### 3.8.2.3 Gas Phase Synthesis of Nanomaterials

Nanostructures with well-controlled sizes, shapes and chemical compositions can be obtained via gas phase synthesis. This route to NSMs is therefore of increasing interest. Other advantages include controlling the reaction mechanisms, obtaining high purity products, and the ability to produce multi-component systems. Table 3.3 lists the gas phase methods under chemical vapour deposition (CVD) and physical vapour deposition (PVD):

**CVD**: Either homogeneous or heterogeneous reactions of gaseous products.

**PVD**: There are several methods. Inert gas condensation (IGC) of nanoparticles from the vapour phase (not specified as such in Table 3.3) is the most general one.

#### 3.8.2.4 Electrodeposition

This process is the inducement of chemical reactions in an aqueous electrolyte solution with the help of an applied voltage. Deposition of nanostructured materials, including metal oxides and chalcogenides, are in this category. The process may be either anodic or cathodic. It is a relatively cheap technique, allowing film thickness to be controlled via the controlled delivery of charge or potential.

#### 3.8.2.5 Summary for Subsection 3.8.2

It is important to note that for all the synthesis methods briefly discussed in Sect. 3.8.2. and listed in Table 3.3, there are some common features, namely that the following conditions should be fulfilled:

- Control of nanoparticle sizes, size distributions, shapes, crystal structures and compositions.
- Control of aggregation and agglomeration.
- Maintaining or improving the purity of nanoparticles.

- High (or higher) reproducibility.
- Physical properties, stabilization of structures.
- Higher mass production, upscaling, and lower costs (Guizard et al. 1992; Herlin-Boime et al. 2004; Iijima 1991; Hofmann 2009; Pokropivny et al. 2007; Arivalagan et al. 2011; Tiwari et al. 2012; Wang 2013; Sarikaya et al. 2003a, 2003b; Ikkala and Brinke 2002; Rao and Nath 2003; Jensen 1989; Kroto et al. 1985; Lakshmi et al. 1997; Yang et al. 2010, 2015; Pang et al. 2015; Luther 2004; Madou 2002; Mao et al. 2004; Mooney and Radding 1982; Oh et al. 2005; Piner et al. 1999; Salata 2004; Sun et al. 2000; Ye et al. 2004; Ensafi et al. 2016; Ghaedi et al. 2016; Lu et al. 2013; Li et al. 2016; Liu et al. 2012; Dai et al. 2016, 27).

## 3.9 Characterization Techniques

There are several characterization techniques available to assist in ensuring the quality of the synthesized nanoparticles. Table 3.4 gives a summary of these techniques.

**Table 3.4** Summary of characterization techniques for nanoparticle synthesis (Umer et al. 2012)

| Technique[a] | Measurement Parameters | Sample Type | Sensitivity |
|---|---|---|---|
| TEM | Particle size and characterization | Dispersed sample on grid | Down to 1 nm |
| SEM | Particle size and characterization | Conductive or sputter coated | Down to 1 nm |
| AFM | Particle size and characterization | | 1 nm to 8 μm |
| XPS | Elements and functionalization ratio | Solid | 3 to 92 nm |
| XRD | Amorphous/crystalline, average particle size in bulk | Solid | Down to 1 nm |
| FTIR | Functional groups | Solution/ powder | 20 Å to 1 μm |
| RAMAN | Purity/defect | Solid/powder | 0.2 to 10 μm |
| UV-Visual | Degree of functionalization | Solution | Scanned and visible regions are 200–400 nm and 400–800 nm, respectively |

[a] TEM = transmission electron microscopy; SEM = scanning electron microscopy; AFM = atomic force microscopy; XPS = X-ray photoelectron spectroscopy; XRD = X-ray diffraction; FTIR = Fourier transform infrared spectroscopy; RAMAN = Raman spectroscopy; UV = ultraviolet

## 3.10 Applications of NSMs

### *3.10.1 Civil and Industrial Applications*

A list of some key civil and industrial applications of nanomaterials is given in this subsection. Most current applications represent evolutionary developments of existing technologies (Guizard et al. 1992; Herlin-Boime et al. 2004; Iijima 1991; Hofmann 2009; Pokropivny et al. 2007; Arivalagan et al. 2011; Tiwari et al. 2012):

(1) Catalysis: Because of high surface area, nanoparticles provide higher catalytic activity, which in turn means less catalyst required to serve the purpose: e.g. platinum *nanoparticles* for the next generation of catalytic converters in combustion engine vehicles.
(2) Medicine: Nanotechnology drugs containing nanoparticles could be delivered to specific cells, thereby reducing side effects and overall drug consumption.
(3) Sunscreens and cosmetics: Nanosized titanium dioxide and zinc oxide are optimally used for the purpose of absorbing and reflecting ultraviolet (UV) rays, while being transparent to visible light.
(4) Construction: Construction could be faster, cheaper and safer, and nanotechnology has the *potential* to achieve these goals. The important fields are *Concrete, Steel, Glass and Coatings.*
(5) Paints: Nanoparticle-containing paints are lighter and have improved performance. This has huge potential for aircraft painting, owing to significant weight reductions.
(6) Displays: Nanomaterials in television screens and computer monitors is one of the much awaited applications.
(7) Food: Anti-microbial agents can be applied directly to the surfaces of coated films used in the production, processing, safety and packaging of foods.

### *3.10.2 Military (Defence) Applications (Tiwari 2012; Altmann 2004)*

Nanotechnology is multidisciplinary, and may be used to create materials with properties that will revolutionize military technology. The military fields in which nanotechnology will have major the impacts are (i) strong lightweight materials; (ii) adaptive, multifunctional materials; (iii) novel detection and protection schemes against the threats of bio/chemical warfare (iv) remote and local soldier monitoring systems; (v) wound and injury triage and emergency treatment systems; and (vi) novel, noncombat, and combat performance enhancement systems, all to improve soldier survivability.

## 3.11 Defence Materials & Stores Research & Development Establishment (DMSRDE) Contributions

DMSRDE was the first laboratory in the Defence Research & Development Organization (DRDO) to begin working on nanotechnology. Since 1992, DMSRDE has been involved in the laboratory-scale synthesis of fullerene and carbon nanotubes by the arc-discharge method. In 1999, synthesis of carbon nanotubes (CNTs) by Catalytic Chemical Vapour Deposition (CCVD) was begun. Currently the CCVD capability includes not only CNTs but also carbon micro/nano coils, inorganic nanotubes, metallic nanowires/nanorods, nanoferns, and bottlebrush structures.

For device and sensor applications, thin films of hybrid nanomaterials have been deposited on various wafers using sputtering, electron beam, and thermal deposition units. The topologically controlled hierarchical nanostructures of graphene-CNT hybrids have been developed by solution-phase chemical methods.

Over the last fifteen years a significant amount of R&D has been done to supply nanomaterials with required tailor-made properties to other DRDO laboratories and establishments. The expertise developed by DMSRDE in synthesizing need-based hybrid nanomaterials is now considered to be suitable and viable for various applications, provided that certain critical features are incorporated. Also, it is technically feasible to incorporate such features into the hybrid nanomaterials developed by DMSRDE.

The basic R&D, technological developments, and product realizations catering for the specific needs of various DRDO laboratories are presented in the following subsections.

### *3.11.1 Synthesis Processes*

#### 3.11.1.1 Carbon Nanotubes (CNTs)/coiled Carbon/carbon Micro-flowers

The important synthetic procedures for CNTs production are: (i) arc discharge of graphite electrodes or laser vaporization of graphite electrodes in the presence of a metal catalyst; (ii) pyrolysis of hydrocarbon vapour by catalytic chemical vapour deposition (CCVD) or pyrolysis of metallocenes; (iii) CVD) on silicon wafers containing lithographically deposited metal dots; and (iv) laser-assisted production of CNTs from acetylene.

Among the production methods reported, only the CCVD method is capable of producing CNTs on a large-scale, because of its lower reaction temperature coupled to the low cost of production. At the DMSRDE the CCVD method has been used to synthesize carbon nanotubes and fibres (Altmann 2004; Mukhopadhyay et al. 1999, 2005; Fonseca et al. 1998; Tiwari et al. 2016; Agarwal et al. 2017). The as-synthesized product was then analyzed using transmission electron microscopy (TEM), scanning

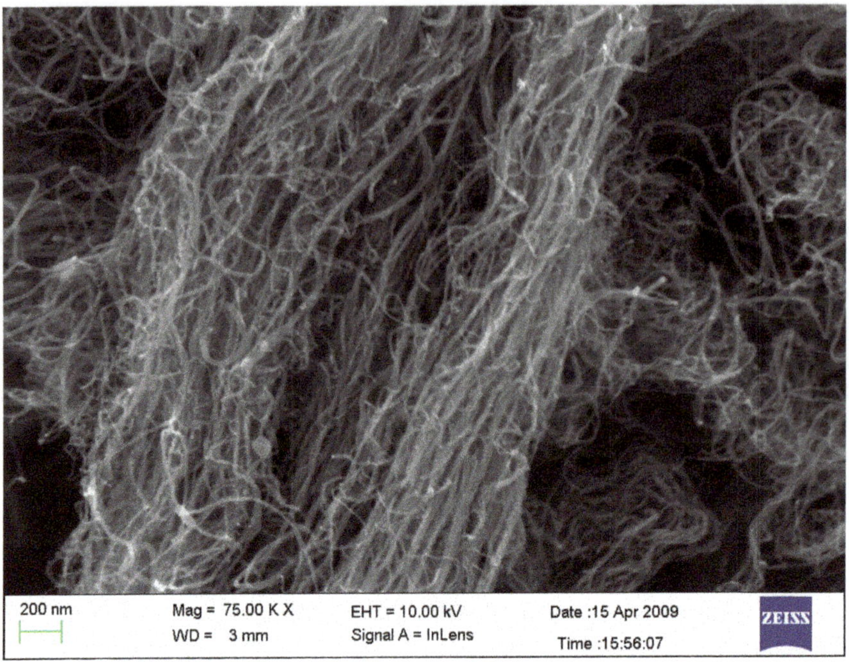


**Fig. 3.1** SEM image of bundles of multi-walled carbon nanotubes (MWCNTs)

electron microscopy (SEM), thermal analysis and Raman spectroscopy. An example SEM image, in this case bundles of multi-walled carbon nanotubes (MWCNTs), is shown in Fig. 3.1.

#### 3.11.1.2 Metallic Nanowires

Nanowire synthesis schemes have been extensively studied, and many methods are reported in the literature. With respect to synthesis methods, the bottom-up approach is favoured since it is less infrastructure-intensive. The following bottom-up approaches are the most used ones (Srivastava et al. 2015):

(1) Electrochemical deposition: Uses polymeric membranes with vertical pores as templates for electrodeposition. This is a versatile technique and a variety of nanowires can be grown depending on the electrolyte solution, although there is a limitation in the variety of suitable conducting substrates. Figure 3.2 is an SEM image of silver nanofern structures obtained by electrodeposition.
(2) Vapour–liquid–solid (VLS) growth: VLS growth was first demonstrated by Wagner et al. (Srivastava et al. 2015). This is a catalyst-assisted growth method that can be used to grow a variety of nanowires. The process basically involves evaporation of the source material followed by its transportation by a carrier gas

**Fig. 3.2** SEMimage of silver nanofern structures obtained by electrochemical deposition

and final condensation on a substrate. A low supersaturation point is needed for the growth of 1D structures. Also, the dimensions of 1D structures depend on the pressure, temperature and the substrate.

(3) Solution phase method: The solution phase synthesis scheme involves the reduction of precursor materials in the liquid phase, resulting in atoms which eventually nucleate clusters and grow, leading to the formation of nanomaterials. This is a well-used technique for synthesis of nanowires.

#### 3.11.1.3 Hierarchical Tailor-Made All-Carbon Nanostructures

The potential for improvements in device performances is based on tailoring the materials properties at the nano level. The main challenges for nanomaterials include the design and tailoring of complex hybrid nanoparticles and nanomaterials (e.g. nanotubes, functionalized surfaces, multi-layers, novel thin films and interfaces) having multiple functions, and then to be followed by device integration. The potential applications include catalysis, spintronic, optical, magnetic and electronic devices (Mukhopadhyay et al. 2005; Tiwari et al. 2016; Agarwal et al. 2017).

At the DMSRDE an easy method has been derived by utilizing iron nanoparticle–graphite hybrid interfaces to control derivatization and delamination of graphite. By using hydroxyl radicals produced catalytically in Fenton's reagent, we not only tailored the manipulation of defect sites in the graphite lattice but also self-exfoliation of graphite layers. The side edges of the graphite contained more functional groups, and this helped not to disturb the honeycomb lattice structure (Agarwal et al. 2017).

### *3.11.2 Applications*

During the last decades, some of DMSRDE's in-house R&D has been on the development of carbon and other hybrid ID/2D nanomaterials. The fundamental building blocks with salient features were successfully synthesized and developed. The intended properties of these materials were also extensively assessed at laboratory level and found to meet the applicable standards. As mentioned at the beginning of this Sect. (3.10), these materials are now considered to be suitable for various applications after incorporating necessary critical features (Altmann 2004; Mukhopadhyay et al. 1999, 2005; Fonseca et al. 1998; Tiwari et al. 2016; Agarwal et al. 2017; Srivastava et al. 2015).

DMSRDE has already developed advanced nanotechnology materials to cater for the specific needs of various DRDO laboratories, including the Naval Science and Technological Laboratory (NSTL), the Defence Laboratory Jodhpur (DLJ), the Defence Research & Development Establishment (DRDE), the Defence Bioengineering and Electromedical Laboratory (DEBEL), and the Naval Materials Research laboratory (NMRL).

Because these nanotechnology materials are subject to 'technology denial', i.e. they are not freely available, the indigenous (Indian) capabilities will immensely benefit the DRDO system laboratories, leading to self-reliance in the technology. We have focussed on the following applications:

- Nanocomposites with high thermal and low electrical conductivity (NSTL requirement).
- Mesoporous nanocomposite membranes for decontamination and disinfection of drinking water (DLJ requirement).
- Silver nanoferns for surface-enhanced Raman spectroscopy (SERS) applications (DMSRDE/ DRDE) requirements.
- Novel nanostructures for gas sensors, biomedical sensors, solar cells, and supercapacitor applications (DMSRDE / DRDE / DEBEL / NMRL) requirements.

## 3.12 Conclusions and Outlook for Future

Nanotechnology is already beginning to transform our existence with new technologies and products. For millions of people in developing countries, nanotechnology has brought the hope of new solutions to problems in particularly water, energy, health care, and education. There is a steep rise in global funding for nanotechnology, but at the same time it is necessary to make sure that its benefits will become more widely recognised. Similarly, the technology challenges must become better appreciated.

We believe that in the future many more nanotechnology products will be developed via simpler and less time-consuming methods than those currently available. Even nanomaterials have great impact on new "*green*" technologies, e.g. fuel cells and catalysts. More specifically, the following issues have to be considered for NSM developments:

- Further development of synthesis and/or fabrication methods for the raw materials (powders) as well as for the nanostructured materials themselves.
- Better understanding of the influence of the sizes of building blocks in NSMs as well as the influence of microstructure on the physical, chemical and mechanical properties of these materials; and also the influence of interfaces on the properties.
- Investigation of catalytic applications of mono- and pluri-metallic nanomaterials.
- Most importantly, the transfer of development technologies into industrial applications, including industrial scale development of the various required synthesis methods.

**Acknowledgements** The authors gratefully acknowledge the motivation from several senior scientists (past as well as present) from the DMSRDE. Some of the inputs are from several of their colleagues – namely Dr. Santosh Kr. Tripathi, Mohd. Imamuddin, Mr. Subhash Mandal, Mr. Sanjay Kanojia, Mrs. Neha Agarawal and other DNMAT members of DMSRDE, and they are all gratefully acknowledged. They are also thankful to the DRDO for the funding and support for these studies. The valuable suggestions made by Dr. RJH Wanhill and his editing efforts for this chapter are also gratefully acknowledged.

## References

Adams, A.C.: VLSI Technology. Ed. S. M. Sze. New York: McGraw-Hill (1988)

Agarwal, N., Bhattacharyya, R., Tripathi, N.K., Kanojia, S., Roy, D., Mukhopadhyay, K., Prasad, N.E.: Derivatization and interlaminar debonding of graphite–iron nanoparticle hybrid interfaces using Fenton chemistry. Phys. Chem. Chem. Phys. **19**, 16329–16336 (2017)

Akerman, M.: Nanocrystal targeting in vivo. Proc. Natl. Acad. Sci. **99**, 12617–12621 (2002)

Allred, D., Sarikaya, M., Baneyx, F., Schwartz, D.: Electrochemical Nanofabrication Using Crystalline Protein Masks. Nano Lett. **5**(4), 609–613 (2005)

Altmann, J.: Military uses of nanotechnology: perspectives and concerns. Secur. Dialogue **35**, 61–79 (2004)

Alwitt, R.S.: Electrochemistry Encyclopedia. In: Cleveland Ohio (http://electrochem.cwru.edu/encycl/) (2002)

Arivalagan, K., Ravichandran, S., Rangasamy, K., Karthikeyan, E.: Nanomaterials and its potential applications. Int. J. Chem. Tech. Res. **3**, 534–538 (2011)

Asoh, H.N.K., Nakao, M., Tamamura, T., Masuda, H.: Conditions for fabrication of ideally ordered anodic porous alumina using pretextured Al. J. Electrochem. Soc. **148**, B152–B156 (2001)

Brodie, I., Muray, J.J.: The physics of microfabrication, Plenum Press, New York. Chan W, Nie S (1998) quantum dot bio conjugates for ultrasensitive nonisotopic detection. Science **281**, 2016–2018 (1982)

Dai, H., Wang, N., Wang, D., Ma, H., Lin, M.: An electrochemical sensor based on phytic acid functionalized polypyrrole/graphene oxide nanocomposites for simultaneous determination of cd(II) and Pb(II). Chem. Eng. J. **299**, 150–155 (2016)

Ensafi, A.A., Rezaloo, F., Rezaei, B.: Electrochemical sensor based on porous silicon/silver nanocomposite for the determination of hydrogen peroxide. Sens. Actuators B Chem. **231**, 239–244 (2016)

Feynman, R.: Plenty of room at the bottom. In. Pasadena: Caltech (1959). https://en.wikipedia.org/wiki/There%27s_Plenty_of_Room_at_the_Bottom

Fonseca, A., Hernadi, K., Piedigrosso, P., Colomer, J.F., Mukhopadhyay, K., Doome, R., Lazarescu, S., Biro, L.P., Lambin, P., Thiry, P.A., Bernaerts, D., Nagy, J.B. Synthesis of single- and multi-wall carbon nanotubes over supported catalysts. Appl. Phys. A **67**, 11–22 (1998)

Ghaedi, H., Afkhami, A., Madrakian, T., Soltani-Felehgari, F.: Construction of novel sensitive electrochemical sensor for electro-oxidation and determination of citalopram based on zinc oxide nanoparticles and multi-walled carbon nanotubes. Mater. Sci. Eng. C **59**, 847–854 (2016)

Guizard, C., Julbe, A., Larbot, A., Cot, L.: Nanostructures in sol-gel derived materials: application to the elaboration of nanofiltration membranes. J. Alloy. Compd. **188**, 8–13 (1992)

Herlin-Boime, N., Mayne-L'Hermite, M., Reynaud, C.: Synthesis of covalent nanoparticles by CO2 laser. Encycl. Nanosci. Nanotechnol. **10**, 301–326 (2004)

Hofmann, H.: Advanced Nanomaterials Course Support; Version 1 (2009). https://ltp.epfl.ch/files/content/sites/ltp/files/shared/Teaching/Master/03-IntroductionToNanomaterials/LectureSupportAll.pdf

Iijima, S.: Helical microtubules of graphitic carbon. Nature **354**, 56–58 (1991)

Ikkala, O., Brinke, G.T.: Functional Materials Based on Self-assembly of polymeric supramolecules. Science, New Series **295**, 2407–2409 (2002)

Jensen, K.: Chemical Vapor Deposition, in Microelectronics Processing: Chemical Engineering Aspects. American Chemical Society, Washington DC (1989)

Kroto, H., Heath, J., O'Brien, S., Curl, R., Smalley, R.: C60: buckminsterfullerene. Nature **318**, 162–163 (1985)

Lakshmi, B.B., Dorhout, P.K., Martin, C.R.: Sol-gel template synthesis of semiconductor nanostructures. Chem. Mater. **9**(3), 857–862 (1997)

Li, H., Wang, W., Lv, Q., Xi, G., Bai, H., Zhang, Q.: Disposable paper-based electrochemical sensor based on stacked gold nanoparticles supported carbon nanotubes for the determination of bisphenol. Electrochem. Commun. **68**, 104–107 (2016)

Liu, S., Lu, F., Liu, Y., Jiang, L.-P., Zhu, J.-J.: Synthesis, characterization, and electrochemical applications of multifunctional Fe3O4@C–au nanocomposites. J. Nanopart. Res. **15**, 1–8 (2012)

Lu, F., Liu, S., Jiang, L.-P., Zhu, J.-J.: Sonochemical fabrication of CdSexTe1−x/au nanotubes and their potential application in biosensing. J. Nanopart. Res. **15**, 1–9 (2013)

Luther, W.: System Model for Nanocomposites Produced by Sol-Gel. Published by Future Technologies Division of VDI Technologiezentrum GmbH, Dusseldorf, Germany (2004)

Madou, M.: Fundamentals of Microfabrication: the science of miniaturization. CRC Press, Taylor and Francis Group, Boca Raton, ISBN 9780849308260 (2002)

Mao, C., et al.: Virus-based toolkit for the directed synthesis of magnetic and semiconducting nanowires. Science **303**(5655), 213–217 (2004)

Mooney, J., Radding, S.: Spray pyrolysis processing. Ann. Rev. Mater. Sci. **12**, 81–101 (1982)

Mukhopadhyay, K., Koshio, A., Sugai, T., Tanaka, N., Shinohara, H., Konya, Z., Nagy, J.B.: Bulk production of quasi-aligned carbon nanotube bundles by the catalytic chemical vapour deposition (CCVD) method. Chem. Phys. Lett. **303**, 117–124 (1999)
Mukhopadhyay, K., Ram, K., Lal, D., Mathur, G.N., Rao, K.U.B.: Double helical carbon microcoiled fibers synthesis by CCVD method. Carbon **43**, 2400–2402 (2005)
Oh, S., Finones, R., Daraio, C., Chen, L.-H., Jin, S.: Growth of nano-scale hydroxyapatite using chemically treated titanium oxide nanotubes. Biomater **26**, 4938–4943 (2005)
Pang, P., Li, H., Liu, Y., Zhang, Y., Feng, L., Wang, H., Wu, Z., Yang, W.: One-pot facile synthesis of platinum nanoparticle decorated reduced graphene oxide composites and their application in electrochemical detection of rutin. Anal. Methods **7**, 3581–3586 (2015)
Piner, R.D., et al.: "Dip pen" nanolythography. Science **283**, 661–665 (1999)
Pokropivny, V., Lohmus, R., Hussainova, I., Pokropivny, A., Vlassov, S., Frantsevich: Introduction to nanomaterials and nanotechnology, ISBN: 978-9949-11-741-3, Tartu University Press, www.tyk.ee, Tartu 2007 (2007)
Rao, C.N.R., Nath, M.: Inorganic nanotubes, Dalton Trans., 1–24 (2003)
Salata, O.: Applications of nanoparticles in biology and medicine. J. Nanobiotechnol. **2**, 3–10 (2004)
Sarikaya, C., Tamerler, A.K.Y., Jen, K.S., Baneyx, F.: Molecular biomimetics: nanotechnology through biology. Nat. Mater. **2**, 577–585 (2003)
Sarikaya, M., Tamerler, C., Jen, A.-Y., Schulten, K., Baneyx, F.: Molecular biomimetics: nanotechnology through biology. Nat. Mater. **2**, 577–585 (2003)
Srivastava, A.K., Bhartia, B., Mukhopadhyay, K., Sharma, A.: Long term biopotential recording by body conformable photolithography fabricated low cost polymeric microneedle arrays. Sens. Actuators, A **236**, 164–172 (2015)
Sun, S., Murray, C., Weller, D., Folks, L., Moser, A.: Monodisperse FePt nanoparticles and ferromagnetic FePt nanocrystal superlattices. Science **287**, 1989–1992 (2000)
Tiwari, A.: Military nanotechnology. Int. J. Eng. Sci. & Adv. Technol. **2**, 825–830 (2012)
Tiwari, J.N., Tiwari, R.N., Kim, K.S.: Zero-dimensional, one-dimensional, two-dimensional and three-dimensional nanostructured materials for advanced electrochemical energy devices. Prog. Mater. Sci. **57**, 724–803 (2012)
Tiwari, N., Pandey, N., Roy, D., Mukhopadhyay, K., Prasad, N.E.: Engineering the physical parameters for continuous synthesis of fullerene peapods. Nanotechnology **27**, 205604–205611 (2016)
Umer, A., Naveed, S., Ramzan, N., Rafique, M.S.: Selection of a suitable method for the synthesis of copper nanoparticles. NANO: Brief Rep. Rev. **7**, 1230005–1230018 (2012)
Wang, Z.L.: Nanowires and Nanobelts: Materials, Properties and Devices. Volume 1: Metal and Semiconductor Nanowires. Springer Science & Business Media (2013)
Yang, S., Qu, L., Li, G., Yang, R., Liu, C.: Gold nanoparticles/ethylenediamine/carbon nanotube modified glassy carbon electrode as the voltammetric sensor for selective determination of rutin in the presence of ascorbic acid. J. Electroanal. Chem. **645**, 115–122 (2010)
Yang, J.-H., Yang, H., Liu, S., Mao, L.: Microwave-assisted synthesis graphite-supported Pd nanoparticles for detection of nitrite. Sensors Actuators B Chem **220**, 652–658 (2015)
Ye, X.-R., Lin, Y., Wang, C., Engelhard, M.H., Wang, Y., Wai, C.M.: Supercritical fluid synthesis and characterization of catalytic metal nanoparticles on carbon nanotubes. J. Mater. Chem. **14**, 908–913 (2004)

# Chapter 4
# A New Frontier in Functional Fluids: Nano Lubricating and Thermally Conducting Fluids

**Jyoti P. Singh, Ashok Kr. Gautam, Jyoti Srivastava, Tandra Nandi, and Eswara Prasad Namburi**

**Abstract** Functional fluids are formulated from a range of premium quality base oils using advanced additive technologies. Nanoparticles dispersed in liquid media are known as Nanofluids. In this chapter we give an overview of the present international scenario of Nanofluids research, followed by synthesis methodologies for a few selected Nanofluids. We also focus on the primary objectives of current Nanofluids research. The chapter also describes (i) the novel methods of nanoparticles synthesis and their stable dispersion in fluids to prepare Nanofluids; (ii) the potential applications of Nanofluids in lubrication and thermal management and some of the key challenges in this field viz. high heat transfer efficiency and lubricity; and finally (iii) the probable mechanism for enhanced thermal conductivity and lubricity.

## 4.1 Introduction

Nowadays, the need for specialty lubricants and thermal fluids for critical applications are fulfilled by specifically formulated functional fluids. The lubricant serves many purposes to protect interacting surfaces and this is discussed in detail in the present

J. P. Singh · A. Kr. Gautam · J. Srivastava
Defence Materials and Stores Research and Development Establishment (DMSRDE), Defence Research and Development Organisation (DRDO), Kanpur, Uttar Pradesh, India

T. Nandi
Formerly With Defence Materials and Stores Research and Development Establishment (DMSRDE), Defence Research and Development Organisation (DRDO), Kanpur, Uttar Pradesh, India

E. P. Namburi (✉)
Former Outstanding Scientist and Ex-Director, Defence Materials and Stores Research and Development Establishment (DMSRDE), Defence Research and Development Organisation (DRDO), Kanpur, Uttar Pradesh, India
e-mail: neswarap@gmail.com

 
E. P. Namburi et al. (eds.), *Novel Defence Functional and Engineering Materials (NDFEM) Volume 1*, Indian Institute of Metals Series,
https://doi.org/10.1007/978-981-99-9791-6_4

chapter. Nowadays, increasing demand for high performance fluids for fast growing industries propel the newer research for development of functional fluids.

Synthetic fluids can be defined as molecules/chemical compounds that are specifically engineered in the laboratory by standard synthetic methodologies. A large number of such synthetic substances are now commercially for applications as lubricants. Silicone based functional fluids are the basis for developing many products with enhanced benefits.

Modern fabrication and synthesis techniques pave the way to produce fine nanoparticles for several advance technological applications. The various physio-chemical, electrical and magnetic properties of nanomaterials is much better than bulk materials (Gleiter 2000). Hence, nanomaterials have attracted tremendous attention of material scientists and researchers across the world.

Nanoparticles dispersion in the liquid may cause a huge enhancement or total change in the existing properties of the base fluid. For this new class of solid–liquid suspensions, the term NANOFLUIDS was first coined in 1995. Typicaly Nanofluid contains very low concentration of nanoparticles, <1% (v/v) and may or may not contain surfactants or other dispersing chemicals. The term was primarily used to describe heat transfer fluids. However, over the years, the term has been extended to other systems like FERROFLUIDS and Nano lubricating fluids.

## 4.2 Background on Nanofluids

As mentioned in Sect. 4.1, modern fabrication techniques pave the way to produce fine nanoparticles for several advance technological applications. The need of the miniaturization technology has propelled the materials development from the early 1950s, micrometre scale to the atomic scale (Sohn 1998; Roco 1998). Hence Nanophase materials have received considerable attention from materials scientists and engineers.

Nanoparticles suspended in solution provide the following advantages:

(i) They avoid rapid settling of the particles, whereas larger particles require the fluid to be circulated, which would cause wear and tear in the vessels containing them. Contrary to above, the suspension stability of nanophase materials are quite long and hence reduce the sedimentation and erosion.
(ii) Due to minute size and larger *overall* surface area of nanoparticles, these materials are quite promising for many applications like that of coolants (Choi and Eastman 1995; Kim et al. 2001) and magnetic dampers (Berkovsky et al. 1993).

However, a major disadvantage of nanoparticles is their high agglomeration tendency and this leads to the emergence of new concept for Nanofluid preparation.

## 4.3 Synthesis of Nanoparticles

Nanoparticles, as the name suggests, are particles with sizes in the Nano regime, i.e. $10^{-9}$ m. These are the smallest human-made objects and have unique properties that are significantly different from the bulk properties. There are two major approaches to synthesizing nanoparticles:

(a) The bottom-up approach: This approach synthesizes nanoparticles from the atomic or ionic scale. The particles may be synthesized by a soft chemical route, "Chime Deuce" or by physical methods. There are several methods for this approach. A few examples are:

   (i) Micro-emulsion based techniques
   (ii) Sol–gel method
   (iii) Dendrimer-encapsulated techniques
   (iv) Hydrothermal synthesis.

(b) The other approach to the synthesis of nanoparticles is the Top Down method in which bulk-sized particles are broken down to the nanoscale by physical methods like mechanical attrition or milling. As is understandable, this method is difficult to achieve and gives less control over the size, shape and monodispersity of the particles, and hence is less popular.

However, the most common method for synthesis of nanoparticles is by physical synthesis techniques such as Inert Gas Condensation (IGC) or by chemical synthetic techniques like Chemical Vapour Deposition (CVD). The reverse micelle technique is also used, as in our laboratory. Here we mention the important aspects of these three techniques.

### *4.3.1 IGC Process*

In this process of preparation of nanomaterials, the material is evaporated into the atmosphere of low density inert gas and this technique is used primarily to produce metallic and ceramic nanoparticles (Pfund 1930; Kimoto et al. 1963; Birringer 1989). During the synthesis process, the material vapour condense to form nanoparticles when they diffuse hot surface to the surrounding gas. The yield of this technique was initially poor but later on the gas condensation technique has also been used to scale up the nanoparticles production (Chang et al. 1994).

### *4.3.2 CVD Process*

In this process a chemical precursor is converted to the gas phase which subsequently dissociates by thermal energy or plasma to generate nuclei (Tavakoli et al. 2007).

The plasma used in CVD is generated by microwave radiation. The metal nuclei are carried out of the plasma reaction zone and then allowed to coagulate. By this method, 5–50 nm particles of Mn, Fe, Co, Zn have been produced (Koch 2006). There are several other examples of preparation of metals and metal oxide/ceramic nanoparticles by this method (Leslie-Pelecky and Rieke 1996; Ying et al. 1995).

### *4.3.3 Reverse Micelle Process*

In this process, micro-emulsions are made by dissolving a suitable surfactant in a mixture of a polar organic solvent and water. Surfactants are molecules consisting of both a hydrophilic and a hydrophobic end. When a surfactant is dissolved into a solvent, nanometer-sized water droplets are formed within the volume and stabilized by the surfactant molecules. The hydrophilic end of the surfactant lies along the outer surface of the water droplet and the hydrophobic end is attached to the non-polar solvent, stabilizing the water droplet and providing mechanical separation from the solvent. The radius of the water droplet depends on $W_0$, where $W_0 =$ (molar number of water)/(molar number of surfactant).

The droplet concentration is proportional to the water volume fraction of the micro-emulsions. Both the micro-emulsion size and concentration can be varied by adjusting the composition of the micro-emulsion (Lade et al. 2000; Schmidt et al. 1999). Brownian motion of the particles causes them to collide and coalesce, and thus allow a rapid exchange of their contents causing them to react. Often external reactants like $NaBH_4$are added to the micro-emulsions as reducing agents. This method has the following advantages:

(a) Good control of the shape and size of the particles
(b) Particle monodispersity
(c) Easy to synthesize
(d) Economically viable.

## 4.4 Preparation of Nanofluids from Nanoparticles

Techniques for the preparation of nanoparticles are crucial to making stable Nanofluids. Some techniques followed are covered below.

### *4.4.1 The Two-Step Method*

In the present method, firstly the nanoparticles are prepared followed by which they are dispersed into a suitable base fluid. The most important factor for making a stable dispersion is given by the Stokes–Einstein equation (Rajagopalan and Hiemenz 1997), which gives the rate of particle sedimentation as

$$v = 2r_p\left|\rho_p - \rho_f\right|g/9\eta_f \quad (4.1)$$

The above equation shows that the for achieving a lower sedimentation rate and stable particle dispersion the size of the particles should be small. Apart from this the smaller difference in the density of the particle and the base fluid ($\rho_p - \rho_f$) and higher fluid viscosity also favours the lower rate of sedimentation. But alteration of the properties of fluids e.g. viscosity, density has its limitation and cannot be changed beyond certain limits. Apart from above key properties other properties also affect the sedimentation rate e.g. surface charge on particle, rheology, chemical interactions, energy inputs etc.

A successful dispersing technique should be able to prevent primary particles from aggregating into larger ones or to make larger particles disintegrate into smaller ones. Researchers use either physical or chemical methods to disperse nanoparticles into a fluid (Rajagopalan and Hiemenz 1997; Ross 1988).

#### 4.4.1.1 Physical Methods

In the physical method, two types of stabilization techniques are mentioned below.

(i) *Mechanical*: Mechanical dispersion is high shear mixing using a high-speed mixer, homogenizer, micro fluidizer, and a colloid mill.
(ii) *Ultrasonication assisted*: Ultrasonics cause elastic waves (vibrations) to propagate through a liquid medium. These vibrations cause cavitation, which stirs the dispersion and breaks up any agglomerates.

#### 4.4.1.2 Chemical Methods

Apart from the physical method, chemical methods are also used as aid to the physical method for the preparation of nanoparticles. The Long-range attractive forces among nanoparticles are because of van der Waals' attractions. In order to stabilize the nanoparticles in the fluid, the attractive forces between particles should be reduced and repulsive forces should be enhanced. The best known stabilization methods are, (i) Electrostatic Method and (ii) Steric Method.

In the electrostatic stabilization method, the particles repel each other due to static surface charge present on the particles. The numerical electrostatic stabilization can be well explained by DLVO (Derjaguin-Landau-Verwey-Overbeek) theory (Verwey and Overbeek 1955; Reerink and Overbeek 1954). The concentration of various components plays a crucial role for stabilization. Apart from concentration, adjustment of pH may also cause electrostatic repulsion or addition of an acid or alkali may also cause the formation of finely dispersed nanoparticles.

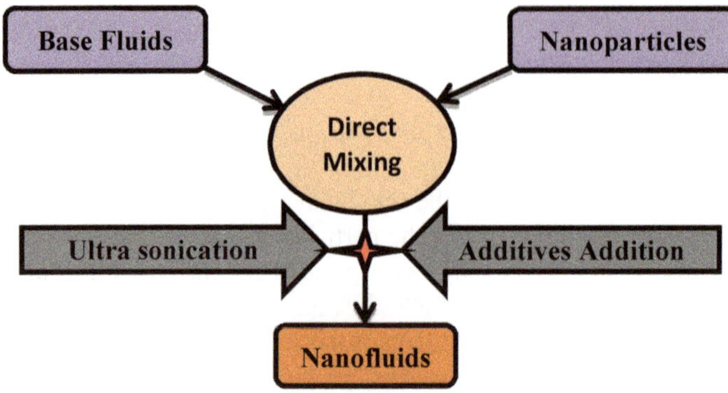


**Fig. 4.1** Two-step preparation process of nanofluids

Steric repulsion also plays an important role in preventing the particles to coalesce. In order to enhance dispersion stability and prevent sedimentation and precipitation, different surfactants and polymeric molecules are used. These molecules consist of an anchor moiety which is attached to the particle surface and a long chain stabilizing moiety, which create hindrance and prevent particles to come closer.

The two-step process is used widely for the preparation of Nanofluids. Various reports shows that the copper and $Al_2O_3$ nanoparticles have been dispersed in base fluids like water and ethylene glycol with or without ultrasonication (Eastman, et al. 1996; Lee et al. 1999; Xie et al. 2002). Figure 4.1 shows a schematic of a two-step process.

## 4.4.2 *One-Step Method*

### 4.4.2.1 Microwave Treatment

Microwave irradiation can also be used for the preparation of certain nanofluids. In an interesting report, copper nanoparticles based nanofluid have been prepared by microwave induced reduction of $CuSO_4{\cdot}5H_2O$ with $NaH_2PO_2{\cdot}H_2O$. The base fluid taken for this single step preparation was ethylene glycol (Zhu et al. 2004). This fast and efficient one-step method resulted in non-agglomerated, crystalline Cu nanoparticles of average diameter 10 nm. To avoid the formation of clusters, PVP (polyvinyl pyrrolidone) was used to retard agglomeration by the steric effect.

### 4.4.2.2 Direct Evaporation–Condensation Method (DEC)

Although the stability of nanoparticles in a dispersion can be controlled using chemical and physical dispersion techniques, the longevity of such a dispersion is still

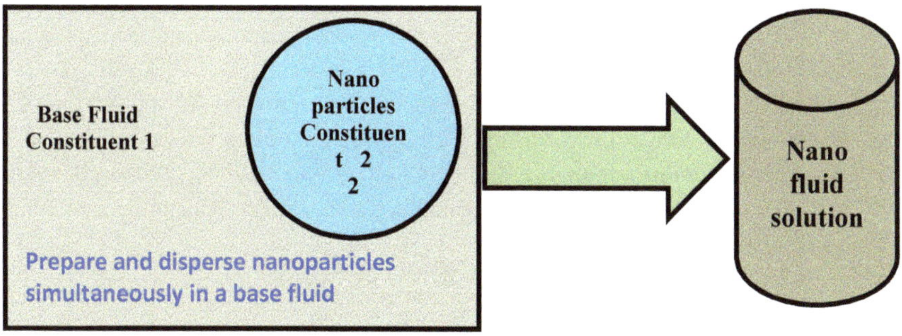


**Fig. 4.2** One-step preparation process of nanofluids

limited. Another method which is used is the Direct evaporation–condensation method (DEC). In this method the vapour from a heated source is passed on to the flowing low vapour pressure liquid, where it starts condensing. This method is called "vacuum evaporation onto a running oil substrate" (Akoh et al. 1978).

The reports reveal nanofluid preparation via the single step Direct Evaporation–Condensation method over gas condensation method when applied for highly conducting metal based preparations. It has been observed that copper nanoparticle based nanofluid prepared the through Direct Evaporation–Condensation method has shown better result than that prepared by other similar methods (Eastman et al. 2001; Asirvatham et al. 2009). Figure 4.2 shows a schematic of a single step process.

The nanofluid preparation via nanoparticles dispersion by the Direct evaporation–condensation method produces the nanofluid with good suspension stability but this method has limitations e.g. it can be used for the liquids with low vapour pressure and the another issue which is very frequently observed is the low oxidation stability of these nanoparticles. Despite these drawbacks, the one-step method uses size control alone for stabilizing the nanoparticle dispersions, and does not use electrostatic control or steric control; this avoids the use of surfactants and acids or alkalis. In a report, the copper nanofluid was prepared via the Direct evaporation–condensation method followed by preparation, and a very small amount of thioglycolic acid was added to the nanofluid to enhance the thermal properties. When the nanofluid was analysed for its properties, not only the thermal properties but a more stable dispersion was also obtained (Eastman et al. 2001). This means that a good electrostatic control was obtained by using the acid in such small amounts. It may be noted that thioglycolic acid is a relatively weak acid when compared, for example, to HCl. Hence, such a small quantity should not give significant problems in practical systems.

## 4.5 Nano Lubricating Fluids

The lubrication is the process of reduction of friction and wear by application of a suitable material known as the lubricant. Lubrication not only reduces wear, but also excessive heat generation, which in turn results in enhanced mechanical efficiency. The lubricant materials may be in any form, e.g. solid, liquid or gas.

### *4.5.1 Principles of Lubrication*

There are two well-defined and extreme conditions of lubrication: (i) fluid film or hydrodynamic lubrication; and (ii) boundary lubrication, which is an intermediate or transition condition of lubrication also known as quasi-hydrodynamic or mixed lubrication. In the first type of lubrication the lubricant film thickness is the key parameter and is the best way to minimize the wear and surface damage. The main feature is that the fluid film is considerably thicker than the dimensions of surface irregularities. The viscosity of the lubricant is the most important factor since it determines the friction, temperature rise, and rate of flow of lubricants.

In the case of boundary lubrication the chemical properties of the lubricant and the bearing surfaces play an important role. A lubricant having polar molecules forms adsorbed layers that may prevent a direct metal-to-metal contact. The extent to which an adsorbed film can prevent the metal-to-metal contact is limited. Under certain load and temperature conditions, the boundary film will fail, so that direct contact *will* take place. This condition is called extreme boundary lubrication. Under these conditions seizure of metallic surfaces and the destruction of one or both of the surfaces begins. The intermediate condition of the above two is called mixed lubrication, and this occurs when both the above mechanisms work together at the same time in the same machine element.

The other type of lubrication is the elsatohydrodynamic lubrication, which refers to highly loaded contacts. In this case, the load is on a very small contact area in the hydrodynamic lubrication. Due to load on the very small area this load can lead to change in the viscosity and elastic deformation of the surface in contact. The increased viscosity leads to increased film thickness and due to larger pressure on the contacts the surface gets flattened locally which results in enhanced load carrying area and finally results in enhancement of film thickness. The result is called elastohydrodynamic lubrication.

Figure 4.3 shows the "Stribeck Curve" named after Richard Stribeck. The stribeck curve shows the graph between the coefficient of friction (CoF) and Hersey number. The hersey number can be defined as $\eta N/P$, Where, $\eta$ = dynamic **viscosity**; N = sliding speed and P = load.

The stribeck curve shows the nature of variation of coefficient of friction and different lubrication regimes. The Hersey number is a dimensionless parameter.

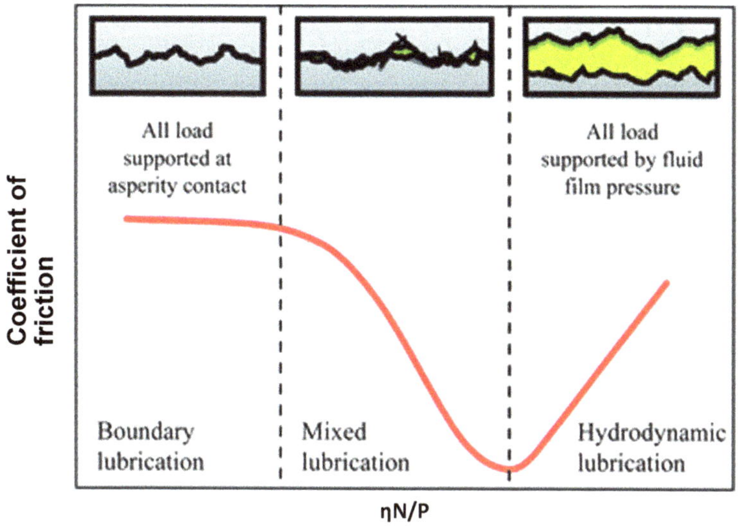


**Fig. 4.3** Stribeck curve

### 4.5.2 *Structure of Surfaces*

The requirement/application of the lubricants depends on the surface texture of the rubbing surfaces. Any engineering surface can have different types of surface roughness, and the surface roughness level differs widely from surface to surface. Therefore, to understand the nature of roughness present on the surface there are many surface characterization techniques available, e.g. profilometry, AFM, and SEM examination. Based on these characterizations the surfaces can principally have three types of roughness and these three types of roughness together form the profile. These are all shown in Fig. 4.4.

### 4.5.3 *Salient Features of Lubricants*

There are four important points related to lubricants and their application:

(i) The periodicity of application of lubricants
(ii) Application methodology
(iii) Proper quality
(iv) Proper quantity.

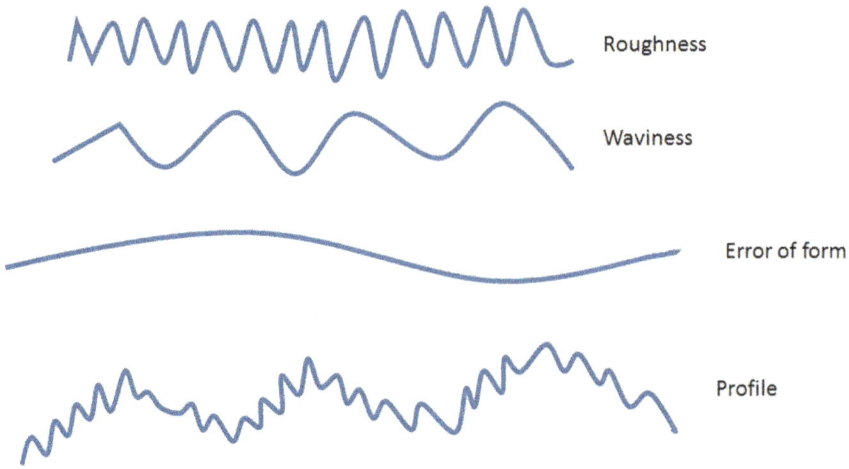


**Fig. 4.4** Surface structure of rubbing surfaces

### *4.5.4 Purposes of Lubrication*

There are many purposes as to why lubricants are applied. The most important purposes are:

1. Reduction of friction and mechanical power required to run the system.
2. Reduction of wear.
3. Reduction in excessive heat generation, which may result in serious component failures.

Apart from these three important and most crucial points, there are many other purposes that are also important, and lubrication may help in improving the efficiency of the machine, these are:

1. Removal of wear debris.
2. Rust prevention.
3. Extension of useful life of a machine and its components.
4. Reduction in breakdowns and unscheduled maintenances.
5. Prevention of inner environment from outer environment.

### *4.5.5 Types of Lubricants*

Based on the physical state of the lubricant, it is classified into three main categories.

1. Liquid lubricants: All oils come into this category.
2. Solid lubricants: Examples are graphite, molybdenum disulphide and tungsten disulphide. All these materials exist in the form of solid powders.
3. Semi-solids: All types of greases come into this category of lubricants.

#### 4.5.5.1 Liquid Lubricants

(A) **Crude oil**: Crude petroleum product was formed millions of years ago and is believed to be formed from various types of vegetable and animal remains that settled on seabeds. Owing to pressure and temperature, these materials underwent various changes and were converted into the different types of hydrocarbons and other constituents of crude oil. Crude oils are very complex mixtures consisting mainly of different types of hydrocarbons with 50 carbon atoms per molecule, or more. Crude oils are found in a variety of types, e.g. light color oil consists mainly of gasoline, and black colored oils of solid asphalts. The oils have many types of impurities and undesired products, and need refining by various techniques. Nearly all lubricating oils are obtained from refining crude oils and are called mineral oils.

(B) **Mineral and blended oils**: Mineral oils are blended with certain types of additives to obtain some special features/properties, and these oils are called blended oils. However, blended oils do not possess various properties like high-temperature sustainability and application/use under high pressure.

(C) **Synthetic oils**: Synthetic oils have unique combinations of properties like high-temperature stability, extended temperature range, and longer lifetime even in an extreme operating condition. Unlike mineral oils, no refining is done. Types of synthetic oils are:

  (i) Diester based
  (ii) Phosphate ester based
  (iii) Silicone oil
  (iv) Poly-glycidyl ethers
  (v) Poly-alkylene glycols.

All of these liquid lubricants should be characterized for important properties e.g. viscosity, viscosity index, pour point, flash point, fire point, neutralization number and insoluble rating or precipitation number.

### *4.5.6 Lubricant Additives*

Lubricant additives are generally organic materials and sometimes inorganic materials. These additives are always added in low concentrations to improve the overall performance of the lubricant in terms of physical, chemical and tribological properties. Based on the type of property to be enhanced, there are three types of additives:

a. **Physical additives**: These additives improve the physical property of the oil, e.g. pour point depressants, dispersants, viscosity index (VI) improvers, demulsifies, foam inhibitors, and tackiness agents.

b. **Chemical additives**: Examples are detergents, seal swell agents, metal deactivators, antioxidants, corrosion inhibitors, anti-rust agents, biocides and peroxide decomposers.
c. **Tribological additives**: These are friction modifiers, antiwear agents, extreme pressure additives and lubricity improvers; for example Zinc diethyl dithiophosphate (ZDDP), magnesium sulphonates, aromatic amines, phenols, silicone oils.

Different types of additives are used in combinations known as additives packages. Depending upon the uses/application, these additive packages differ for different types of lubricating oil.

## 4.6 Nanomaterial-Based Tribological Additives

Nanolubricant is the lubricant containing a stable suspension of nanoparticles intended to enhance the tribological properties. Different types of nanoparticles are reported which enhance the tribological properties of base oil, when stably suspended in small quantity. Table 4.1 summarises the most used synthesis methods for different nanoparticles that can be used in Nano lubricant formulations.

Development of new and efficient Nano lubricating fluids is driven by the automobile industry with the objective to develop high-efficiency automobile systems (Li et al. 2006). As discussed previously, solid lubricants (e.g. molybdenum disulphide, tungsten disulphide and graphite) are used as friction modifiers. In many recent studies it has been shown that nanoparticles of these materials and many other materials are much more efficient lubricant additives compared to their bulk materials with larger flake size (Lee et al. 2009a; Wu et al. 2007; Chinas-Castillo and Spikes 2003; Hu et al. 2002; Rapoport et al. 2003; Zhang et al. 2014; Park et al. 2011; Chou and Lee 2008; Shenoy et al. 2012; Hwang et al. 2011).

Nanoparticles then, when used as lubricant additives, follow different mechanisms for their lubrication action. Various groups have proposed many mechanisms for the lubricating behaviorof nanoparticles. Wu et al. and Chinas-Castillo et al. have shown in their work the rolling/sliding effect (Hwang et al. 2011). Hu et al. explored the mechanism of the lubricating behaviour of Nano magnesium borate as a lubricating oil additive that follows the protective film mechanism (Hwang et al. 2011). Rapoport et al. worked on tungsten disulphide nanoparticles under mixed lubrication and showed the mechanism of third body effect (Hwang et al. 2011). Liu et al. investigated the mechanism of copper Nanoparticles on a tribologically stressed surface and showed the mending effect (Hwang et al. 2011).

Figure 4.5 shows that in the case of Nano lubricants there is a shift of the mixed film lubrication regime owing to micropolishing. This happens because asperity contacts are reduced by nanoparticles filling-in valleys between the asperities, thereby smoothing the surface, as shown in Fig. 4.6.

**Table 4.1** Nanoparticles synthesis methods

| Particles | Methods of synthesis | Particle sizes or size range (nm) | Shape | Reference |
|---|---|---|---|---|
| $MoS_2$ | Sono chemical | 15 | Nanoparticles | Mdleleni et al. (1998) |
| $MoS_2$ | Wet chemical | 21 | Nanoparticles | Betke and Kickelbick (2014) |
| $MoS_2$ | Solvothermal | 10 to 50 | Nano sheets and Nano flowers | Vattikuti and Byon (2015) |
| $MoS_2$ | Hydrothermal | Spherical and lamellar hollow multilayer (7–9 layers) 40–70 | Spherical and lamellar hollow multilayer nanoparticles | Shyyko et al. (2016) |
| $MoS_2$ | Chemical vapour deposition (CVD) | 0.632 | Nanoparticles | Gao and Zhang (2014) |
| BN | Milling | 100 | Nano sheets | Greco et al. (2011) |
| $TiO_2$ | Co-precipitation | 20–25 | Nanoparticles | Ingole et al. (2013) |
| $WS_2$ | Solvothermal method | 50 to 200 | Nanoparticles | Sun et al. (2011) |
| $WS_2$ | Micro emulsion | 7–12 | Nanoparticles | Ghoreishi et al. (2012) |
| $WS_2$ | Mechanical activation | 10 | Nano sheets | Wu et al. (2010) |
| $WS_2$ | High-energy ball milling | 10–15 | Nano rods | Zhang et al. (2007) |
| CuO | Hydrothermal microwave | 5 | – | Alves et al. (2016) |
| CuO | Chemical | 61, 91,78 | Nanoparticles, Nano rods, Nanorhombics | Farbod (2015) |
| Nano carbon (Nano diamond) | Detonation of high energy explosives | 5–10 | – | Nunn et al. (2015) |
| Graphite | Ball milling | 500 | Nano sheets | Huang et al. (2006) |
| Graphite | Chemical oxidation and microwave puffing | 18 | Nano sheets | Yu and Qiang (2012) |
| Graphite | Chemical | 3–30 | Nano sheets | Hargreaves and Cooper (2016) |

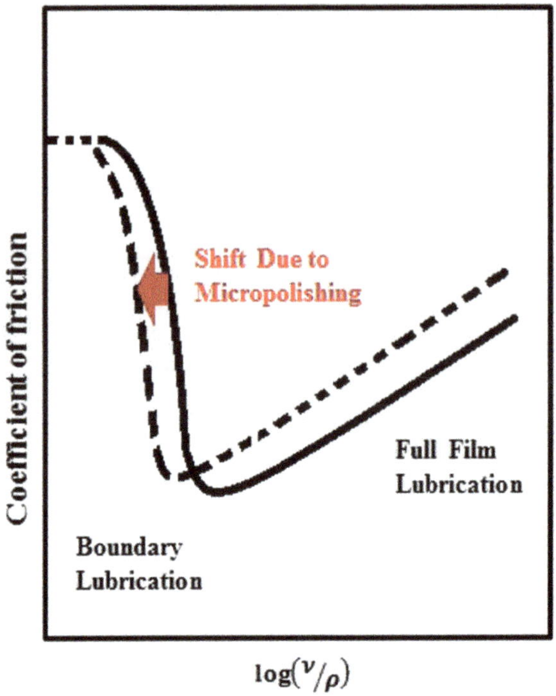


**Fig. 4.5** Shifting of the stribeck curve in the mixed film regime

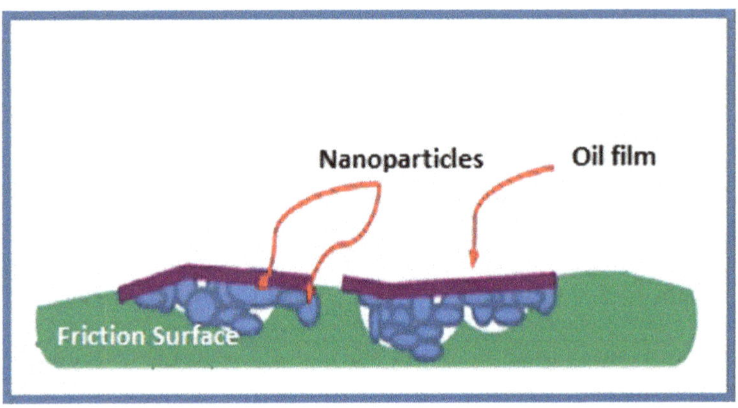


**Fig. 4.6** Filling-in of valleys between asperities by nanoparticles

In the case of automobile engines, 5% of the total energy generated is wasted as frictional loss. The present technological need is to understand the surface topography at the nano-level and develop new surfaces and coatings with low friction and wear. In recent years, world-wide groups have done tremendous work on the tribological application of carbon nanomaterials, but no review has covered all these. In this chapter, we shall discuss in Sect. 4.6.1 the tribological application of all types of carbon nanomaterials. Carbon-based lubricant additives are especially important because of their exceptional properties and low shear strength, which is duc to weak Van Dcr Waals' forces acting between two consecutive layers of the lubricant.

## *4.6.1 Carbon Nanostructures and Nano Molybdenum Disulphide Based Lubricant Additives*

### 4.6.1.1 Carbon Nanotube (CNT) Based Lubricant Additives

Carbon nanotubes (CNTs) are one-dimensional nanomaterial with cylindrical shapes, e.g. Fig. 4.7. They can be both single-walled and multi-walled (MWNTs), and have been extensively studied for their antiwear, friction modifying and lubricating properties for the last two decades (Zhang et al. 2014), and are potential tribological additives owing to some exceptional properties, e.g. high stiffness, high strength, high-chemical inertness, high thermal conductivity, and an $sp^2$ bonded carbon network. Zhang et al. have well covered the recent work in their review (Zhang et al. 2014).

CNTs could be used for tribological application in three different ways: as additives; CNT-Polymer composites and CNT-Metal/ceramic composites. Many mechanisms have been proposed for an understanding of the tribological action of CNTs, but for real applications there are issues and challenges (Zhang et al. 2014; Park et al. 2011). For example, when CNTs are used as additives in any oils, the stability of the suspension is very low and preparation of stable suspensions remains difficult.

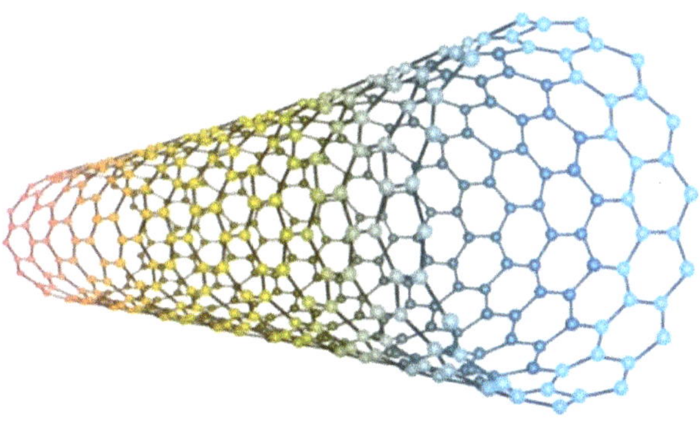

**Fig. 4.7** Single-walled carbon nanotube

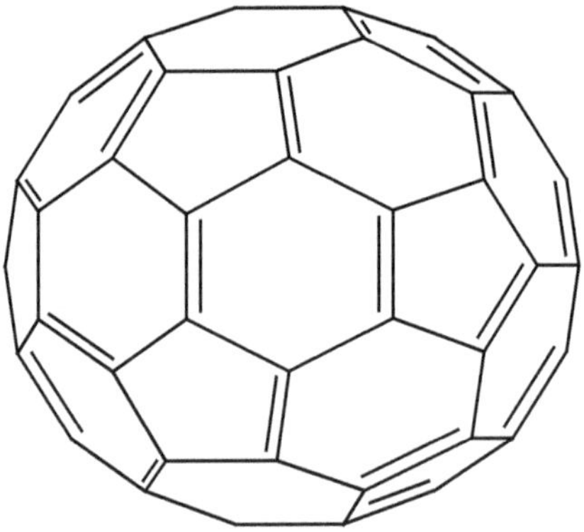

**Fig. 4.8** Fullerene (C60) molecule example

#### 4.6.1.2 Fullerene-Based Lubricant Additives

Fullerenes, which have 60 carbon atoms in a spherical structure, see Fig. 4.8, and are zero-dimensional materials, have shown much potential as lubricant additives. During the last two decades various groups have worked on fullerenes for exploring their tribological properties. Gupta et al. have experimentally demonstrated the role of fullerenes as lubricant additives for oils and greases (Gupta and Bhushan 1994). These zero-dimensional materials not only reduce friction but also show wear reduction when used as an additive in oils (Ginzburg et al. 2002). Such results demonstrate the potentials of fullerenes as lubricant additives.

#### 4.6.1.3 Graphene-Based Lubricant Additives

Graphene is a two-dimensional, atomically thick solid material that has the potential for many technological applications. These include its tribological properties, which, however, have not been much investigated. Berman et al. have briefly covered the work done on all aspects of graphene tribology in a recent review (Berman et al. 2014). Graphene can easily be coated on any substrate at nano to micro levels by solution phase coating. Owing to this easy coating technique, graphene is becoming increasingly popular as a solid lubricant and lubricant additive. Based on many computational studies, various mechanisms have been proposed for the frictional mechanism of graphene on the nanoscale (Smolyanitsky et al. 2012; Guo et al. 2007).

#### 4.6.1.4 Nano Graphite-Based Lubricant Additives

Graphite is another carbon-based solid lubricant and the most widely used among all of them. The lubrication property of graphite is due to its lamellar structure and low shear strength. Application of Nano graphite instead of micron size graphite is preferred for many reasons, e.g. better dispersability and capability of filling nano asperities. Nano graphite can be used for high-temperature applications in various engine components and other machinery tools, e.g. open gear, ball bearings, and machine-shop works. Graphite is also commonly used for lubricating locks, since a liquid lubricant allows the particles to get stuck in locks, thereby worsening the problem. Another advantage of Nano graphite is that it is best suited for lubrication under atmospheric conditions, since water vapour is a necessary component for graphite lubrication: water reduces the bonding energy between the hexagonal planes of graphite to a lower level than that of the adhesion energy between a substrate and the graphite. Also, in an oxidising atmosphere graphite is effective at high temperatures up to 450 °C continuously, and can withstand much higher temperature peaks.

#### 4.6.1.5 Nano Molybdenumdisulphide Based Lubricant Additives

$MoS_2$ belongs to a family of the nearly forty different layered transition metal dichalcogenides (TMD). These offer exciting opportunities for various applications primarily due to their crystal structure. The crystal structure of $MoS_2$ consists of molybdenum atoms loosely bound by Vander Waals' interactions between two layers of sulphur atoms. The S-Mo-S layered structure shows easy sliding contact that results in a low coefficient of friction.

$MoS_2$ has many technologically important applications in various fields e.g. semiconductors, sensors, and catalysis. However, its applicability for a positive outcome is very dependent on the purity with respect to the stoichiometry and phase. Therefore, there is a need to study the synthesis methods in order to obtain better yields with high purity. Generally, there are two synthesis approaches for obtaining $MoS_2$ nanoparticles: physical methods and chemical methods.

## 4.7 Thermally Conducting Fluids

Cooling and efficient heat dissipiation is one of the most frequently occurring challenges faced by various industries. Diesel engine components are evolving rapidly due to continuing demands for more horsepower and better fuel efficiency. Most aircraft and missiles face the problem of heating of the engines and aerodynamic heating of other parts. Hence cooling systems must sometimes operate at high temperatures while transferring heat to the surrounding environment. Water or a conventional heat transfer fluid suffice for many applications. But there is need for enhancement of the efficiency of cooling for hi-tech supersonic and subsonic engine systems, and supercomputer systems. Also, efficient cooling systems is one of the most important technology resulting in manufacturing of various electric vehicles, which operate at comparatively much lower temperatures than in IC engines. Thus there is a demand for the development of a new, highly efficient coolant.

Conventional coolants have poorer heat transfer properties than most crystalline solids. Therefore, it has been predicted that fluids containing suspended solid particles should have more effective transfer than conventional heat transfer fluids. Maxwell's model (Maxwell 1873) explains the effect of volume fraction of solid spherical particles dispersed in any base fluid on effective thermal conductivity of solid fluid suspension. Heat transfer is affected by the effective surface area of particles and increases with increasing surface area. Therefore, for an efficient heat transfer fluid it is desirable to have a large surface area/volume ratio to gain the effective heat transfer. Hamilton and Crosser (Hamilton and Crosser 1962) focussed on the influence of an enhancement in the particle surface area by controlling the shape of micron-sized particles. However, these gave only one order of magnitude improvement. Also, the micron-sized particles settle out of the suspension, and they are unsuitable for emerging miniaturized technologies because they cause clogging of the tiny channels in these devices. These factors coupled with the fact that nano-sized particles afford the maximum surface area to volume ratio, provide motivation for the engineering fluids with nanoparticles suspended in it (Hamilton and Crosser 1962). Several types of solid particles were tried, like oxide ceramic particles, metal nanoparticles, and multiwall nanotubes. Here we review some results obtained with each of these types.

### 4.7.1 Oxide Ceramic Nanofluids

The oxide ceramic based Nanofluids have been studied for their heat transfer properties at lower particles concentration (1–5%). Lee et al. conducted experiments to assess the thermal conductivity of Nanofluids containing copper oxide and aluminum oxide nanoparticles in water and ethylene glycol. They discovered that these Nanofluids exhibited significantly higher thermal conductivities compared to the same liquids without nanoparticles (Lee et al. 1999). The greatest enhancement in thermal conductivity (20%) was observed in ethylene glycol when it was mixed with 4% CuO nanoparticles with an average diameter of 35 nm. Wang et al. also investigated the effective thermal conductivities of Nanofluids with $Al_2O_3$ nanoparticles measuring 28 nm and CuO nanoparticles measuring 23 nm (Wang et al. 1999).

The effective thermal conductivities were seen to increase as the average particle size decreased and was seen to depend on the dispersion technique. Xie et al. (2002) concluded from their studies on alumina nanoparticles The thermal conductivity of Nanofluids is influenced by both the particle size and the pH of the aqueous suspension, as well as the thermal conductivity of the base fluid.

Das et al. (2003) conducted experiments using the temperature oscillation technique to investigate the thermal conductivities of Nanofluids containing CuO or $Al_2O_3$ nanoparticles in water. They made a significant finding that within a small temperature range of 21–51 °C, there was a two- to four-fold increase in thermal conductivity. Moreover, Nanofluids with smaller CuO nanoparticles demonstrated the highest enhancement. Das and colleagues also proposed that the strong temperature dependence of thermal conductivity is attributed to nanoparticle movement. This discovery suggests that Nanofluids could be considered smart fluids, as they can sense their thermal environment, making them desirable coolants for high energy density devices.

### 4.7.2 *Metallic Nanofluids*

Copper nanoparticles with an average diameter <10 nm were dispersed in ethylene glycol at loadings of up to 0.5% by volume, resulting in minimal agglomeration. The addition of less than 1% vol. of thioglycolic acid further improved the dispersion of nanoparticles. Thermal conductivity was measured at room temperature (Eastman et al. 2001; Asirvatham et al. 2009). These studies demonstrated that metallic nanoparticles significantly enhance thermal conductivity compared to the same volume fraction of oxide nanoparticles. Furthermore, the inclusion of thioglycolic acid as a stabilizing agent showed a remarkable improvement compared to Nanofluids without acids. A mixture of approximately 0.3% vol. Cu nanoparticles dispersed in ethylene glycol with 1% thioglycolic acid increased effective thermal conductivity by 40%. Notably, fresh fluids exhibited higher conductivities than those stored for up to two months.

### 4.7.3 *MWNT Nanofluids*

The two-step method was used to produce oil suspensions of MWCNTs with loadings up to 1% by volume. The thermal conductivity of these suspensions was examined at room temperature (Choi et al. 2001). Figure 4.9 demonstrates a significant enhancement in thermal conductivity (150%) due to the presence of nanotubes, which is the highest ever observed in a liquid medium. Comparable results have been obtained for polymer/nanotube composites as well.

The measured thermal conductivities are non-linear with nanoparticle loadings. However, all existing theories predict a linear relationship (Bonnecaze and Brady 1990, 1991).

### 4.7.4 *Summary for Sects. 4.7.1–4.7.3*

From the information presented in Sects. 4.7.1–4.7.3 it is observed that oxide Nanofluids have the lowest thermal conductivity enhancements amongst the three types of Nanofluids studied. In addition, because of the abrasive, nature of ceramic oxides, these materials are not used for heat transfer applications. In addition, compared to oxide Nanofluids the solid concentrations in copper or nanotube Nanofluids are one order of magnitude lesser for a similar thermal conductivity

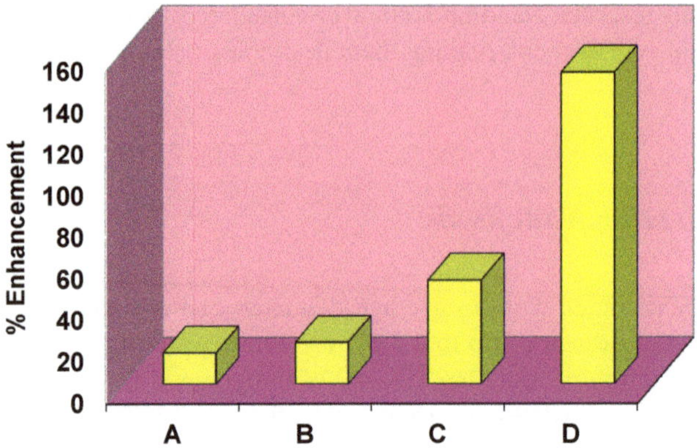


A: 4.3 vol % $Al_2O_3$; 33 nm size B: 4 vol % of CuO; 35 nm size
C: 0.5 vol % Cu ~ 10 nm size D: 1 vol % MWNTs

**Fig. 4.9** Enhancement of thermal conductivity by different types of Nanofluids

enhancement. Hence, it is possible to produce metallic or nanotube Nanofluids that are less viscous than oxide Nanofluids. Table 4.2 lists the dispersion methods and maximum thermal conductivity ratios of the Nanofluids covered in this chapter.

**Table 4.2** Dispersion methods and maximum thermal conductivity ratios for a wide range of nanofluids covered in this chapter

| Nanomaterials | Base fluids | Dispersion methods | Loading vol% | Max $k_e/k_f$ | Refs. |
|---|---|---|---|---|---|
| Cu | Ethylene glycol | One step method Thioglycolic acid | 0.3 | 1.4 | Eastman et al. (2001) |
| MWNT | Synthetic PAO | Two step method Polymer | 1.0 | 2.57 | Bonnecaze and Brady (1991) |
| $Al_2O_3$ | Water<br>Ethylene glycol | Two step method | 4.0<br>5.0 | 1.09<br>1.19 | Lee et al. (1999) |
| CuO | Water<br>Ethylene glycol | Two step method | 3.4<br>4.2 | 1.12<br>1.22 | Lee et al. (1999) |
| $Al_2O_3$ | Ethylene glycol | Two step method Mechanical/ultra sonication | 8.0 | 1.4 | Das et al. (2003) |
| $Al_2O_3$ | Water | Two step method Filtration | 3.0 | 1.12 | Das et al. (2003) |
| Soot particles | Diesel engine oil | Motor oil dispersant | 1.7 | 1.14 | Bonnecaze and Brady (1990) |

$k_e$: effective thermal conductivity of Nanofluids
$k_f$: thermal conductivity of the base fluids
$k_e/k_f$: Increment in thermal conductivity with respect to the base fluid

## 4.8 Potential Mechanisms for Enhanced Conductivity

While classical thermal conductivity models have been validated for mixtures with low concentrations (Bonnecaze and Brady 1991), the experimental data used in these validations only considered particles of micron sizes (Meredith and Tobias 1961; Turner 1976). Consequently, these theories, which were designed for fluids with particles significantly larger than nanoparticles, cannot accurately forecast the thermal characteristics of Nanofluids.

Keblinski et al. (2002) investigated four potential mechanisms explaining the enhanced heat transfer in Nanofluids: Brownian motion of particles, molecular-level layering at the liquid/particle interface, heat transport within the nanoparticles, and nanoparticle clustering effects. Similar to Wang et al. (1999), they found that Brownian motion has minimal impact on energy transport. However, they demonstrated that interface layering of liquid molecules and nanoparticle clustering can facilitate rapid heat transport. Through molecular dynamics simulations, they observed that heat moves in a ballistic manner rather than through diffusion inside nanoparticles. They noted that phonons oscillate within the nanoparticles with a period corresponding to the time it takes for a phonon to travel across a particle, which increases with particle size.

**N.B**: All classical models predict heat transport through diffusion, whereas the model of Keblinski et al. (2002) uses a ballistic approach. This suggests that macroscopic heat transfer may not be a relevant parameter for the thermal transport properties of Nanofluids. On the other hand, the interfacial transmission coefficient might be, i.e. it may determine how much heat can go through the solid/liquid interface, rather than getting reflected back within particles.

In Nanofluids, the presence of solid-like liquid layers creates an intermediate physical state between a solid and bulk liquid. These solid-like nanolayers play a crucial role in enhancing thermal conductivity as they act as a thermal bridge between the solid nanoparticles and the bulk liquid. Hence, the inclusion of these layers is essential for improving heat transfer in Nanofluids.

Non-equilibrium molecular dynamics (MD) simulations in which a temperature gradient is imposed (Xue et al. 2003) also studied the solid–liquid interfacial resistance. These simulations reveal that the bonding strength between liquid and solid atoms plays a key role in determining the interfacial thermal resistance. Thus the thermal conductivity of the solid particles might be much less important than the thermal resistance at the solid–liquid interface. The thermal resistance in a wetting liquid for nanometer-sized particles is small and gives rise to enhanced thermal conductivities. Without going into the mathematical details, we can simply say that these findings indicate that for nanoparticles in a non-wetting liquid the particle–matrix interfaces will exhibit a high thermal resistance, while for wetting liquids the thermal resistance will be small and equivalent to the temperature drop over several atomic distances (1 nm).

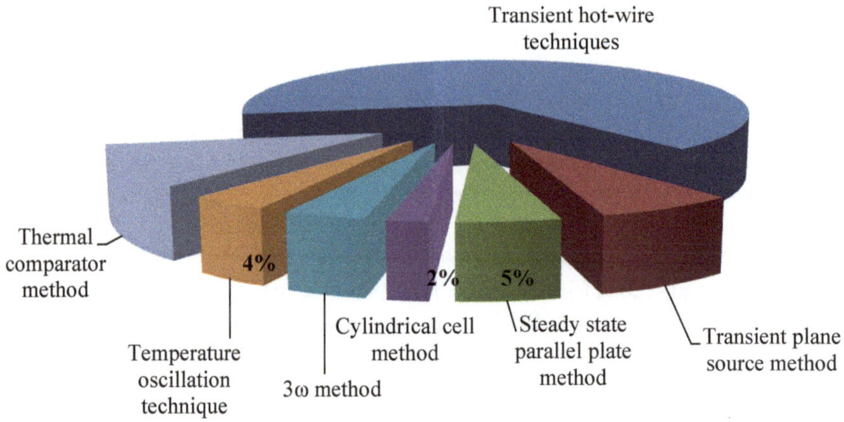


**Fig. 4.10** Shares of the use of different thermal conductivity measurement techniques (Paul et al. 2010; Lee et al. 2009b)

It is important to note that measurement of thermal properties has been a challenge for a long time. Different methods and techniques have been used for measuring the thermal conductivity of Nanofluids. These are:

(i) Transient hot-wire techniques
(ii) Thermal constants analyzer techniques
(iii) Steady-state parallel-plate method
(iv) Cylindrical cell method
(v) Temperature oscillation technique
(vi) 3ω method
(vii) Thermal comparator method.

The Transient hot-wire techniques (ASTM D5930 and D7896) and Thermal comparator methods (ASTM E289, E207, and E967)are the major techniques used, see Fig. 4.10.

## 4.9 Stability of Nanofluids

The stability of Nanofluids is one of the most important parameters influencing their properties. The stability is achieved by several methods. Generally, the common techniques used for the preparation of stable Nanofluids are:

- Addition of surfactants or surface modification
- pH control
- Ultrasonic vibration (agitation).

### 4.9.1 Effects of Surfactants or Additives

Surfactants are chemical compounds and reduce the surface tension of the base fluid and increase absorption of the particles. Another effect is to stop clustering and agglomeration. The following are some popular surfactants:

- Oleic acid
- Cetyltrimethylammoniumbromide (CTAB) (Zhu et al. 2009)
- Dodecyl trimethylammonium bromide (DTAB)
- Sodium octanoate (SOCT)
- Sodium dodecyl benzene sulphonate SDBS (Zhu et al. 2009)
- Sodium dodecyl sulphonate (SDS) (Turner 1976)
- Gum Arabic (Betke and Kickelbick 2014)
- Polyvinylpyrrolidone (PVP) (Choi et al. 2001)
- Hexadecyltrimethylammoniumbromide (HCTAB)
- Salt.

It should be pointed out that addition of surfactants is not effective for Nanofluids working at high temperatures. This is probably because of damage to the bonding between the surfactant and nanoparticles (Zhu et al. 2007; Assael et al. 2005). One investigation found that 60 °C is the critical temperature (Zhu et al. 2009).

### 4.9.2 pH Control

It is reported that the stability of Nanofluids increased when the pH changed, owing to strong repulsive forces (Zhu et al. 2009; Assael et al. 2005). Xie et al. (2003) observed this after a simple acid treatment of CNTs in a water-based solution. Lee et al. (2009b) have investigated the effect of agglomeration on $Al_2O_3$ based Nanofluids by changing the pH. Wang et al. have found two optimum pH values of 8 for $Al_2O_3$/water and about 9.5 for Cu/water (Wang and Zhu 2009). Fovet et al. (2001) studied the effect of pH on stability and thermal conductivity of titanium dioxide based Nanofluids.

### 4.9.3 Ultrasonic Agitation (Vibration)

Ultrasonic agitation is one of the most frequently used techniques for stabilization of Nanofluids. There are various reports where ultrasonication has been used for the preparation and stabilization of Nanofluids. Munson et al. (2006) have provided a good description of the effect of sonication on various Nanofluids.

### *4.9.4 Functionalization of Nanoparticles*

This technique is also useful for the enhancement of long-term stability of Nanofluids. This is a surfactant-free technique, whereby direct injection of functional Nanoparticles into the base fluid can provide better stability (Yu and Xie 2012). The stability of CNT-containing Nanofluids can be improved by introducing hydroxyl groups onto the surfaces of CNTs (Chen and Xie 2010).

## 4.10 Nano Lubricating Fluid Developments at DMSRDE, DRDO

DMSRDE, Kanpur has worked for the development of nanolubricating fluids. The different types of naoparticles have been synthesized and used for the development of nanolubricating fluids for enhancement of the tribological properties of existing lubricants. Two major types of nanoparticles which have been synthesized and studied in detail are Molybdenum disulphide Nanoparticles and Carbon Nanomaterials.

### *4.10.1 Molybdenumdisulphide-Based Nano Lubricating Fluids*

Ultrafine molybdenum disulphide particles ($MoS_2$) were successfully developed by a wet chemical synthesis method. The synthesis is carried out in an aqueous solution using ammonium molybdate, citric acid, and ammonium sulphide as the reactants. The quantities of citric acid, as well as molar concentration of the reactants, play a significant role in synthesis. In the present investigation, a dispersion unit has been used at the reduction step, since stirring speed plays an important role to obtain $MoS_2$ nanoparticles. Characterizations of $MoS_2$ were carried out by Fourier Transformation Infrared Spectroscopy (FTIR), Raman spectroscopy, Electron Microscopy (TEM), X-ray Diffraction XRD, and Thermo-Gravimetric Analysis (TGA) studies. The obtained nanoparticles were spherical, amorphous, monodispersed and with a size of $\leq$5 nm, see Fig. 4.11. The thermal stability of the $MoS_2$nanoparticle samples was retained up to 400 °C.

$MoS_2$ nanoparticles (<20 nm) synthesized were used for preparation of the nanolubricating fluid, in a base oil (SAE-30 engine oil). The tribological properties were studied by Four ball tester and Pin-on-Disk instruments. The Four ball tester wear study confirmed a wear reduction of 20%, and the Pin-on-Disc test showed a 27% reduction of the coefficient of friction, both with respect to commercially available nanoparticles of size 70 nm. The above enhancement in tribological properties is because of particles size reduction from commercial Nano $MoS_2$ (70 nm) to synthesized nanoparticles of size <20 nm. See also Fig. 4.12.

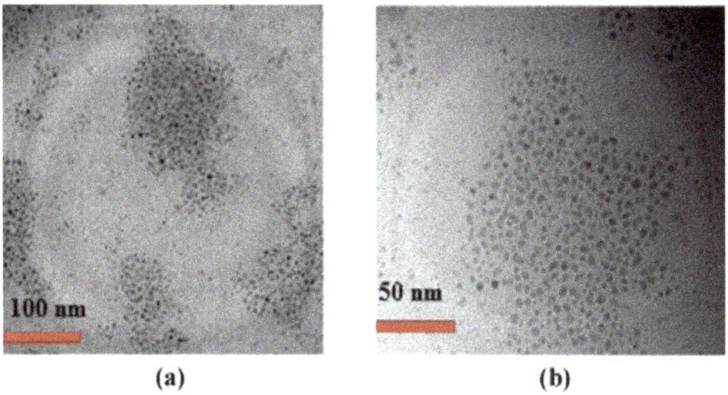


**Fig. 4.11** **a** TEM image of molybdenum disulphide nano particles (<20 nm), **b** magnified view

## *4.10.2 Carbon Nanomaterials-Based Nano Lubricating Fluids*

Carbon nanostructures-based nanomaterials, which can be used as lubricant additives are discussed in Sect. 4.6.1. Graphene, fullerenes and CNTs are expensive and difficult to produce on a large scale. Therefore they are not used for large scale production of Nanolubricants. The carbon nanomaterials that DMSRDE is using for nanolubricant preparation have low production costs and the production is easy to scale-up. These materials are discussed in the following subsections.

### 4.10.2.1 Nano Graphite

Nanographite has been reported to be used as a lubricant additive, but the major challenge with nanographite is its size and poor dispersability in Nanofluids. The smaller the size, the better its dispersion stability. In the present work, Nanographite of size <50 nm has been synthesized and characterized, and a nanolubricating fluid based on it has been developed. This fluid has good dispersion stability and greatly enhanced tribological properties compared to the neat lubricant.

Nano graphite was synthesized starting from micrographite by a simple route, Fig. 4.13. This procedure of nanographite synthesis results in graphitic nanoparticles of size ~50 nm, as confirmed by TEM and Dynamic Light Scattering (DLS) analysis, Fig. 4.14a–c. The synthesized material had a very uniform particle size distribution which was evident from Dynamic Light Scattering (DLS) analysis which showed a very low value of the PolyDispersity Index (PDI) i.e. (<1) Fig. 4.14c. This was also evident from the TEM analysis. The material was further characterized by different characterization techniques e.g. Selected Area Electron Diffraction (SAED), FTIR spectroscopy, Ultraviolet–Visible (UV–Vis) spectroscopy and Raman spectroscopy. All characterization results were well correlated.

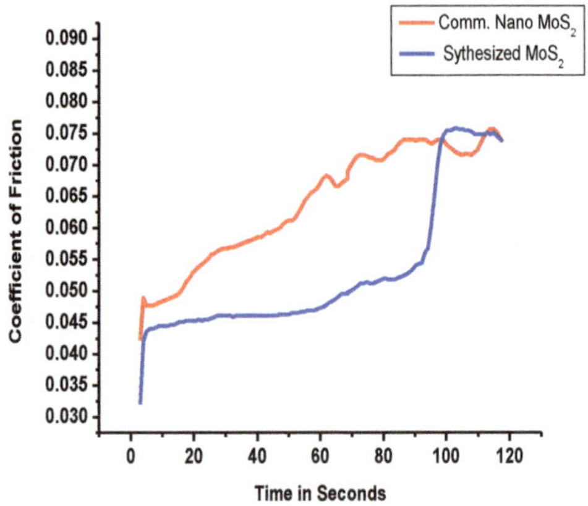


(a)

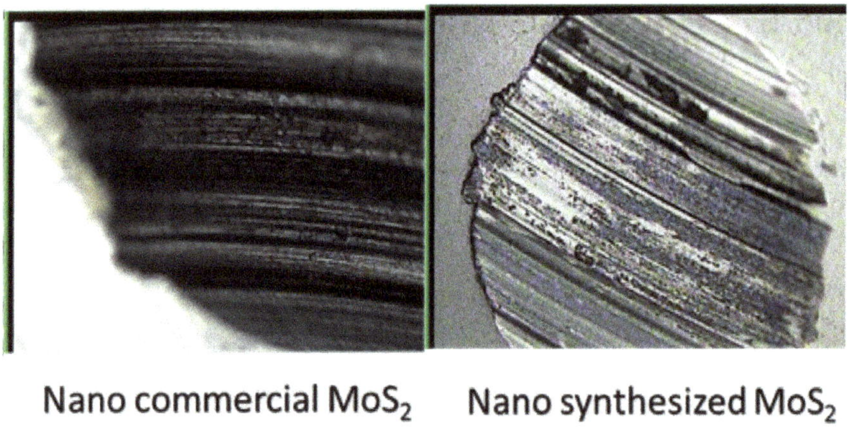


(b1) (b2)

**Fig. 4.12** **a** Variation of coefficient of friction for commercial and synthesized nanoparticles at 50 N load; **b1** and **b2** wear scar images

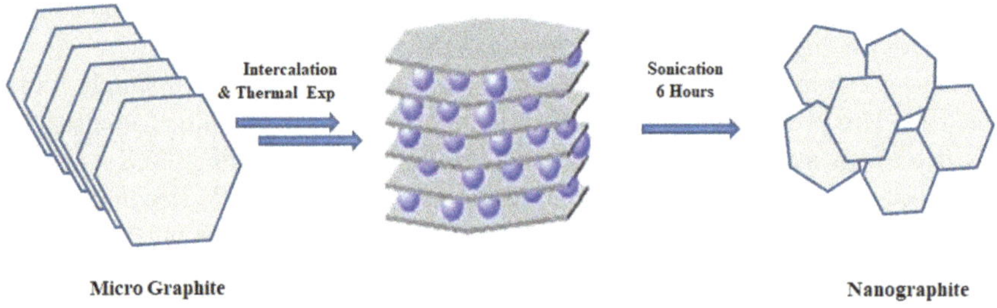


**Fig. 4.13** Nanographite preparation scheme, starting from micrographite (<20 μm)

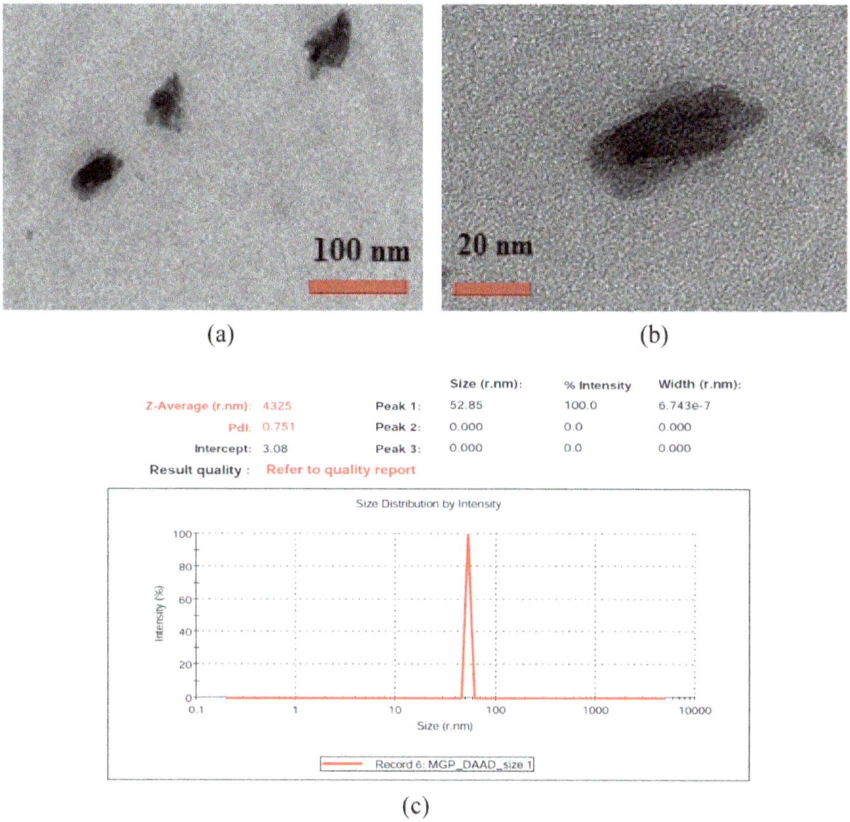


**Fig. 4.14** **a** TEM of nanographite, **b** magnified view of nanographite: magnified view, **c** DLS analysis result of nanographite

#### 4.10.2.2 Graphitic Carbonnanoparticles

Graphitic carbon nanoparticles are a new member in the lubricant family, first prepared by Singh et al. (2018). These materials are mostly spherical in shape, and not only help in stable dispersion of the materials in Nanofluids, but also enhance the tribological properties. The probable mechanism for enhanced tribological properties is suggested to be the ball-bearing mechanism. Carbon nanoparticles were synthesized from candle soot and subjected to fractional isolation and characterization. Highly thermally stable (onset of degradation 697 °C) and crystalline carbon nanoparticles (size 20–25 nm) of superior purity were isolated by a multi-stage separation and isolation process. During this process various other fractions with sizes less and more than the size 20–25 nm were isolated and characterized. The TEM image and DLS analysis result of carbon nanoparticles are shown in Fig. 4.15 (Singh et al. 2018).

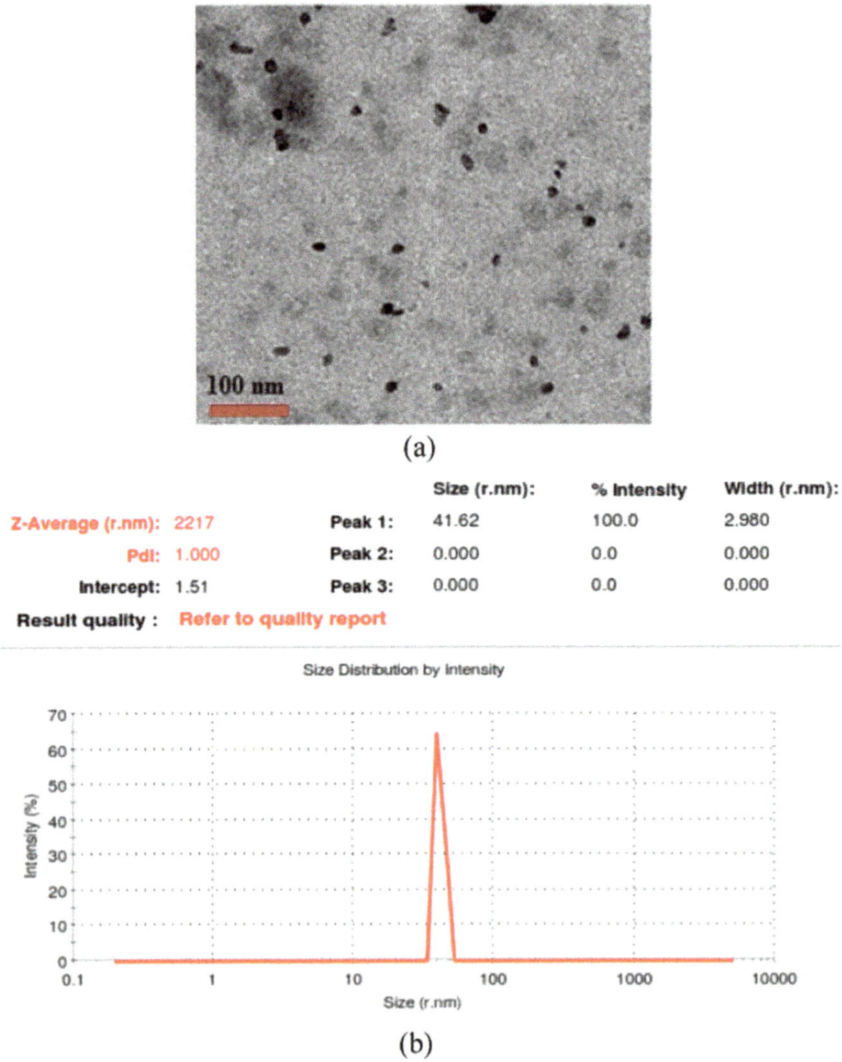


**Fig. 4.15** **a** TEM image of carbon nanoparticles and **b** dynamic light scattering (photon correlation spectroscopy) results for carbon nanoparticles

Nano lubricating fluids based on graphite, nanographite, and graphitic carbon nanoparticles were prepared and their tribological and thermal properties evaluated:

(1) The friction and wear properties were evaluated in the first instance using a Pin-on-Disk instrument, and the wear properties were additionally checked using a Four-ball tester. The results are shown in Fig. 4.16 and Tables 4.3 and 4.4. The nanographite-containing oil demonstrated a 7X reduction in wear and 13X lower coefficient of friction compared to the neat engine oil. The tribological

properties of the carbon nanoparticles-containing oil also showed significant enhancement, but less than those of the nanographite-containing oil.

(2) The thermal conductivity measurements of all the Nanofluids showed there was upto 5% enhancement in thermal conductivity.

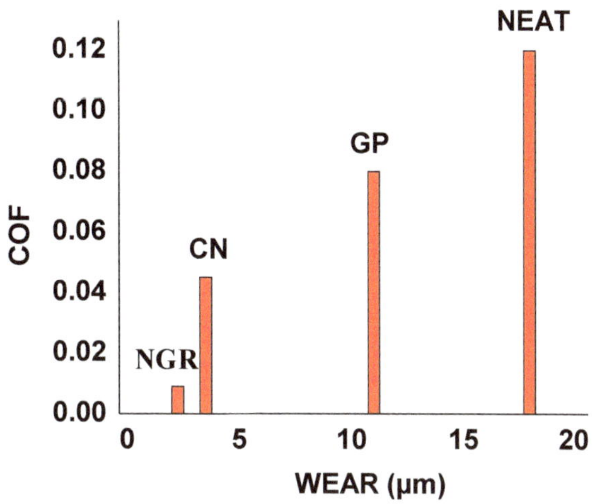


**Fig. 4.16** Comparison of coefficient of friction (COF) and wear at 49 N load (ASTM G 99). KEY: NGR = Nanographite particles; CN = Carbon nanoparticles; GP = micrographite particles; NEAT = neat engine oil

**Table 4.3** The tribological property at 49 N load (ASTM G 99)

| NLF | Coefficient of friction | Wear (μm) |
|---|---|---|
| NGR | 0.009 | 2.6 |
| CN | 0.045 | 3.5 |
| GP | 0.08 | 11 |
| Neat | 0.12 | 18 |

**Table 4.4** Wear test results of nanofluids by four-ball testing (IP 239)

| ISL: Initial seizure load | | | WSD: Wear scar diameter | | | |
|---|---|---|---|---|---|---|
| Sample | Additives % (g/ml) | Surfactant % | Sonication time (min) | ISL (N) | Wear test WSD (mm) (Load: 490 N) | Percentage reduction in wear |
| Nanographite based NLF | 0.3 | 0.10 | 90 min | 736 | 0.51 | 33 |
| CN based NLF | 0.3 | 0.10 | 90 min | 710 | 0.58 | 24 |
| Graphite based NLF | 0.3 | 0.10 | 90 min | 638 | 0.0.71 | 7 |
| Neat SAE-30 oil | … | … | … | 589 | **0.76** | … |

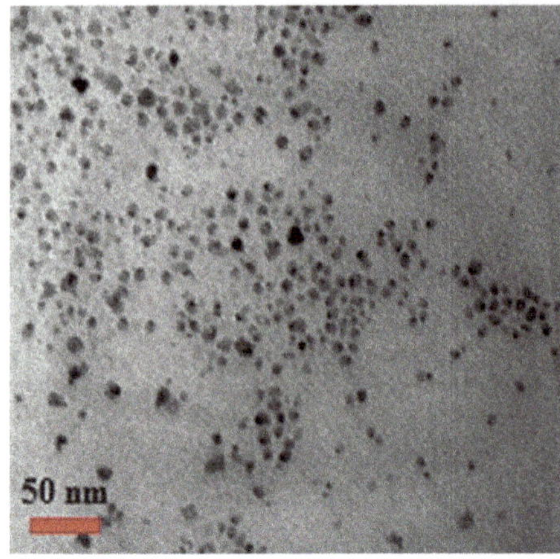


**Fig. 4.17** TEM of Cu nanoparticles

## *4.10.3 Thermally Conducting Fluids*

### 4.10.3.1 Cu Nanoparticle-Based Thermally Conducting Fluid

The synthesis of highly crystalline and well-dispersed Cu nanoparticles has been carried out based on a micro-emulsion mediated method. The reaction parameters of synthesis have been optimized to 5 g-batch level. As of yet particles in the size range of 5–12 nm have been obtained as has been observed by TEM analysis, Fig. 4.17. These particles were crystalline in nature, as shown by Powder X-ray Diffraction (PXRD) analyses, Fig. 4.18. The crystallinity of these nanoparticles is very important because crystalline solids have much larger thermal conductivities than those of fluids.

A mini heat exchanger, see Fig. 4.19, with 90% heat exchange effectiveness, was designed and fabricated in-house to measure the heat transfer rates of Nanofluids. Theoretical heat transfer rates were calculated using reported relationships in the literature for conventional fluids and Nanofluids. For a DMSRDE-developed copper nanoparticle Nanofluid (1% of 5–12 nm particles in ethylene glycol) there was a 27% increase in heat transfer rate at 45 °C as compared to neat ethylene glycol.

### 4.10.3.2 Silver Nanoparticles Decorated Reduced Graphene Oxide-Based Thermally Conducting Fluids

Reduced graphene oxide decorated with silver nanoparticles was synthesized in two steps on a laboratory scale and used for the preparation of a Nanofluid. The graphite oxide was prepared by the Hummers method. Chemical reduction along with decoration of silver nanoparticles was done using sodium borohydride as the reducing agent. This changed the colour from brown to black. After complete addition the mixture was stirred by a magnetic stirrer (1000–12000 rpm) at 100 °C for 24 h to

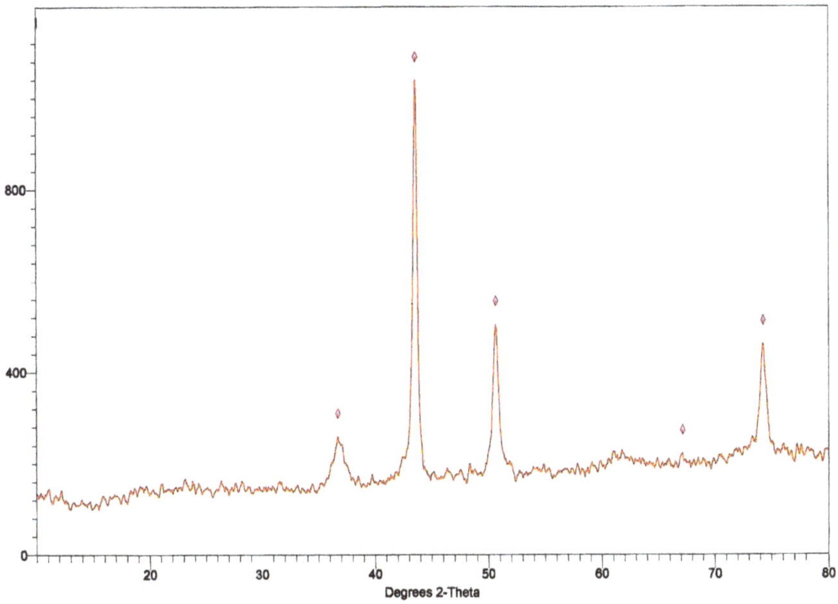


**Fig. 4.18** PXRD of Cu nanoparticles

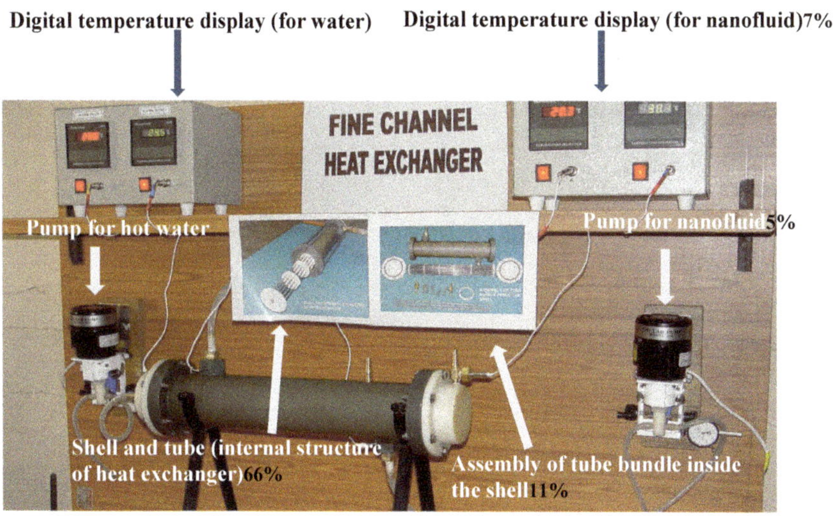


**Fig. 4.19** DMSRDE fabricated heat exchanger

ensure reduction of the graphene oxide. The product was filtered and washed until neutral and stored in suspension form.

Figure 4.20a shows the silver decorated reduced graphene oxide SAED image, in which bright spots are observed without any specific pattern: this infers a semi-crystalline nature of the material. Figure 4.20b is the TEM image: the silver nanoparticles on the graphene sheets vary in size from 10 to 20 nm, together with some lumps. Figure 4.20c is the Raman spectroscopy plot. The presence of a peak at 595.6 $cm^{-1}$ further confirms the presence of nano silver.

These silver nanoparticles were then suspended in a 1:1 ethylene glycol/water mixture. Performance evaluation was done with a 300 cc, 50 hp rotary engine at speeds from 4000 to 7000 rpm at the DRDO VRDE Institute Ahmed Nagar. Figure 4.21 shows the engine rig.

When the rig test was carried out at 5500 rpm speed, the DMSRDE developed Nanofluid resulted in 55 kg thrust, and at that time the fuel flow rate was 9.5 lit/h. This compares with the conventional coolant, where the same thrust was obtained at 6500 rpm and the fuel flow was 13.14 lit/h. Thus the DMSRDE-developed Nanofluid

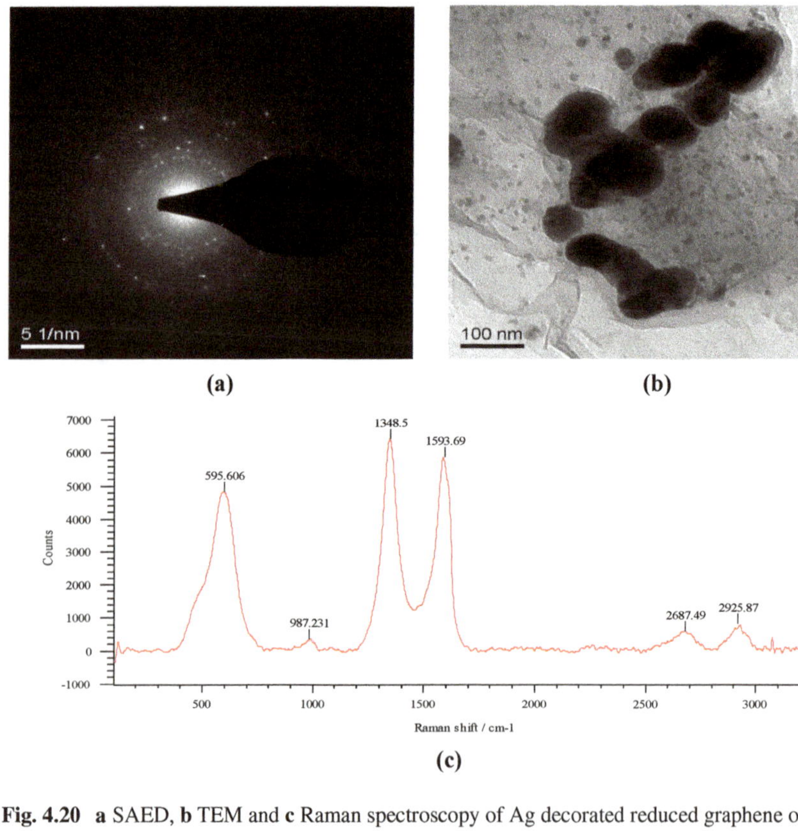


**Fig. 4.20** **a** SAED, **b** TEM and **c** Raman spectroscopy of Ag decorated reduced graphene oxide

**Fig. 4.21** 55 hp rotary engine rig

resulted in a fuel saving of 3.6 lit/h for the same thrust. Table 4.5 gives more details of the engine rig results.

**Table 4.5** Performance evaluation results of the DMSRDE-developed Nanofluid

| Engine speed (rpm) | Thrust (kg) | | Fuel flow (litres/hour) | | Engine–in temperature (°C) | | Engine–out temperature (°C) | |
|---|---|---|---|---|---|---|---|---|
| | DMS fluid | Existing fluid | DMS fluid | Existing fluid | DMS fluid | Existing fluid | DMS fluid | Existing fluid |
| 4000 | 29.7 | 21.2 | 4.4 | 4.55 | 43.4 | 45.5 | 72.0 | 71.4 |
| 4500 | 39.0 | 26.0 | 6.3 | 6.15 | 51.8 | 45.8 | 73.4 | 72.0 |
| 5000 | 45.7 | 32.2 | 7.4 | 6.67 | 56.5 | 46.0 | 74.4 | 73.0 |
| 5500 | 55.1 | 39.7 | 9.5 | 8.69 | 58.7 | 48.0 | 75.8 | 73.5 |
| 6000 | 65.1 | 48.6 | 11.6 | 10.7 | 61.2 | 50.0 | 76.5 | 73.6 |
| 6500 | 74.1 | 57.0 | 14.1 | 13.1 | 62.4 | 55.5 | 78.8 | 74.0 |
| 7000 | 89.4 | 67.0 | 19.0 | 15.6 | 63.1 | 60.0 | 81.9 | 75.0 |
| 7500 | 91.2 | 76.0 | 20.8 | 19.5 | 64.2 | 64.0 | 84.3 | 76.0 |

### *4.10.4 Future Directions for Nanofluid Developments*

Applied research on Nano fluids is a newly emerging field. Nanoparticles can be added to motor oils, antifreeze coolants, automatic transmission fluids, greases, metal-working fluids and refrigerants. However, the applied research is challenging owing to technical problems such as stability, toxicity, corrosion particulate formation, emission, and fouling. Since nanoparticles offer several interesting properties that the bulk materials do not possess, such as thermal enhancement, lubrication improvement, super para magnetism, etc., they offer a wide spectrum of potential applications. Nanofluids are being developed mainly for applications as dampers, sealants, automobile engine oils, and coolants.

## 4.11 Summary

Nanofluid research is an emerging interdisciplinary field where nanoscience and industrial engineering meet, and together could provide a key for new technologies. We are still in a research stage, mainly because of the need to develop high-volume low-cost production methods for industrial applications. Some of the key challenges in this field are to understand the mechanisms of the unique properties of Nanofluids and to develop models that can link particle size, polydispersity index, zeta potential, surface hydrophobicity, nanostructure etc., directly to the properties of the Nanofluids. Hence Nano Fluids offer enormous opportunities to solve interesting and basic scientific questions as well as technological issues.

**Acknowledgements** The authors wish to acknowledge the guidance provided by Dr. Dipak Kumar Setua, DRDO fellow and the help rendered by Mr. Dev Roop Arya in preparing this chapter. The authors also wish to acknowledge VRDE, DRDO Ahmadnagar for performance evaluation of the Nanofluids developed by DMSRDE and Mr. A.K. Srivastava, DFL DMSRDE and IIT (ISM) Dhanbad for tribological characterization.

## References

Akoh, H., et al.: Magnetic properties of ferromagnetic ultrafine particles prepared by vacuum evaporation on running oil substrate. J. Cryst. Growth **45**, 495–500 (1978)

Alves, S., et al.: Nanolubricants developed from tiny CuO nanoparticles. Tribol. Int. **100**, 263–271 (2016)

Asirvatham, L.G., et al.: Experimental study on forced convective heat transfer with low volume fraction of CuO/water nanofluid. Energies **2**(1), 97–119 (2009)

Assael, M., et al.: Thermal conductivity enhancement in aqueous suspensions of carbon multi-walled and double-walled nanotubes in the presence of two different dispersants. Int. J. Thermophys. **26**(3), 647–664 (2005)

Berkovsky, B., Medvedev, V.F., Krakov, M.S.: Magnetic Fluids: Engineering Applications. Oxford University Press (1993)
Berman, D., Erdemir, A., Sumant, A.V.: Graphene: a new emerging lubricant. Mater. Today **17**(1), 31–42 (2014)
Betke, A., Kickelbick, G.: Bottom-up, wet chemical technique for the continuous synthesis of inorganic nanoparticles. Inorganics **2**(1), 1–15 (2014)
Birringer, R.: Nanocrystalline materials. Mater. Sci. Eng. A **117**, 33–43 (1989)
Bonnecaze, R., Brady, J.: A method for determining the effective conductivity of dispersions of particles. Proc. R. Soc. Lond. Ser. A Math. Phys. Sci. **430**(1879), 285–313 (1990)
Bonnecaze, R., Brady, J.: The effective conductivity of random suspensions of spherical particles. Proc. R. Soc. Lond. Ser. A Math. Phys. Sci. **432**(1886), 445–465 (1991)
Chang, W., et al.: Chemical vapor condensation of nanostructured ceramic powders. Nanostruct. Mater. **4**(3), 345–351 (1994)
Chen, L., Xie, H.: Surfactant-free nanofluids containing double-and single-walled carbon nanotubes functionalized by a wet-mechanochemical reaction. Thermochim. Acta **497**(1–2), 67–71 (2010)
Chinas-Castillo, F., Spikes, H.: Mechanism of action of colloidal solid dispersions. J. Tribol. **125**(3), 552–557 (2003)
Choi, S., et al.: Anomalous thermal conductivity enhancement in nanotube suspensions. Appl. Phys. Lett. **79**(14), 2252–2254 (2001)
Choi, S.U., Eastman, J.A.: Enhancing Thermal Conductivity of Fluids with Nanoparticles. Argonne National Lab., IL (United States) (1995)
Chou, C., Lee, S.: Rheological behavior and tribological performance of a nanodiamond-dispersed lubricant. J. Mater. Process. Technol. **201**(1–3), 542–547 (2008)
Das, S.K., et al.: Temperature dependence of thermal conductivity enhancement for nanofluids. J. Heat Transfer **125**(4), 567–574 (2003)
Eastman, J.A., et al.: Anomalously increased effective thermal conductivities of ethylene glycol-based nanofluids containing copper nanoparticles. Appl. Phys. Lett. **78**(6), 718–720 (2001)
Eastman, J., et al.: Enhanced thermal conductivity through the development of nanofluids. MRS Online Proc. Lib. Arch. **457** (1996)
Farbod, M.: Morphology dependence of thermal and rheological properties of oil-based nanofluids of CuO nanostructures. Colloids Surf. A **474**, 71–75 (2015)
Fovet, Y., Gal, J.-Y., Toumelin-Chemla, F.: Influence of pH and fluoride concentration on titanium passivating layer: stability of titanium dioxide. Talanta **53**(5), 1053–1063 (2001)
Gao, B., Zhang, X.: Synthesis of $MoS_2$ inorganic fullerene-like nanoparticles by a chemical vapour deposition method. S. Afr. J. Chem. **67**(1), 6–11 (2014)
Ghoreishi, S., et al.: Nanoparticles synthesis of tungsten disulfide via AOT-based microemulsions. Mater. Res. Bull. **47**(6), 1438–1441 (2012)
Ginzburg, B., et al.: Antiwear effect of fullerene C 6 0 additives to lubricating oils. Russ. J. Appl. Chem. **75**(8), 1330–1335 (2002)
Gleiter, H.: Nanostructured materials: basic concepts and microstructure. Acta Mater. **48**(1), 1–29 (2000)
Greco, A., et al.: Friction and wear behaviour of boron based surface treatment and nano-particle lubricant additives for wind turbine gearbox applications. Wear **271**(9–10), 1754–1760 (2011)
Guo, Y., Guo, W., Chen, C.: Modifying atomic-scale friction between two graphene sheets: a molecular-force-field study. Phys. Rev. B **76**(15), 155429 (2007)
Gupta, B., Bhushan, B.: Fullerence particles as an additive to liquid lubricants and greases for low friction and wear. Lubr. Eng. **50**(7) (1994)
Hamilton, R.L., Crosser, O.: Thermal conductivity of heterogeneous two-component systems. Ind. Eng. Chem. Fundam. **1**(3), 187–191 (1962)
Hargreaves, N.J., Cooper, S.J.: Nanographite synthesized from acidified sucrose microemulsions under ambient conditions. Cryst. Growth Des. **16**(6), 3133–3142 (2016)
Hirtzel, C., Rajagopalan, R.: Colloidal Phenomena, Advanced Topics, pp. 15–23. Noyes. Park Ridge, NJ (1985)

Hu, Z.S., et al.: Preparation and tribological properties of nanometer magnesium borate as lubricating oil additive. Wear **252**(5–6), 370–374 (2002)
Huang, H., et al.: An investigation on tribological properties of graphite nanosheets as oil additive. Wear **261**(2), 140–144 (2006)
Hwang, Y., et al.: Effect of the size and morphology of particles dispersed in nano-oil on friction performance between rotating discs. J. Mech. Sci. Technol. **25**(11), 2853–2857 (2011)
Ingole, S., et al.: Tribological behavior of nano $TiO_2$ as an additive in base oil. Wear **301**(1–2), 776–785 (2013)
Keblinski, P., et al.: Mechanisms of heat flow in suspensions of nano-sized particles (nanofluids). Int. J. Heat Mass Transf. **45**(4), 855–863 (2002)
Kim, P., et al.: Thermal transport measurements of individual multiwalled nanotubes. Phys. Rev. Lett. **87**(21), 215502 (2001)
Kimoto, K., et al.: An electron microscope study on fine metal particles prepared by evaporation in argon gas at low pressure. Jpn. J. Appl. Phys. **2**(11), 702 (1963)
Koch, C.C.: Nanostructured Materials: Processing, Properties and Applications. William Andrew (2006)
Lade, M., et al.: On the nanoparticle synthesis in microemulsions: detailed characterization of an applied reaction mixture. Colloids Surf. A **163**(1), 3–15 (2000)
Lee, S., et al.: Measuring thermal conductivity of fluids containing oxide nanoparticles. J. Heat Transfer **121**(2), 280–289 (1999)
Lee, K., et al.: Understanding the role of nanoparticles in nano-oil lubrication. Tribol. Lett. **35**(2), 127–131 (2009a)
Lee, K., et al.: Performance evaluation of nano-lubricants of fullerene nanoparticles in refrigeration mineral oil. Curr. Appl. Phys. **9**(2), e128–e131 (2009b)
Leslie-Pelecky, D.L., Rieke, R.D.: Magnetic properties of nanostructured materials. Chem. Mater. **8**(8), 1770–1783 (1996)
Li, B., et al.: Tribochemistry and antiwear mechanism of organic–inorganic nanoparticles as lubricant additives. Tribol. Lett. **22**(1), 79–84 (2006)
Maxwell, J.C.: A Treatise on Electricity and Magnetism, vol. 1. Clarendon Press, Oxford (1873)
Mdleleni, M.M., Hyeon, T., Suslick, K.S.: Sonochemical synthesis of nanostructured molybdenum sulfide. J. Am. Chem. Soc. **120**(24), 6189–6190 (1998)
Meredith, R.E., Tobias, C.W.: Conductivities in emulsions. J. Electrochem. Soc. **108**(3), 286–290 (1961)
Munson, B.R., Young, D.F., Okiishi, T.H.: Fundamentals of Fluid Mechanics. Wiley, USA (2006)
Nunn, N., et al.: Tribological properties of polyalphaolefin oil modified with nanocarbon additives. Diam. Relat. Mater. **54**, 97–102 (2015)
Park, K.-H., Ewald, B., Kwon, P.Y.: Effect of nano-enhanced lubricant in minimum quantity lubrication balling milling. J. Tribol. **133**(3), 031803 (2011)
Paul, G., et al.: Techniques for measuring the thermal conductivity of nanofluids: a review. Renew. Sustain. Energy Rev. **14**(7), 1913–1924 (2010)
Pfund, A.: Rev. Scient. Instrum. **1**, 397 (1930)
Rajagopalan, R., Hiemenz, P.C.: Principles of colloid and surface chemistry. Marcel Dekker, New-York **8247**, 8 (1997)
Rapoport, L., et al.: Tribological properties of $WS_2$ nanoparticles under mixed lubrication. Wear **255**(7–12), 785–793 (2003)
Reerink, H., Overbeek, J.T.G.: The rate of coagulation as a measure of the stability of silver iodide sols. Discuss. Faraday Soc. **18**, 74–84 (1954)
Roco, M.: Reviews of national research programs in nanoparticle and nanotechnology research-nanoparticle and nanotechnology research in the USA. J. Aerosol Sci. **29**(5), 749–751 (1998)
Ross, S., Morrison, E.: Colloidal Systems and Interfaces (1988)
Schmidt, J., Guesdon, C., Schomäcker, R.: Engineering aspects of preparation of nanocrystalline particles in microemulsions. J. Nanopart. Res. **1**(2), 267–276 (1999)

Shenoy, B., et al.: Effect of nanoparticles additives on the performance of an externally adjustable fluid film bearing. Tribol. Int. **45**(1), 38–42 (2012)
Shyyko, L.O., Kotsyubynsky, V.O., Budzulyak, I.M.: The importance of surfactant and its type on $MoS_2$ nanoparticles formation. J. Nanosci. Nanotechnol. **16**(8), 7792–7796 (2016)
Singh, J.P., et al.: Carbon nanoparticle synthesis, separation, characterization, and tribological property evaluation. Sep. Sci. Technol. **53**(14), 2314–2326 (2018)
Smolyanitsky, A., Killgore, J., Tewary, V.: Effect of elastic deformation on frictional properties of few-layer graphene. Phys. Rev. B **85**(3), 035412 (2012)
Sohn, L.L.: Nanotechnology: a quantum leap for electronics. Nature **394**(6689), 131 (1998)
Sun, S., Li, Z., Chang, X.: Synthesis and structural characterization of tungsten disulfide nanomaterials. Mater. Lett. **65**(19–20), 3164–3166 (2011)
Tavakoli, A., Sohrabi, M., Kargari, A.: A review of methods for synthesis of nanostructured metals with emphasis on iron compounds. Chem. Pap. **61**(3), 151–170 (2007)
Turner, J.: Two-phase conductivity: the electrical conductance of liquid-fluidized beds of spheres. Chem. Eng. Sci. **31**(6), 487–492 (1976)
Vattikuti, S., Byon, C.: Synthesis and characterization of molybdenum disulfide nanoflowers and nanosheets: nanotribology. J. Nanomater. **2015**, 9 (2015)
Verwey, E.J.W., Overbeek, J.T.G.: Theory of the stability of lyophobic colloids. J. Colloid Sci. **10**(2), 224–225 (1955)
Wang, X.-J., Zhu, D.-S.: Investigation of pH and SDBS on enhancement of thermal conductivity in nanofluids. Chem. Phys. Lett. **470**(1–3), 107–111 (2009)
Wang, X., Xu, X., Choi, S.U.S.: Thermal conductivity of nanoparticle-fluid mixture. J. Thermophys. Heat Transfer **13**(4), 474–480 (1999)
Wu, Y., Tsui, W., Liu, T.: Experimental analysis of tribological properties of lubricating oils with nanoparticle additives. Wear **262**(7–8), 819–825 (2007)
Wu, Z., et al.: Synthesis of $WS_2$ nanosheets by a novel mechanical activation method. Mater. Lett. **64**(7), 856–858 (2010)
Xie, H., et al.: Thermal conductivity enhancement of suspensions containing nanosized alumina particles. J. Appl. Phys. **91**(7), 4568–4572 (2002)
Xie, H., et al.: Nanofluids containing multiwalled carbon nanotubes and their enhanced thermal conductivities. J. Appl. Phys. **94**(8), 4967–4971 (2003)
Xue, L., et al.: Two regimes of thermal resistance at a liquid–solid interface. J. Chem. Phys. **118**(1), 337–339 (2003)
Ying, J.Y., Tschöpe, A., Levin, D.: Synergistic effects and catalytic properties tailored by nanostructure processing. Nanostruct. Mater. **6**(1–4), 237–246 (1995)
Yu, X., Qiang, L.: Preparation for graphite materials and study on electrochemical degradation of phenol by graphite cathodes. Adv. Mater. Phys. Chem. **2**(02), 63 (2012)
Yu, W., Xie, H.: A review on nanofluids: preparation, stability mechanisms, and applications. J. Nanomater. **2012**, 1 (2012)
Zhang, L., et al.: $WS_2$ nanorods prepared by self-transformation process and their tribological properties as additive in base oil. Mater. Sci. Eng. A **454**, 487–491 (2007)
Zhang, W., Ma, G., Wu, C.: Anti-friction, wear-proof and self-lubrication application of carbon nanotubes. Rev. Adv. Mater. Sci. **36**, 75–88 (2014)
Zhu, H.-T., Lin, Y.-S., Yin, Y.-S.: A novel one-step chemical method for preparation of copper nanofluids. J. Colloid Interface Sci. **277**(1), 100–103 (2004)
Zhu, H., et al.: Preparation and thermal conductivity of suspensions of graphite nanoparticles. Carbon **45**(1), 226–228 (2007)
Zhu, D., et al.: Dispersion behavior and thermal conductivity characteristics of $Al_2O_3$–$H_2O$ nanofluids. Curr. Appl. Phys. **9**(1), 131–139 (2009)

# Chapter 5
# Electrically, Magnetically and Strain Field Assisted Smart/Functional Nano Fluids

**Sandeep Kumar, Ajay Katiyar, Nizamuddin Khan, Jyoti Srivastava, Tandra Nandi, and Eswara Prasad Namburi**

**Abstract** Smart Nano fluids are special class of nano materials having ability to reversibly change their deformation and flow characteristics significantly with a response time of millisecond under the ordinance of magnetic field and/or stress-strain field. In this chapter we have discussed various ways of synthesising magnetic and shear thickening fluids, its characterisation aspects keeping the objective of achieving high yield stress and high viscosity without increasing the off-state viscosity. A noble approach of synthesising magnetic particles of high saturation magnetisation and a multidisperse suspension to enhance the packing fraction to fill the gaps and avoid agglomeration is discussed in detail. We have also discussed various ways of characterising the nanofluids using rheometry. We have mainly discussed the types of flow which are being generally used in application, for example, oscillatory and rotational flow behaviour characterisation methods which have been explained.

## 5.1 Introduction

Every living system gives response to stimuli and we have learnt this from nature to develop or design materials which can give response to external stimuli like magnetic field, electric field, stress or stain etc. The materials that can response reversibly to external stimuli are called smart materials. Studies on smart material, also referred to as Intelligent Material, began with increasing pace since 1990. Smart materials

S. Kumar · A. Katiyar · N. Khan · J. Srivastava · T. Nandi
Defence Materials and Stores Research and Development Establishment (DMSRDE), Defence Research and Development Organisation (DRDO), Kanpur, Uttar Pradesh, India

E. P. Namburi (✉)
Former Outstanding Scientist and Ex-Director, Defence Materials and Stores Research and Development Establishment (DMSRDE), Defence Research and Development Organisation (DRDO), Kanpur, Uttar Pradesh, India
e-mail: neswarap@gmail.com

 
E. P. Namburi et al. (eds.), *Novel Defence Functional and Engineering Materials (NDFEM) Volume 1*, Indian Institute of Metals Series,
https://doi.org/10.1007/978-981-99-9791-6_5

have now become a reality for potential commercial applications. Smart behavior is manifested by materials due to one or more factors such as changes in shape, viscosity, storage modulus, loss modulus, yield stress, conductivity, optical changes etc. For example, photochromic polymers change their color reversibly on exposure to light, magnetic fluids change its viscosity under magnetic field and Shear thickening fluid (STF) has the tendency of thickening at critical shear rate. The structures made out of those materials are called 'smart structures'.

Rheology of functional fluids is an extremely important processing parameter. It fills the gap between the developing a material and taking it to stage of a successful product. This is achieved by understanding the flow behavior of these materials under various stress and strain field. In this chapter, we have gone through all the complex phenomena a functional fluid shows compared to the simple Newtonian fluids. We have also seen the types of rheometer and various rheological tests that we can perform to characterize the functional fluids for its rheological properties like yield stress, viscosity, storage modulus, loss modulus, loss tangent, complex viscosity etc.

The present chapter is organized considering the following areas:

i. Magneto Sensitive Nano Fluids.
ii. Shear Thickening Fluids.
iii. Rheology of complex fluids.

## 5.2 Magneto Sensitive Nano Fluids

Magneto sensitive fluid is a stable colloidal suspension of magnetic particles in a carrier fluid. Many liquid bases including hydrocarbons, fluorocarbons, esters, organo-metallics, polyphenyl ethers etc. and several types of magnetic particles like metallic oxide ($Fe_3O_4$, $Fe_2O_3$), metals and their alloys (Fe, Co, Ni) can be selected. Brownian motion keeps the particles suspended indefinitely, so that there is no precipitation under gravity or magnetic field. Since concentration of the particles is only about 3–10% by volume, there is little effect on physical and chemical characteristics of the fluid. Particles are stabilized with an absorbent coating. Various proprietary fluids have since been tailored for specific particles and liquid bases.

Ferrofluids are stable mixtures of surfactant coated magnetic nanoparticles in a liquid medium. S. S. Papel made the first ferrofluid by using a ball mill grinder for magnetic particles of magnetite in the presence of oleic acid (Papel 1965). These small particles in a magnetic fluid behave as single-domain soft magnets. When a ferrofluid is placed in a magnetic gradient the bulk shape of the fluid will distort (Fig. 5.1) (Phule and Ginder 1998). Figure 5.1 shows formation of spikes under a magnetic field.

Thermodynamically, these ferro colloids behave as solute molecules. The magneto sensitive response of these fluids has a origin due to polarization of suspended particles in presence of an external magnetic field (Grinder 1998). In addition, they display increased viscosity and increased apparent density on application of a magnetic gradient (Ginder and Davis 1994).

**Fig. 5.1** Distortion in bulk shape of ferrofluid under magnetic field

## *5.2.1 Applications*

There are unlimited applications of MSFs (Scherer et al. 2005; Ginder 1996; Carlson and Charzan 1994; Johnson 1998; Jolly et al. 1998; Popplewell 1984; Raj et al. 1995; Tang et al. 2000). Various people target specific applications and develop a fluid for themselves. There are various tests specific to each application and if we go by specific applications most devices that use MSFs can be classified into three groups. The first one is having fixed poles, also called as the Pressure Driven Flow mode (PDF), for example, servo valves, actuators, dampers and vibration/shock absorbers which work on the PDF mode. The second one is the relatively moveable or transferable poles also called as the Direct Shear (DS) mode. Some examples are clutches, brakes and locking devices. The last one is the Squeeze Film (SF) mode. The third is rarely used which is the biaxial elongated mode for high force and slow-motion applications. Semi-active dampers for the automotive and aircraft industry are most likely the biggest application of MR fluids. The benefit of using MR fluids instead of conventional oil bases from their large dynamic forces combined gives very rapid mechanical responses. Furthermore, MR dampers do not require mechanical valves to restrict flow contrary to hydraulic dampers. The automotive suspension system, Magne Ride, first incorporated MR technology in the Cadillac Seville STS as early as 2002. An improved version of that vehicle's primary suspension system is currently used as a standard suspension or an option in many models for Cadillac, Buick, Chevrolet, Holden Special Vehicles, Ferrari and Audi. MR fluids are also used in rotary brakes to be used in exercise equipment, pneumatic actuators and steer-by-wire systems. The vehicle's secondary suspension is also benefitted from MR devices. An example is the Rheonetic RD-1005–3 MR damper manufactured by Lord Corporation. MR dampers are also used in civil engineering applications, in seismic protection, cable stayed bridges, and biomedical applications (e.g. prosthetic leg developed by Biedermann Motech GmbH). MR fluid is also relevant in the polishing of optical lenses (QED technologies) and impact or shock loading applications (e.g. active recoil of large caliber gun systems and bullet-proof vests).

The MSFs developed by us has a unique application in surface finishing at nanometer level of optical surfaces using a magnetorheological nanofinishing instrument. We have used water as a suspending medium which has got several advantages for nano finish upto 10 nm level of the optical surface as it is reusable, redispersable and has a self- cleaning characteristic by taking away the debris itself.

### *5.2.2 Mechanism*

Particles in the suspension remain randomly oriented in the absence of an applied magnetic field. An applied magnetic field induces a dipole in each drop. The interaction of the applied field between the resultant dipoles causes the particles to form columnar structures, parallel to the applied magnetic field. Figure 5.2 shows the change in the internal structure of the suspension on application of the magnetic field.

These columnar structures restrict the motion of the fluid so increase viscosity (Tang et al. 2000) and yield stress (Huang and Lai 2000; Luan et al. 1996; Miao et al. 2009a, 2009b; Shorey et al. 2001; Kordonski et al. 2009) known as the magnetorheological effect (Miao et al. 2009a). Improvement of the MR effect and/or reduction of the size of the magnetorheological devices are important challenges for these applications. One of the possible solutions is to use anisotropic particles like rod/cube-like producing higher MR Response compared to spherical ones.

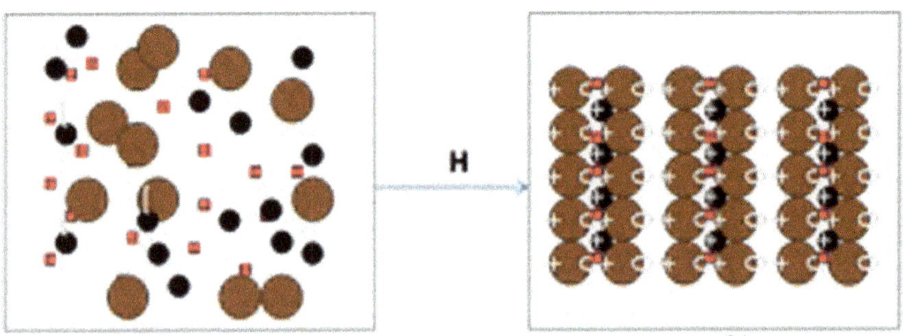


(a) (b)

**Fig. 5.2** **a** Suspension droplets without external magnetic field, where the droplets are randomly oriented **b** Long chains formed by the droplets when subjected to a magnetic field

## 5.3 Shear Thickening Fluids

Body armour is an essential equipment for the police and military. In past few decades, scientists are interested to develop modern armour with improved impact resistance to protect against high impact. Although, besides the above-mentioned advancements, modern body armour still needs improvement to meet modern body armour requirements. The current body armour made of metal plates or from layers of fabric is generally heavy and bulky. In general, most of these armours are rigid and so it is difficult to use on arms, legs and necks. The medieval suits of plate armours have been provided gaps and joints for comfort, and the present body armour generally protects the head and the torso of a person (Ashcroft et al. 2001).

In present era, the body armours for chest protection have been developed with multiple layers of either Kevlar [poly(phenylenediamine terephthalate)], or Spectra or Dyneema [ultrahigh molecular weight polyethylene] or Zylon [poly(benzobisoxazole)] or poly(phenyleneterephthalamide) fibres with ceramic tile inserts (Lee et al. 2003; Ahmad et al. 2007; Forster et al. 2011; Medvedovski 2010; Hazell et al. 2008). When the bullet/projectile hits the body armour, the kevlar fabric absorb and scatter the impact energy to the consecutive layers to reduce the impact energy of the projectile and stop the projectile from penetrating (Ahmad et al. 2007, 2008; Mylvaganam and Zhang 2007). However, present materials used for body armour are too bulky and stiff for application in extremities protection. Thus, an alternate material is required which can replace the present body armour materials with same or better ballistic performance.

One of the newest types of body armour that is presently under extensive research is the one that could give both flexibility and light weight. This property of flexibility together with light weight comes from the addition of a liquid to the existing armour materials. Such "liquid armour" is not entirely ready for combat, but it is suggested that liquid body armour technology has the potential to produce a novel thin, flexible, light weight and cheaper material with similar or better ballistic properties compared to the present bulkier vests (Lee et al. 2003; Wagner et al. 2007, 2009; Hassan et al. 2010). Furthermore, this material can also be extended for other uses in defence like protective clothing, flexible protective devices and shields and stab resistant clothing and devices. Such liquids, when added to the fabric, have special property of shear thickening that gives instant hardness to the material upon impact and reverts back to normalcy upon release of impact energy. At a first instance, this material could be used in a soldier's sleeves and pants which are the key areas that are not protected by ballistic vests but needs flexible vests.

Shear thickening fluid is a non-Newtonian type fluid with higher concentration of solid dispersion. Generally, the viscosity of these types of fluids increases abruptly after a critical value of shear stress (Hoffman 1974; Barnes 1989). The shear thickening behaviour in these types of fluids is due to the development of jamming clusters (known as hydro-clusters). The hydrodynamic lubrication forces between particles are responsible for the formation of hydro-clusters (Bossis and Brady 1989; Foss and Brady 2000). Extensive studies have been made in the past to understand the concept of shear thickening using experiments like rheo-optics (Bender and Wagner 1995), neutron scattering (Bender and Wagner 1996; Maranzano and Wagner 2001a; Lee and Wagner 2006) stress-jump rheological measurements (Kaffashi et al. 1997) and computer simulations (Bossis and Brady 1989). Even many reports exit which quantitatively forecast the onset of shear thickening behaviour for concentrated and electrostatically stabilized dispersions of hard spheres (Maranzano and Wagner 2001a, 2001b).

## 5.4 Smart Properties of Shear Thickening Fluids

The recent development in the properties of STF and its composition for the improvement of stab (resistance to the force exerted by means of an edged blade) and ballistic resistance as well as its dampening behaviour has led to many innovative applications.

### *5.4.1 Response against Stab and Puncture*

Generally, stab threats have two different mechanisms: cut and puncture. Puncture denotes the impact due to the instruments like ice awls, or picks with sharpened tip without cutting edge. These types of threats are of major concern to correctional officers, since sharp-pointed objects are somewhat easy to manage. Cut refers to impact with knives having continuous cutting edge. Knife threats are usually more challenging to impede compared to the puncture, because the continuous cutting edge presents a longer sharpened region more capable of compromising protective materials during a stab-type impact. Some of the advanced developments in the study relevant to resistance against stab and puncture of nano colloidal suspensions such as STFs are as follows:

The stab and puncture resistance of STFs of various composition impregnated in fabrics like Kevlar shows that the inter yarn friction is the most important contributor to the puncture resistance while the particle hardness plays the significant role in stab resistance of the impregnated fabric (Haris et al. 2015). The loading of the polymer in STFs also affects the thickening behaviour and it should be optimized for the special regards for the best performance. The best possible composition of Poly-Ethylene Glycol (PEG) and nano silica is highly effective in protective clothing purposes (Cwalina et al. 2015). Hypodermic needle puncture tests predict the role of mobility of yarn and yarn friction in impact resistance. The simple intercalation of yarn with STF improved the yarn friction and puncture resistance, while inter fabric layer friction plays a more complex role both of which depend also upon other process parameters like needle size, weave structure of fabric etc. (Petel et al. 2015). The clear demonstration of the neat and STF impregnated fabric against stab resistance is shown in Fig. 5.3a and b. It is clear from the figures that STF impregnated Kevlar shows the high impact resistance under high load.

**Fig. 5.3** Demonstration of neat and STF impregnated kevlar against stab resistance

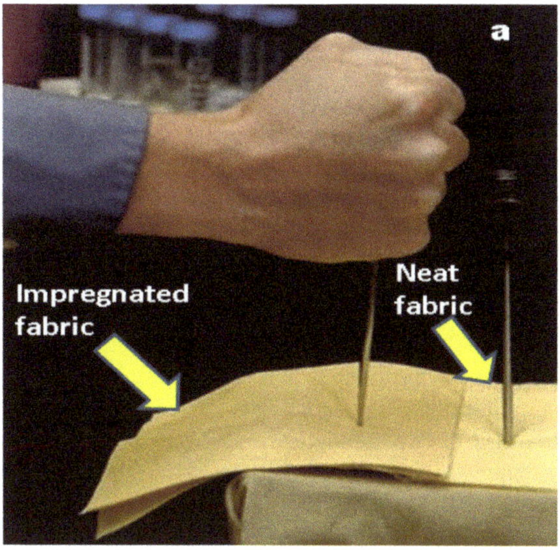


(a)

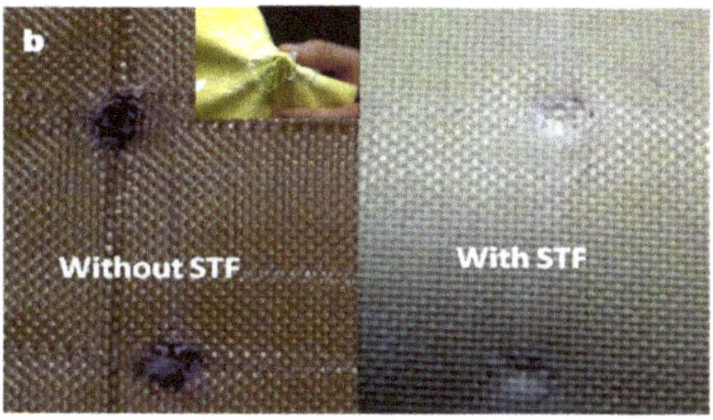


**(b)**

## *5.4.2 Ballistic Impact Response*

The particle strength and fluid density influence the depth of penetration and ballistic resistance of STF. Particle strength plays the significant role in determining the response to the impact. The effect of particle strength in ballistic response of various Newtonian and non-Newtonian solid–liquid dispersions demonstrates that high particle strength such as silica in STFs alone shows major disparity from the expected behaviour. Moreover, the influence of particle strength vanishes under very high velocity impacts (Wang 2015).

Impact based applications of STF Kevlar fabric composites under various parameters such as shock absorption, cushioning, energy dissipation etc. are low velocity impact response. The particle size also influences the ballistic performance i.e. reducing the particle size increased the ballistic performance of the impregnated fabric. Moreover, the smaller particle size enhances the adhesion characteristic of the particle with the yarn, leading to better coupling/bridging in the inter yarn level (Soutrenon and Michaud 2014).

### *5.4.3 Energy Absorption Characteristic*

Various parameters such as particle concentration, viscosity etc. of STF influence the damping behavior and show advantage under high strain applications (Leonowicz et al. 2014). Changing the oligomer chemistry and the molecular weight of dispersed particles and the medium helps to customize the impact characteristic/energy absorption calibre of STFs according to the requirements (Kordani and Vanini 2014). Generally, STF shows the highest energy absorption capability when the ratio of ethanol to Silica is 3:1 (Lu et al. 2015).

## 5.5 Influence of Process Parameters on STF/STF-Kevlar Composites

The major concern with the STF is to optimize the conditions/process parameters which play an important role in the thickening behaviour. The influence of variation of process parameters such as temperature, padding pressure, pH, surfactant, coating of particles etc. affect the final properties of STF impregnated Kevlar and thus affect the potential applications of the same in various fields.

### *5.5.1 Effect of Temperature*

The effect of temperature on shear thickening fluids can be determined in terms of critical shear rate i.e. critical shear rate is shifted towards the higher magnitude as the temperature is increased, while the shear thickening ratio decreases under the same temperature. However, the shear thickening ratio increases with the increase in temperature of suspension.

Further, research reveals the gelation effect of nano-silica at elevated temperatures when the loading of silica particles exceeded beyond a critical value. The changes in hydrogen bonds as well as in the solvation layer at the surface of $SiO_2$ particles are responsible for the gelation characteristics (Obradovic et al. 2014).

### 5.5.2 Effect of pH

The pH of STF significantly affects the critical shear rate, if the pH of STF is brought near the solution's iso-electric point, and the magnitude of the critical shear rate is reduced. Therefore, the pH plays an important role in the preparation/synthesis of STF according to the requirements.

### 5.5.3 Effect of Additives on the Properties of STF

The dilatant properties of STFs can be extensively altered by adding certain additives such as graphene oxide, metal powder, carbides etc. that change the inter particle behaviour or the particle medium interactions. Therefore, the selection of an additive is very important in the synthesis of STF.

### 5.5.4 Effect of Functionalization of Nano Silica Particles

Functionalization of silica particles by using the silane coupling agent shows better performance of STF against ballistic resistance compared to the fumed nano silica based STF due to the increased number of bonds between Kevlar fabric and nano silica.

In recent years, STF is utilized in making the liquid body armour to protect JAWANs/soldiers against high impact attacks such as knife, bullet etc. and save their life (Lee and Wagner, 2003). The liquid body armour provides the flexibility at all the joints of the body of the soldiers which make them comfortable during the impacts. The shear stress vs. shear rate diagram and the utility of the thickening effect for astronauts has been demonstrated in Fig. 5.4 (N. J. Wanger).

STF has shown attraction of scientific community FOR towards the nano-polishing of rough surfaces (Li et al. 2015). The basic principle of STF in the finishing process is illustrated in Fig. 5.12. However, some abrasive particles are added to the STF (known as shear thickening polishing (STPF) to make suitable for finishing process which contribute to the finishing process. When the relative motion is made to polish the surface, the STPF thickened and shear stress increased sharply in the contact area & the abrasive particles play the role in removing the material from the surface to be finished.

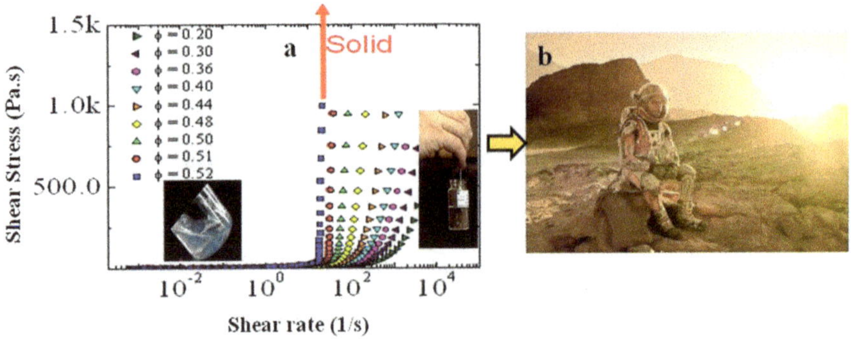


**Fig. 5.4** **a** Shear stress versus shear rate diagram **b** utility of STF technology for the astronaut (N. J. Wanger)

The detailed mechanism of finishing by STPF is illustrated in Fig. 5.6. Figure 5.6a shows the condition without shear thickening i.e. there is no relative motion between the surface and STPF. As the relative motion between the STPF and surface starts the abrasives inside the STPF becomes active (as illustrated in Fig. 5.6b). As the relative motion increases, the shear force tries to balance the resistance force provided by roughness peaks (which further depends on the area of shear and work piece material yield stress) and the small material removal by shear off action from the surface starts (Fig. 5.6c). When the relative motion between the polishing surface and STPF crossed a limiting value sufficient to overcome the resistance force of the roughened peaks the larger chips removal occurs due to the thickening effect in STPF (Fig. 5.6d). The completion of material removal is demonstrated in Fig. 5.6e when the strength of shear thickening reduces as the abrasive passed over the peak due to the reduction in shear stress.

The previous works focused on evaluating the viscosity and other rheological properties such as thixotropy, viscoelasticity, shear thinning etc. of STF under the steady shear as well as oscillatory shear flows (Yoon, et al. 2011; Dennis, et al. 2013; Katiyar et al. 2012, 2015). The linear viscoelastic response of STFs typically shows solid-like behaviour. However, the viscoelastic response under nonlinear oscillatory conditions, with large strain amplitudes, demonstrates some amazing characteristics. The complex viscosity of STFs abruptly jumped to the higher levels when the strain value exceeds a certain magnitude. Such type of phenomenon is designated as strain thickening or flow-blocking (Khan and Raghvan 1997). The non-linear oscillatory rheology is essential to recognize the onset of shear thickening behaviour as well as to estimate the time scale required to spawn shear thickening response. The time scale and minimal magnitude of strain amplitude are essentially required to produce hydro-clusters responsible for thickening effect and, of interest, for the technological utilities (Florida 2009).

## 5.6 Mechanism of Shear Thickening Fluid

Generally, there are three major mechanisms responsible for shear thickening behaviour of STF.

- Hydro-clustering
- Order–disorder transition
- Dilatancy

### 5.6.1 *Hydro-Clustering*

The hydro-cluster mechanism was first investigated by Brady and Bossis in 1985 (Brady and Bossis 1985). The science behind the mechanism is that particles are pushed into each other by shear flow and try to move away from each other leads to overcome the viscous forces from the lubrication gaps between the neighbouring particles. This suggests the critical shear rate beyond which the particles stick together transiently due to lubrication forces and grow in to large clusters known as hydro-clusters. These hydro-clusters increase the effective viscosity of the fluids onset of critical shear rate. The demonstration of the mechanism has been shown in Fig. 5.5.

Figure 5.7a shows the equilibrium state of the fluid under normal shear load. When the shear load increase beyond the critical magnitude the hydro-clusters are formed and increase the effective viscosity against shear flow (Fig. 5.7b).

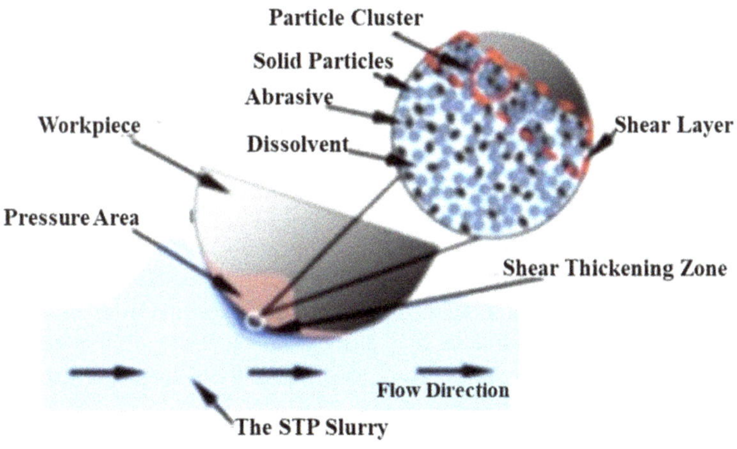


**Fig. 5.5** Mechanism of STF finishing [BinghaiLyu et al. 2015]

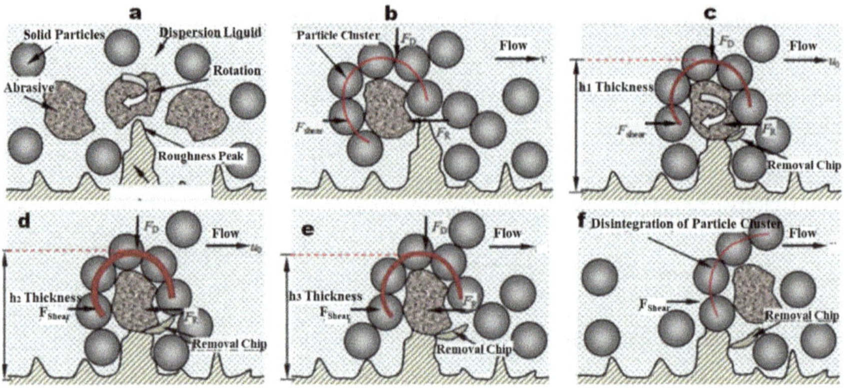


**Fig. 5.6** The complete mechanism of material removal process from the surface by STPF **a** Without shear thickening **b** With shear thickening **c** With a small size chip removal by shear off action **d** Integration of particle cluster **e** Material removal complete **f** Disintegration of particle cluster [BinghaiLyu et.al 2015]

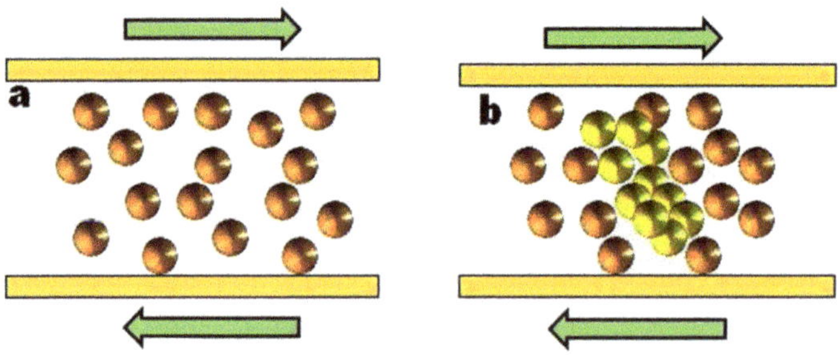


**Fig. 5.7** **a** Equilibrium state of fluid under normal shear load and **b** formation of hydro-clusters under high shear load

## 5.6.2 *Order–disorder Transition*

The order disorder transition was introduced by Hoffman in 1974 (Hoffman 1972). Order–disorder transition refers to the microstructure transition from order layers at low shear rate to a disorder state at high shear rates. However, discontinuous shear thickening effect can occur without an order disorder transition (Egres and Wagner 2005).

## 5.6.3 *Dilatancy*

Dilatancy is the characteristic feature of dense granular flow in which particles attempt to go around each other during shearing without following the direct path.

Hence, their packing volume increases (Onoda and Liniger 1990; Reynolds 1885). According to some previous studies, the onset of dilatancy and discontinuous shear thickening coincide, hence, this mechanism cannot be considered as the mechanism of shear thickening.

## 5.7 High Shear Flow Phenomenon in Shear Thickening Fluids

Various factors/parameters affect the viscosity of Non-Newtonian fluids such as time length for which the fluids has been flowing and magnitude of flow. The rheology cannot be defined by a single number due to the various shear rates for different fluid processing and operations. Moreover, the viscosity is of the fluid which may change with time length of continuous deformation. Therefore, it is inadequate to define the flow behaviour of any fluid only by the viscosity at a constant shear rate. Hence, rheological study of fluids is an utmost important study to understand the flow characteristics under different conditions. Most of the suspensions are non-Newtonian in nature and different flow phenomena arise when these fluids are subjected to the shear rate above 10 1/s. These phenomena can be classified into three major categories: (a) Suspended phase characteristics (b) Dispersing phase flow characteristics (c) particle–particle interactions polymeric effects when macromolecular substances are solubilized in the liquid phase.

Generally, most of the dispersing liquid phase such as water, or various solvents show Newtonian behaviour. The non-Newtonian behaviour arises due to the addition of liquid/solid particles in the liquid phase. The momentum transfer altered through the bulk fluid due to the translation and rotation of particles inside the suspension under the superimposed flow in addition to the flow motion, resulting in increase or decrease in shear stress depending on (a) the disperse and dispersing phase interactions (b) mechanical interactions of the dispersed particle due to the hydro-dynamic effects (c) particle–particle interactions deal with the degree of aggregation of particles. All these types of interactions play an important role in determining the flow characteristics.

The net potential energy manifested by secondary forces such as electrostatic, Vander Waals, steric, salvation etc. contribute to determining the bulk rheological properties of the fluids. These secondary interactions develop a complex nature flow structure with in the non-Newtonian fluids. Hence, the association and disassociation of these particle control the rheological parameter under shear flow. The orientation of asymmetric nature of particles contributes to an additional role in the rheology of suspension.

The rheology of low concentrated solutions is mostly described by the interfacial interactions between the particle and liquid. However, enough liquid phase is available between particles for their lubrication under shear flow, and the interaction between particles is mostly controlled by the liquid. With increasing concentration

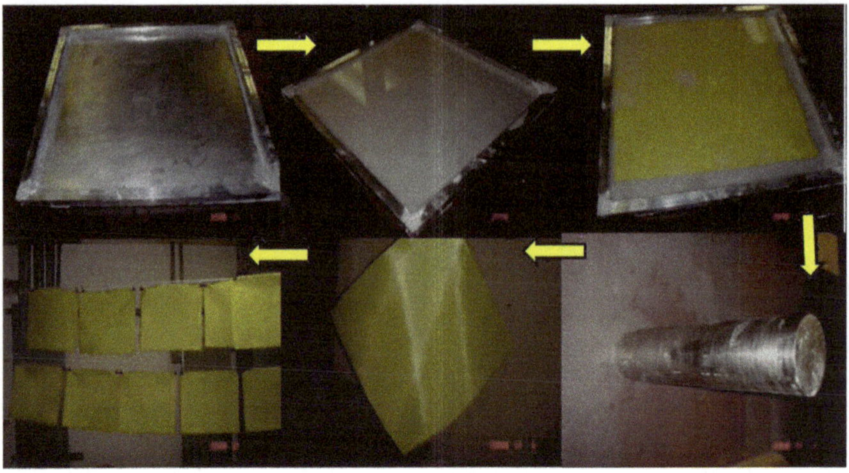

**Fig. 5.8** Steps of impregnation technique of STF in fabric

of solid phase, the mean particles distance reduces and drastic interactions such as particle–particle and hydrodynamic effects affect the flow.

## 5.8 Impregnation of STF in Fabrics

Impregnated Kevlar demonstrates high impact resistance against high impact loads. However, the impregnation technique is also important for making the liquid body armour and is illustrated step by step in Fig. 5.8. Initially, the STF/ethanol should be taken in a plane container, afterwards the fabric should be dipped/soaked for 1 min and subsequently, the soaked fabric should be squeezed by a cylindrical roller to remove the excess STF/ethanol solution from the soaked fabric. The final soaked fabric is hanged for 02 days at room temperature to evaporate ethanol (a solvent) from the solution of STF/ethanol.

## 5.9 Rheology of Complex Fluids

### *5.9.1 Basics of Rheology*

Rheology as the name suggests is the science of deformation and flow of materials. Materials are having two broad classifications, one end is Newtonain viscous liquids following Newton's law of viscosity and the other end is Hookean elastic solids that follow Hooks law of elasticity. Most of the materials come in-between these two extreme behaviours called viscoelsatic materials.

### *5.9.2 Experimental Methods of Rheology*

The flow properties are determined experimentally using rheometers and just for measuring viscosity we use viscometer. Rheometer gives us a complete flow profile of materials under various types of complex loading history generally experienced by any material during its processing in an instrument.

There are two types of rheometer—one is shear type and the other is extensional flow type rheometer. Shear rheometers are further classified based on the mode of generating shear forces on the materials either by rotation of plates in rotational rheometer or by pressure driven mode in capillary rheometer. Extensional rheometers are very rare due to problems in generating homogeneous extensional flow for liquids with low viscosity.

## 5.10 Rheological Test Methods

### *5.10.1 Yield Stress and Stress Ramp*

The maximum stress after which materials start flowing is called yield stress, below this critical value the material is elastic and above this structure breaks and the material flows easily. This is determined by using a shear stress ramp as shown in Fig. 5.9. Determining the yield stress is very important for various applications where we want to know the amount of stress required for a material for its pumping or for storing in a container.

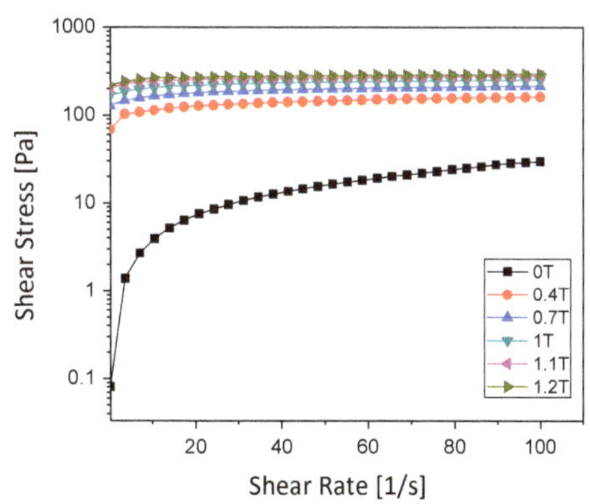


**Fig. 5.9** Yield stress measurement of a magneto sensitive fluid based on the shear stress maximum method in a shear rate ramp

### 5.10.2 Critical Strain and Stress Sweep

Rheological properties of a viscoelastic material are independent of strain or stress and to allow this critical value the material is behaving linearly and there is a trend in various properties like storage and loss modulus with frequency. After reaching this critical value materials starts behaving non-linearly and storage modulus declines abruptly. So, measuring the strain or stress amplitude dependence of the storage and loss moduli is a good first step taken in characterizing viscoelastic behaviour: A strain sweep or stress sweep experiment will establish the extent of the material's linearity.

### 5.10.3 Structure and Frequency Sweep

Once the linear viscoelastic regime (LVR) is determined using the above protocol, we can also do a frequency sweep experiments below those critical value of stress or strain to determine about the effect of colloidal forces, the interactions among particles or droplets. In a frequency sweep, measurements are made over a possible range of oscillation frequencies at a constant oscillation amplitude and temperature.

### 5.10.4 Creep Recovery Experiments

The creep test determines the time-dependent nature of a sample. It tells us about zero shear viscosity and compliance value which measures the elastic recoil of a material. After a sample is allowed to creep under constant load, the material's elastic behaviour can be obtained by performing a recovery test where we abruptly remove the imposed stress and measure the extent the sample recovers.

## 5.11 Research and Technology Development at DMSRDE, Kanpur

### 5.11.1 Properties of Neat Silicone Oil

The properties of neat silicone oil with and without magnetic field are also done to show that there is no MR effect in the neat oil. The flow test was showing that there is almost no change in properties after varying the magnetic field from 0 to 1 T.

### *5.11.2 Oscillatory Time Sweep Experiments on Water Based Nanofluids*

Neat silicone oil and water were also devoid of any magnetorheological effect. We did oscillatory time sweep test to check the efficacy of the magnetic field effect on iron oxide/water suspension. Results were showing very clear effect of the magnetic field on the suspension. Storage modulus was showing a huge increase but was strongly dependent on the strength of the magnetic field as shown in Fig. 5.11a and b for unstable MSFs without adding laponite and for stable MSFs with laponite.

Figure 5.11 shows variation of storage modulus as a function of magnetic field strength. Magnetic field application has two types of suspension, first one is without having any stabilizing agent and another one is laponite stabilized suspension for oscillatory stress of 50 Pa at frequency 0.1 Hz. The figure clearly shows that with increase in the strength of the magnetic field, elastic modulus keeps on increasing and finally attains a saturated value. These phenomena can be easily explained on the basis of magnetization of the magnetic sensitive particles in the presence of the magnetic field to get polarized and form a chain structure in the direction of the magnetic field. Initially time required for magnetization and chain formation decreases with increase in magnetic field strength so there is a saturation effect which can be seen into the storage modulus value at 1.1 and 1.2 T of magnetic field (Fig. 5.10).

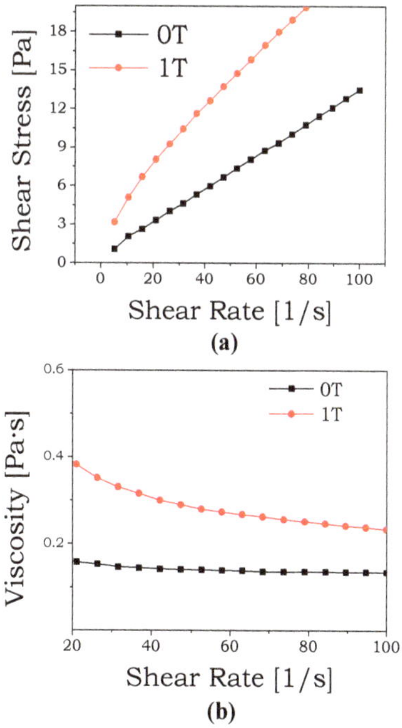


**Fig. 5.10** Shear rate sweep test on virgin silicone oil **a** Newtonian behavior of shear stress with shear rate **b** viscosity with shear rate

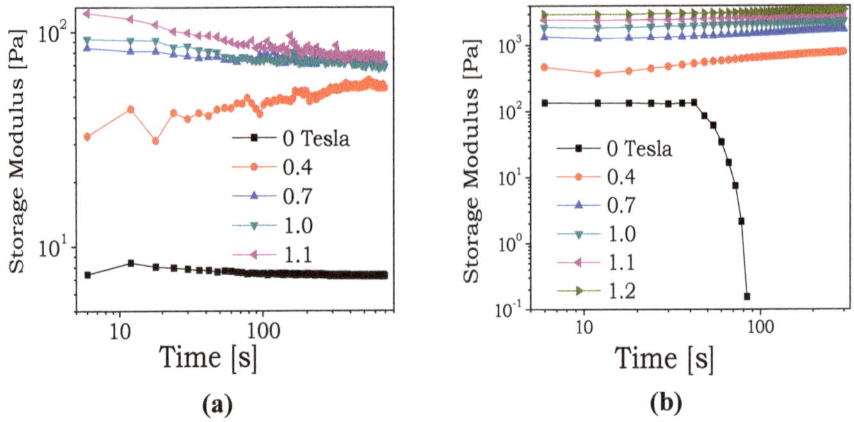


**Fig. 5.11** Storage modulus variation with time under the ordinance of various magnetic fields for **a** unstable composition **b** stable composition

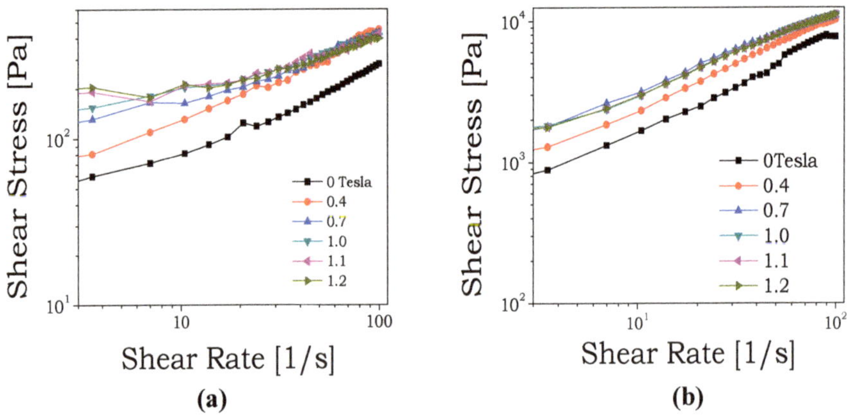


**Fig. 5.12** Yield stress (determined using flow experiment) variation with shear rate at various magnetic fields for **a** unstable composition **b** stable composition

### *5.11.3 Flow Behavior of Water Based Nanofluids*

Defining yield stress in a very simple way is that its lowest stress required for the starting of deformation is breaking the field induced structures. It is of special interest in the application of these fluids in torque transfer type behavior required to have yield stress characterization.

As it can be seen in Fig. 5.12 yield stress increases with magnetic field strength and gets saturated further. This happens because at increased field strength the magnetic particle or the ferrofluid magnetization gets saturated. The yield stress will increase with the external field strength and eventually becomes dependent of magnetic field as the saturation magnetization reaches its maximum.

We can also observe that there is significant amount of increase in yield stress for unstable composition to stable composition. The maximum yield stress we can achieve with 1.2 T of magnetic field is 180 Pa and for the stable composition the yield stress is 1050 Pa. So we can say that stability plays a crucial role in increasing the properties and efficiency of magnetorheological fluids. The increase can also be supported with the yield stress required to move the samples at zero magnetic field. We can see that there is 20 Pa of stress required for unstabilised composition and 115 Pa stress is required to move the stabilised composition at zero fields. Laponite ageing increases the viscosity of the suspension and makes the dispersing medium to hold the heavy iron oxide particles into the suspension to give better results.

## *5.11.4 Creep Behavior of Water Based Nanofluids*

Having analyzed the MR fluid for yield stress using the flow test, elastic storage modulus using oscillatory flow test, creep experiments were carried out depicting the flow deformation at various stresses. Figure 5.13 shows creep flow behavior of unstable and stable water based suspensions at constant stress of 2000 Pa for varying strengths of the magnetic field. It can be seen that the stabilized suspension shows a plateau in creep compliance at 0.7 T magnetic field whereas the unstabilized suspension shows a plateau in 1.1 T of magnetic field. The system undergoes lesser deformation and hence creep compliances decrease with increase in strength of the magnetic field. Eventually at sufficiently high magnetic field, the compliance reaches a plateau and can also be correlated with the yield stress at that particular stress where suspension stops flowing. So we can say that stress of 2000 Pa is yield stress for stabilized suspension at 0.7 T, whereas unstabilized suspension will require lesser stress to show yielding behavior. The unsteady shear flow behavior of MSFs is better investigated using creep experiments at constant stress. These experiments reveal that MSFs only behave in the linear viscoelastic regime when subjected to extremely low stress levels. For very large stresses, the MR fluid flows with a very low viscosity level.

## *5.11.5 Development of Shear Thickening Fluids (STFs)*

### 5.11.5.1 Synthesis of Shear Thickening Fluids

The shear thickening fluids (STFs)are synthesized by using silica nanoparticles (~7 nm) in different weight percentages i.e. 10, 20, 25 and 30%) and dispersed these particles in the base fluid i.e. mixture of ethanol and polyethylene glycol (PEG 400). The ethanol is mixed in PEG in 1:6 ratios for making the silica particles'

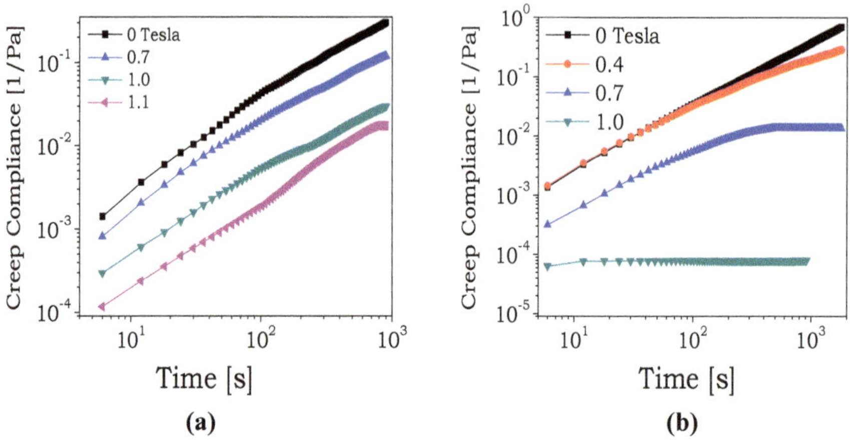


**Fig. 5.13** Creep compliance evolution with time under various magnetic fields at 2000 Pa stress for **a** unstable suspension **b** stable suspension

dispersion easy. For uniform dispersion of silica particles in PEG, a ultrasonication technique was used for 3–4 h. After the sonication, the intermediate solvent (ethanol) was evaporated by heating the samples up to 79–80 ºC on the heating plate and final STF has been collected in sealed bottles for further studies.

#### 5.11.5.2 Characterization

The STF samples are tested for its rheological properties using a rheometer (Physica MCR 301, Anton Paar, Germany). The testing was carried out at 25 °C for all runs. Under rotary mode the shear rate range was used using a linear ramp at 0–3500 $s^{-1}$. A peltier plate and a roughened parallel plate of 25 mm diameter have been used. Each data point has a five second interval and totally twenty points have been taken. The peltier system can control the temperature over a range of −40 to 200 °C with a cooling/heating rate up to 20 ºC/min and a temperature accuracy of ± 0.1 ºC.

Prior to beginning the rheological measurements the effect of wall slip at solid boundaries in the rheometer is examined. In this study, a parallel plate having diameter of 25 mm with roughened surface and a cone plate with cone angle 1º geometries are used to examine the wall slip for suspensions with shear thickening flow profiles. The use of two different geometrical configurations ensured that instrumental artifact, like the wall slip, is present around the maximum viscosity at higher shear rates (above 2000$s^{-1}$), which might result in an error of about 15% in the viscosity value. Reversibility is examined by performing the thixotropic shear test. All measurements presented here were reproducible.

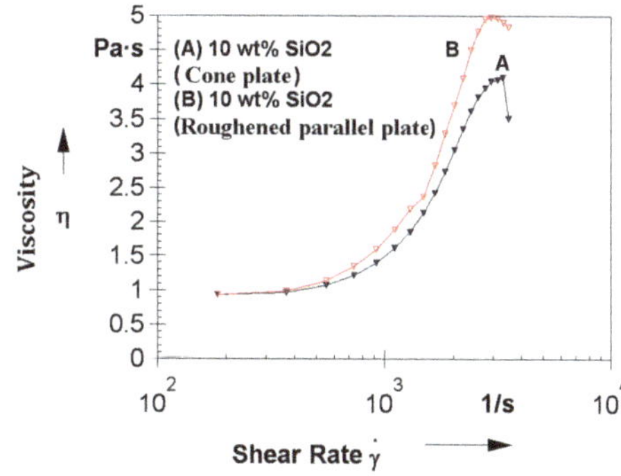


**Fig. 5.14** Comparison of results for cone plate and roughened parallel plate for 10 wt% $SiO_2$

#### 5.11.5.3 Rheological Measurements

Figure 5.14 illustrates the comparison of results for 10 wt% fumed silica concentration with cone plate & roughened parallel plate to test the wall slip. It was observed that the wall slip is present around the maximum viscosity at higher shear rate above $2000s^{-1}$, which could result in an error of about 15% in the viscosity value. The shear thinning regime at high shear rates indicate that the behavior is real and not an artifact of the wall slip.

Figure 5.15a and b demonstrate the viscosity versus shear rate graph. It is found that shear thickening effect is insignificant in the case of pure PEG, however, for the prepared STF, it starts above 200 $s^{-1}$ shear rate and is more prominent at higher shear rate (above 1000 $s^{-1}$). It is found that sharpness of the curves increases by increasing the particles loading i.e. thickening behaviour of STF increases with the increase in silica nanoparticles loading. It is also found that zero shear viscosity jumps dramatically between 0 and10 wt% and not changes much between 10 and 30 wt%. The maximum viscosity ~9 Pa.s for 30 wt% and ~8 Pa.s for 25 wt% of $SiO_2$ at high shear rates is accomplished. This phenomenon can be explained by Hydrocluster theory. At low shear rate, the inter particle interaction forces are balanced with hydrodynamic lubrication force hence viscosity is approximately constant. However, at high shear rate the hydrodynamic lubrication forces are greater than the inter particle interaction forces which causes hydrodynamic clustering, where the hydrodynamic clusters are compact group of particles that stick together due to short range lubrication forces leading to increase in viscosity. It is also observed that fumed silica based STFs show the almost same trends with different viscosity peaks. The STFs based on the modified silica are also tested experimentally for viscosity under steady shear with the same parameters. It is found that thickening starts at lower critical shear rates as compare to fumed silica based STFs. The viscosity changes from 22 to 2000 Pa.s when wt% changes from 20 to 30% (Fig. 5.15b).

Figure 5.16 shows the effect of temperature on the shear thickening behaviour for STF sample containing 25 wt% of fumed silica. It was observed that with the decrease in temperature from 15 to −10 °C, the critical shear rate is shifted from 466 to 16.2 $s^{-1}$ and the magnitude of viscosity increases from 78 to 332 Pa.s. The

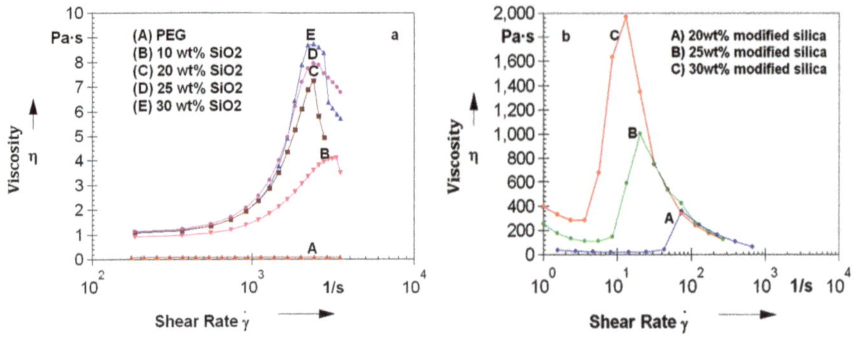


**Fig. 5.15** Viscosity versus shear rate graph with different particles loading of **a** fumed silica **b** modified silica

**Fig. 5.16** Effect of temperature on critical shear rate and viscosity

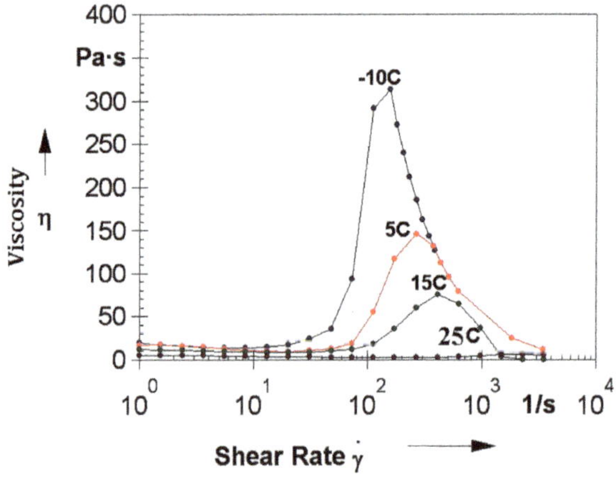


above data demonstrates the influence of temperature on shear thickening behavior of STF. Therefore, the temperature effect should be kept in consideration during the design of soft body armours having specific performance level. Otherwise, the armour designed for low temperature regions may fail, if used in hot locations.

### *5.11.6 Conclusions*

In recent years, smart functional materials are one of the most progressive fields having strategic applications. Magnetorheological and shear thickening material technologies are under development and not established nationally or even disclosed properly on the international platform. Establishing these technologies require novel aspects of nanoscience and technology along with the newly developing science of chemistry. We have presented a comprehensive way of developing these materials practically and get its incredibility checked with the help of the rheometer.

**Acknowledgements** The authors acknowledge the DMCT Division members of DMSRDE, faculty members of IIT Kanpur and IIT Madras for their extensive help in characterization. The authors wish to acknowledge the valuable association of all the members of DFL, DMSDRE for their kind support. They feel greatly indebted to Dr. R.J.H. Wanhill for his painstaking efforts to revise, review and rearrange many parts of the manuscript to improve its presentability.

## References

Ahmad, M.R., Ahmad, W.Y.W., Salleh, J., Samsuri, A.: Performance of Natural Rubber Coated Fabrics under Ballistic Impact. Malays. Polym. J. **2**, 39–51 (2007)

Ahmad, M.R., Ahmad, W.Y.W., Salleh, J., Samsuri, A.: Effect of fabric stitching on ballistic impact resistance of natural rubber coated fabric systems. Mater. Des. **29**, 1353–1358 (2008)

Ashcroft, J., Daniels, D.J., Hart, S.V.: Selection and Application Guide to Personal Body Armour, p. 1. The National Institute of Justice's National Law Enforcements and Corrections Technology Centre, US (2001)

Barnes, H.A.: Shear-thickening ("Dilatancy") in suspensions of non-aggregating solid particles dispersed in Newtonian liquids. J. Rheol. **33**, 329 (1989)

Bender, J.W., Wagner, N.J.: Optical measurement of the contributions of colloidal forces to the rheology of concentrated suspensions. J. Colloid Interface Sci. **172**, 171–184 (1995)

Bender, J.W., Wagner, N.J.: Reversible Shear thickening in monodisperse and bidisperse colloidal dispersions. J. Rheol. **40**, 899 (1996)

Bossis, G., Brady, J.F.: The rheology of Brownian suspensions. J. Chem. Phys. **91**, 1866 (1989)

Brady, J.F., Bossis, G.: The rheology of concentrated suspensions of spheres in simple shear flow by numerical simulation. J. Fluid Mech. **155**, 105–129 (1985)

Carlson, J.D., Charzan, M.J.: Magnetorheological Fluid Dampers. Lord Corporation USA (1994)

Cwalina, C.D, Dombrowski, R.D, McCutcheon, C.J, Christiansen, E.L., Wagner, N.J.: MMOD puncture resistance of EVA suits with shear thickening fluid (STF)–armor tm absorber layers. Procedia Eng. **103**, 97–104 (2015)

Dennis, J.D., et al.: Rheological characterization of monomethylhydrazine gels. J. Propuls. Power. **29**(2), 313–320 (2013)

Egres, R.G., Wagner, N.J.: The rheology and microstructure of acicular precipitated calcium carbonate colloidal suspensions through the shear thickening transition. J. Rheol. **49**(3), 719–746 (2005)

Clements, F.E., Mahfuz, H.: Enhance the stab resistance of flexible body armor using functionalized silica nanoparticles. J. Appl. Phys. **105**, 064307 (2009)

Forster, A.L., Pintus, P., Messin, G.H.R., Riley, M.A., Petit, S., Rossiter, W., Chin, J., Rice, K.D.: Hydrolytic stability of polybenzobisoxazole and polyterephthalamide body armour. Polym. Deg. Stab. **96**, 247–254 (2011).

Foss, D.R., Brady, J.F.: Structure, diffusion and rheology of Brownian suspensions by stokesian dynamics simulation. J. Fluid Mech. **407**, 167–200 (2000)

Ginder, J. M.: Rheology Controlled by Magnetic Fields. Encycl. Appl. Phys. In: Immergut, E. (ed.) VCH, New York (1996)

Ginder, J.M., Davis, C.S.: Shear stresses in magnetorheological fluids: role of magnetic saturation. Appl. Phys. Lett. **65**, 3410 (1994)

Grinder, J. M.: MRS Bull. **26** (1998)

Haris, A., Lee, H.P, Tay, T.E, Tan, V.B.C.: Shear thickening fluid impregnated ballistic fabric composites for shock wave mitigation. Int. J. Impact Eng. **80**, 143–151 (2015)
Hassan, T.A., Rangari, V.K., Jeelani, S.: Synthesis, processing and characterization of shear thickening fluid (STF) impregnated fabric composites. Mat. Sci. Engg A. **527**, 2892–2899 (2010)
Hazell, P.J., Roberson, C.J., Moutinho, M.: The design of mosaic armour: the influence of tile size on ballistic performance. Mater. Des. **29**, 1497–1503 (2008)
Hoffman, R.L.: Discontinuous and dilatant viscosity behavior in concentrated suspensions. I. Observation of flow instability. Trans. Soc. Rheol. **16**(1), 155–173 (1972)
Hoffman, R.L.: Discontinuous and roject viscosity roject in concentrated suspensions. II. Theory and experimental results. J. Colloid Interface Sci. **46**, 491–506 (1974)
Huang, J., Lai, P.: Formation and polarization of dipolar chains. Physica A **281**, 105 (2000)
Johnson, G.L.: Passive magnetorheological clutch. General Motors Corporation, USA, patent no. US5848678A (1998)
Jolly, M.R., Bender, J.W., Carlson, J.D.: Properties and applications of commercial magnetorheological fluids. In: SPIE 5th Annual International Symposium on Smart Structures and Materials. San Diego (1998)
Kaffashi, B., O'Brien, V.T., Mackay, M.E., Underwood, S.M.: Elastic-like and viscous-like components of the shear viscosity for nearly hard sphere, Brownian suspensions. J. Colloid Interface Sci. **187**, 22–28 (1997).
Katiyar, A., Singh, A.N., Shukla, P., Nandi, T.: Rheological behavior of magnetic nanofluids containing spherical nanoparticles of Fe–Ni. Powder Technol. **224**, 86–89 (2012)
Katiyar, A., Dhar, P., Das, S.K., Nandi, T.: Near-field magnetostatics and Néel–Brownian interactions mediated magneto-rheological characteristics of highly stable nano-ferrocolloids. Soft Matter. **11**(8), 1614–1627 (2015)
Khan, S.A., Raghvan, S.R.: Shear thickening response of fumed silica suspensions under steady and oscillatory shear. J. Colloid Interface Sci. **185**, 57–67 (1997)
Kordani, N., Vanini, A.S.: Optimizing the ethanol content of shear thickening fluid/fabric composites under impact loading. J. Mech. Sci. Technol. **28**(2), 663–667 (2014)
Kordonski, W., Sekeres, A., James, R.: Magnetorheological fluid-based high precision finishing technology. US Patent 7 557 566 (2009)
Lee, Y.S., Wagner, N.J.: Rheological properties and small-angle neutron scattering of a shear thickening, nanoparticle dispersion at high shear rates. Ind. Eng. Chem. Res. **45**, 7015–7024 (2006)
Lee, Y.S., Wetzel, E.D., Wagner, N.J.: The ballistic impact characteristics of Kevlar woven fabrics impregnated with a colloidal shear thickening fluid. J. Mater. Sci. **38**, 2825–2833 (2003)
Lee, Wetzel, Wagner.: The ballistic impact characteristics of kevlar woven fabric impregnated with shear thickening fluid. J. Mater. Sci. **38,** 2825–2833 (2003)
Leonowicz, M., Kozłowska, J., Wierzbicki, Ł. Rheological fluids for energy absorbing systems. Appl. Mech. Mater. **440**, 13 – 18 (2014). Trans Tech Publ.
Li, M., Lyu, B., Yuan, J., Dong, C., Dai, W.: Shear-thickening polishing method. Int. J. Mach. Tools Manuf **94**, 88–99 (2015)
Lu, Z., Wu, L., Gu, B., Sun, B.: Numerical simulation of the impact behaviors of shear thickening fluid impregnated warp-knitted spacer fabric. Compos. B Eng. **69**, 191–200 (2015)
Luan, H.M., Korman, C., Willenbacher, N.: Rheometry on magnetorheological (MR) fluids. Rheol. Acta **35**, 417 (1996)
Maranzano, B.J., Wagner, N.J.: The effects of interparticle interactions and particle size on reversible shear thickening hard sphere colloidal dispersions. J. Rheol. **45**, 1205 (2001a)
Maranzano, B.J., Wagner, N.J.: The effects of particle size on reversible shear thickening of concentrated colloidal dispersions. J. Chem. Phys. **114**, 10514 (2001b)
Medvedovski, E.: Ballistic performance of armour ceramics: influence of design and structure–part I. Ceram. Int. **36**, 2103–2115 (2010)
Miao, C., Shafrir, S.N., Lambropoulos, J.C., Mici, J., Jacobs, S.D.: Shear stress in magnetorheological finishing for glasses. Appl. Opt. **48**, 2585 (2009b)

Miao, C., Shafrir, S.N., Lambropoulos, J.C., Jacobs, S.: Conference 7426: Magnetorheological Fluid-based High Precision Finishing Technology Optical Manufacturing and Testing VIII, San Diego, CA, Aug. 4–5, (CD) 7426–11 (2009).
Mylvaganam, K., Zhang, L.C.: Ballistic resistance capacity of carbon nanotubes. Nanotechnol. **18**, 475701 (2007).
Obradovic, V., Stojanovic, D.B, Jancic–Heinemann, R., Zivkovic, I., Radojevic, V., Uskokovic, P.S., Aleksic, R.: Ballistic properties of hybrid thermoplastic composites with silica nanoparticles. J. Eng. Fibers Fabr. **9**(4) (2014)
Onoda, G.Y., Liniger, E.G.: Random loose packings of uniform spheres and the dilatancy onset. Phys. Rev. Lett. **64**(22), 2727 (1990)
Papel, S. S.: Low viscosity magnetic fluid obtained by the colloidal suspension of magnetic particles, U. S. Patent, 3215572 (1965)
Petel, O.E, Ouellet, S., Loiseau, J., Frost, D.L., Higgins, A.J.: A comparison of the ballistic performance of shear thickening fluids based on particle strength and volume fraction. Int. J. Impact Eng. **85**, 83–96 (2015)
Phule, P.P., Ginder, J.M.: The materials science of field-Responsive Fluids. MRS Bull. (1998)
Popplewell, J.: Technological applications of ferrofluids. Phys. Technol. **15**, 150 (1984)
Raj, K., Moskowitz, B., Casciari, R.: Advances in ferrofluid technology. J. Magn. Magn. Mat. **149**, 174 (1995)
Reynolds, O.: LVII. On the dilatancy of media composed of rigid particles in contact. With experimental illustrations. Lond., Edinb., Dublin Philos. Mag. J. Sci. **20**(127), 469–481 (1885)
Scherer, C., Figueiredo Neto, A.M., Braz.: Ferrofluids: properties and applications. J. Phys. 3A **35** (2005)
Shorey, A.B., Jacobs, S.D., Kordonski, W.I., Gans, R.F.: Experiments and observations regarding the mechanisms of glass removal in magnetorheological finishing. Appl. Opt. **40**, 20 (2001)
Soutrenon, M., Michaud, V.: Energy dissipation in concentrated monodisperse colloidal suspensions of silica particles in polyethylene glycol. Colloid Polym. Sci. **292**(12), 3291–3299 (2014)
Tang, X., Zhang, X., Tao, R.: Structure-enhanced yield stress of magnetorheological fluids. J. Appl. Phys. **87**, 2634 (2000)
Wagner, N.J., Wetzel, E.D.: Advanced body armour utilizing shear thickening fluids. US Patent No. 7,226,878 B2 (2007)
Wagner, N.J., Wetzel, E.D.: Advanced body armour utilizing shear thickening fluids. US Patent No. 7,498,276 B2 (2009)
Wang, Y., Li, S.K., Feng, X.Y.: The ballistic performance of multi-layer kevlar fabrics impregnated with shear thickening fluids. Appl. Mech. Mater. **782**, 153–157 (2015)
Yoon, C., et al.: Numerical modeling of injection of shear-thinning gel propellants through plain-orifice atomizer. J. Propuls. Power. **27**(5), 944–954 (2011)

# Chapter 6
# Polymer Precursors for High Technology Applications

**Rakesh Kr. Gupta, Raghwesh Mishra, Suresh Kumar, Ashok Ranjan, and Eswara Prasad Namburi**

**Abstract** This chapter summarizes the development of polymeric materials as precursors for ceramics. Emphasis is on the silicon-based polymeric materials that are potential precursors of silicon carbide (SiC) based ceramics. Different synthetic methods have been discussed and a brief review of the processing of ceramics has been covered. Development of the precursor route to ceramics, in particular polycarbosilane (PCS) to SiC, and a brief account of the contributions of DMSRDE (DRDO) in this field have been described, whereby some of the applications of indigenous PCS have been discussed.

**Keywords** Precursor polymer · Polycarbosilane · Silicon carbide · CMC · C–SiC Composites

## 6.1 Introduction

Metals and alloys have served well as structural materials in aerospace and defence technologies. However, for high temperature applications it is becoming difficult for researchers to further improve their thermal stability, owing to inherent limitations (Reed 2006). Metals have high density, high thermal expansion coefficient and are prone to corrosion at elevated temperatures. Therefore, alternative materials are needed to significantly increase the service temperatures to enhance the performance of structural components in advanced aircraft, spacecraft and weapons.

---

R. Kr. Gupta · R. Mishra · S. Kumar · A. Ranjan
Defence Materials and Stores Research and Development Establishment (DMSRDE), Defence Research and Development Organisation (DRDO), Kanpur, Uttar Pradesh, India

E. P. Namburi (✉)
Former Outstanding Scientist and Ex-Director, Defence Materials and Stores Research and Development Establishment (DMSRDE), Defence Research and Development Organisation (DRDO), Kanpur, Uttar Pradesh, India
e-mail: neswarap@gmail.com



E. P. Namburi et al. (eds.), *Novel Defence Functional and Engineering Materials (NDFEM) Volume 1*, Indian Institute of Metals Series,
https://doi.org/10.1007/978-981-99-9791-6_6

Advanced ceramics generally have low density, low thermal expansion coefficient, high strength, high elastic modulus, high hardness, high thermo-oxidative and thermal stability, and excellent chemical resistance; therefore, have been extensively explored for high temperature structural applications components for gas turbine engines, tiles for re-entry vehicles, and tiles for Tokamak applications.

Ceramic matrix composites (CMCs) and particularly CFCMCs are of high interest for high-temperature structural applications owing to their high strengths and moduli at high temperatures (Hench and Ulrich 1984; Krenkel and Bansal 2005). Future high technology applications, e.g., missiles, aerial vehicles and energy-efficient engines will use high-temperature CMCs for hot-end components. Among these, SiC matrix composites are very important (Kumar et al. 2011a, b; Kohyama et al. 2000).

SiC is a non-oxide ceramic material with outstanding thermal stability, semiconductor properties, chemical inertness, high resistance to oxidation, good mechanical properties, and an extreme hardness approaching those of diamond and boron nitride (Choyke et al. 2004). SiC, in powder form, is industrially produced by reduction of silica with carbon in an electrically heated furnace. SiC prepared in this fashion is an intractable solid and requires further crushing into usable grains. The largest users of silicon carbide are metallurgical, abrasive, and refractory industries. SiC also finds applications in electric heating elements in furnaces, electronic devices, and nuclear radiation damage-resistant areas (Kimoto and Cooper 2014; Feinroth 2013).

The inherent properties of ceramics like brittleness, high hardness and very high softening temperatures make it very difficult to convert ceramics into complex shapes like films and fibres from powders, thus restricting their applications. This difficulty forced researchers to find some polymers that could be processed into the desired shape using conventional polymer forming methods, and whose subsequent pyrolysis gives the needed ceramic. For these polymers, which could be used as precursors and converted into ceramics after pyrolysis, the term "pre-ceramic polymers" has been used (Wynne and Rice 1984).

Dunogues et al. (Birot et al. 1995) have pointed out in their review that a compromise of the following properties should be possessed by an idealized pre-ceramic polymer: (1) sufficiently high molecular weight of the precursor polymers so that volatilization of oligomers is prevented; (2) the presence of cages or rings in polymeric structures to decrease the elimination of volatile fragments owing to backbone cleavage and the latent reactivity in the polymers; (3) viscoelastic properties such as fusibility, malleability and solubility to ensure easy processing; and (4) low organic group content in order to increase ceramic yield and avoid excess free carbon. [N.B: different combinations of substituents and changes in polymer microstructure have dramatic effects on the final ceramic properties].

These precursor polymers have attracted a great deal of attention because otherwise it is difficult to make ceramic forms that can be fabricated with ease at comparatively low temperatures (Birot et al. 1995; Loh et al. 2000). Ceramics obtained through controlled pyrolysis of polymeric precursors are called pre-ceramic polymer derived ceramics. Several advantages, such as the wide variety of precursors, their ease of handling and simple forming methods like polymer infiltration and pyrolysis (PIP), resin transfer moulding (RTM) and injection moulding make precursor route extremely promising for the realization of a new class of ceramics with unique and exceptional properties.

Results of the first experiments of this new synthesis strategy were reported in the early1960s by Ainger et al. (Ainger et al. 1960), and Chantrell et al. (ChantrellPG and Popper 1965), describing the production of non-oxide ceramics from molecular precursors. In the early 1970s Verbeek, Winter and Mansmann (Winter et al. 1974) developed the first practical transformation of polyorganosilicon compounds (polysilazanes, polysiloxanes and polycarbosilanes) to ceramic materials, mainly for manufacturing small-diameter ceramic fibres of $Si_3N_4$/SiC for high temperature applications. Synthesis of SiC materials was further developed by the work of Fritz and Raabe (Fritz and Raabe 1956), who were contemporaries of Yajima (Yajima et al. 1975a, 1978): the so-called Yajima process for the preparation of SiC is mainly based on the thermolysis process of polycarbosilanes (PCSs) in inert argon atmosphere.

The work on silicon-based polymeric precursors for SiC ceramics paved the way for many more ceramics to be synthesized via a precursor route. The most important precursors are polysiloxanes, polysilazanes and PCSs (Birot et al. 1995; Mera and Riedel 2009). Not only binary but ternary ceramics could be synthesized by this route. Increasing the complexity of the starting precursors could lead to further improvements of the properties of the final ceramics, e.g. the introduction of boron atoms inside the starting polymers could produce ceramics in the quaternary SiBCN system. Furthermore, the chemical modification of the starting polymers with precursors of metallic elements like Al, Zr and Ti could provide a further increase of the complexity of the system, and the realization of ceramic components with a wider variety of composition and outstanding thermal properties (Yu et al. 2008; Babonneau and Soraru 1991; Yang et al. 1996).

PCSs are the most common pre-ceramic materials for SiC and have found wide use for making SiC fibres and composites since the pioneering work of Yajima and co-workers (Yajima et al. 1975b; Ly et al. 2001). However, PCS commercially viable synthetic methods are mostly safeguarded by patents and confined to only a few developed countries (Yajima et al. 1981; Curtis et al. 1986).

In order to get SiC fibres and matrices with the required properties and compositions it is essential to have appropriate PCSs. Several reports are available in which PCSs have been modified by incorporating metal, organic moieties or other groups into the main chain. The most widely used process for functionalization is the reaction of Si–H groups of PCSs with metal salts and unsaturated organic moieties. For example, a titanium-incorporated PCS (Ti-PCS) has been prepared by the reaction of Si–H group containing PCS with titanium alkoxides to give easier spinning into fibres (Okamura et al. 1988). Similarly, an iron-containing PCS (Fe-PCS) has been synthesized by the reaction of a PCS with iron acetylacetonate (Mishra et al. 2009). After pyrolysis an Fe-PCS gives magnetic SiC having EMI absorbing properties (Ginzburg et al. 2002). Monolithic SiC structures have been obtained by pyrolysis of epoxy-modified PCSs (Fujiki et al. 2000). A review on precursor derived ceramics has discussed many aspects of precursor technology (Colombo et al. 2010).

The next sections briefly describe important aspects of silicon chemistry, the polymer route for ceramics, various synthesis methods of PCSs, and features of

**Table 6.1** Comparison of various silicon and carbon bond lengths and energies

| Bond type | Bond length (Å) | Bond energy (kJ/mol) |
|---|---|---|
| Si–Si | 2.33 | 226 |
| Si–C | 1.90 | 318 |
| C–C | 1.54 | 334 |
| Si–O | 1.63 | 531 |
| C–O | 1.44 | 340 |
| Si–Cl | 2.11 | 471 |
| C–Cl | 1.81 | 335 |
| Si–H | 1.48 | 318 |
| C–H | 1.09 | 411 |

pre-ceramic polymer processing. The last section of this chapter will cover the high technology applications of precursor polymers.

## 6.2 Silicon Chemistry

Silicon is a second-row element in the periodic table directly under carbon. This suggests similar chemistry and reactivity to that of carbon. Silanes when compared to their analogous carbon compounds, particularly in nucleophilic substitution, exhibit enhanced reactivity and relative rates. This trend can be related to silicon's larger size and lower electronegativity, and the atomic ability to impart low-energy orbital bonding. Table 6.1 shows the bond lengths and bond energies of various carbon and silicon bonds with other elements (Pawlenko 1986).

Correlations to silicon's reactivity or stability can be made and explained using Table 6.1. For example, Si–X (X = halogen) bonds are very prone to nucleophilic attack but resistant to hemolytic cleavage. These observations are explained by the large electronegativity differences and high polarizability of both atoms, thus producing a highly ionic bond (Brook 2000).

## 6.3 Organosilanes

Organosilanes are the basic starting materials for synthesis of (polysilanes) PSs and PCSs and govern the properties and applications of these polymers. A silicon compound that contains at least one C–Si bond in the structure is called an organosilane. None of the organosilanes occur naturally and all are prepared synthetically. The number of types of organosilanes is as large as that of organic compounds. The other groups, along with organic groups, present on Si may be halogens, alkoxy, amino, isocyanate, etc. In general, organosilanes substituted exclusively with saturated hydrocarbon groups (no hydrolysable groups) are called 'non-functional', while organosilanes having hydrolysable groups like alkoxy and halogens are called 'functional' organosilanes. The organosilanes are also 'functional' if they have unsaturated organo groups, substituted organo groups with reactive atoms or groups, or Si–H groups. Silanes having hydrogen, chlorine, nitrogen, methoxy, ethoxy or acetoxy directly bonded to silicon are respectively called silicon hydrides, chlorosilanes, silylamines (silazones), alkoxysilanes and acetoxysilanes.

Organosilanes have been studied since the 1800s and several methods and routes have been established by various workers. The first tentative steps were made by Friedel and Crafts (1863–1880) and later by Ladenburg. Heating dialkylzinc with $SiCl_4$ in a sealed tube at 160 °C gave alkylsilane (Friedel and Ladenburg 1871). Kipping (1904) in 1904 used the Grignard reaction to prepare organosilanes. Stock, Polis, Pape, B. N. Dolnov, K. A. Andrianov and many other researchers have also contributed. Important methods reported for the synthesis of organosilanes include (a) Direct synthesis; (b) Substitution reactions; (c) Elimination reactions; (d) Disproportionation reactions of halosilanes with organosilanes; and (e) Addition reactions/Hydrosilylation reactions which include addition of Si–H group-containing compounds with unsaturated compounds (Colombo et al. 2010; Stark et al. 1982; Price 1947). Of these synthesis methods, the last one (e) is very versatile, easy to use and control in laboratory conditions.

## 6.4 Silicon Polymers

Silicon, similar to carbon, has the unique ability to catenate and form stable long homo-atomic chains. Some polymers having hetero, or homo chains of silicon are polysiloxanes, PSs, PCSs and polysilazanes. PSs and PCSs are important pre-ceramic polymers of SiC-based ceramics (Birot et al. 1995; Allcock 2002; Seyferth 1988; Allcock et al. 1992; Lee et al. 2010).

## 6.5 Organosilicon Polymers as Pre-ceramic Polymers of Si-Based Ceramics

Pre-ceramic polymers may be generally called a class of hybrid materials, consisting of Si-based polymeric chains and side chain functionalities. A simple and schematic representation of a silicon-based polymer monomeric unit is depicted in Fig. 6.1. The main backbone is constituted by silicon atoms bonded to a generic X group, with X = Si, O, NH, BR, $CH_2$ or N=C=N. The nature of X is probably the most important variable, since it determines the class of the final polymer: Si–Si backbone defines the PSs class, while Si–O, Si–NH, Si–BR, Si–$CH_2$ and Si–N = C = N define

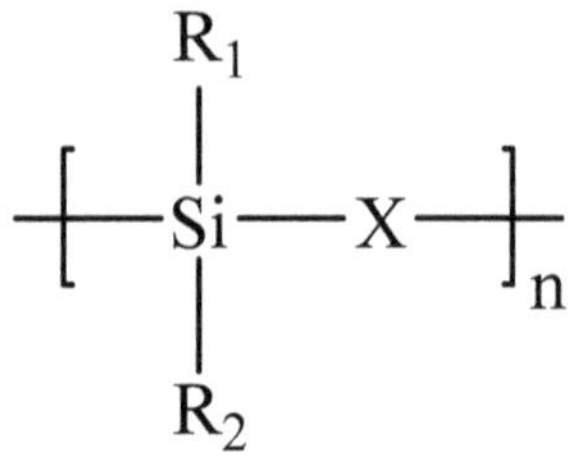

**Fig. 6.1** Representative structure of a silicon-based polymer

**Table 6.2** Main types of silicon-based pre-ceramic polymers, depending on the X group of the polymer backbone

| Group 'X' | Type of polymer |
|---|---|
| O | Polysiloxanes, Polysilsesquioxanes |
| Si | PSs |
| C | PCSs |
| N | Polysilazanes, Polysilsesquiazanes |
| B | Polyborosilanes |

the polysiloxanes, polysilazanes, polyborosilanes, PCSs and polysilylcarbodiimides, respectively, see Table 6.2.

The side chain functional groups $R_1$ and $R_2$ are generally carbon-based (e.g. H, alkyl, aryl, etc.). The nature of these side groups determines the properties and characteristics of the polymers. Modifying the R groups enables properties like thermal stability, solubility, and viscosity-temperature relation to be easily tailored (Colombo et al. 2010). Moreover, the presence of suitable functional side chain groups is essential to achieve cross-linking of the precursor polymer before the pyrolysis step, in order to retain the shape. The mechanism of cross-linking depends on the nature of side chain groups (e.g. condensation reactions if R = –OH functionalities, addition reactions if R = –vinyl groups). The nature and the quantity of the different side groups determine, during the pyrolysis step, the content of residual carbon in the final ceramics.

In principle, since more than one X group could be present at the same time in the polymer backbone, a wider range of possible polymer compositions should be considered (Greil 2000a). For example, polyborosiloxanes are a class of hybrid polymers where both X = B and X = O are present in the main chain. A more comprehensive description and graphical representation of silicon-based pre-ceramic polymers is thus presented in Fig. 6.2, as schematized by Colombo et al. (Colombo et al. 2010).

From the foregoing, it is in principle possible to achieve an extremely wide variety of pre-ceramic polymers. However, not all classes of pre-ceramic polymers have become equally 'popular' and widespread in the recent past. Apart from the properties and performances of the final ceramics, other factors must be considered during the selection of a pre-ceramic polymer. The cost effectiveness of the precursors is one of the most important parameters of a material or a component, especially from an

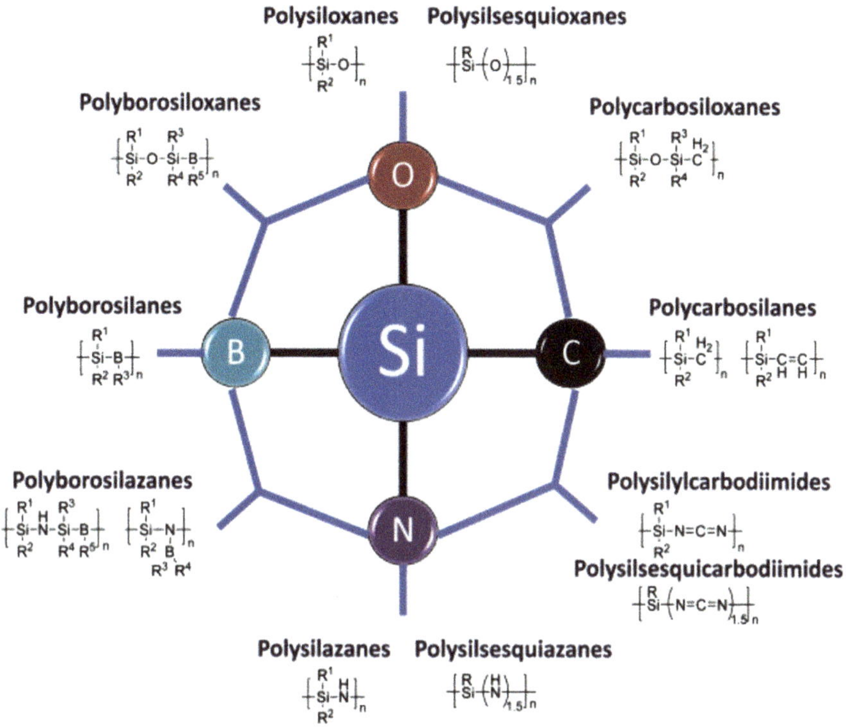


**Fig. 6.2** Main types of silicon-based pre-ceramic polymers in the Si–O–C–N–B system

industrial point of view. Another important factor is the specific handling that every precursor requires, which could be influenced by its chemical durability, thermal stability, toxicity and reactivity towards the atmosphere or other substances.

Polysiloxanes definitely represent the most widely studied class of pre-ceramic polymers (Abe and Gunji 2004), and their low price and high thermo-chemical stability make them the preferred choice for the easy realization of silicon oxycarbide glasses with extremely interesting thermo-mechanical properties up to nearly 1200 °C.

When higher thermal stability is required, other classes of pre-ceramic polymers should be considered, for example, PCS (for ~1400 °C) or polyborosilazanes (for ~1600 °C). The higher performances of polymers (such as polysilazane, PCS, etc.) are accompanied by higher costs of the precursors, more complex synthesis routes, often higher reactivity towards other substances (e.g. moisture and oxygen), and higher toxicity. These factors generally increase the processing complexity and are disadvantageous for their applicability on an industrial scale.

The most important synthesis routes for the production of silicon-based precursor polymers are based on chlorosilanes as the main starting compounds (Colombo et al. 2010; Chojnowski et al. 2000). Other synthesis methods have been proposed during the last decades (e.g. sol–gel), but the chlorosilane route is the most preferred because, in most cases, chlorosilanes are by-products of the silicone industry and offer the advantages of low cost, availability, and easily achievable high purity by distillation. The general formula of chlorosilanesis $R_xSiCl_{4-x}$, where $0 \leq x \leq 3$ and R is a generic side group. Nowadays, the chemistry of organochlorosilanes is very well known, and the reactivity of this class of compounds has been extensively exploited for the synthesis of many different organosilicon polymers such as PSs, PCSs, polysilsesquioxanes, polyborosilazanes, polysilylcarbodiimides, polyorganosilazanes, etc. A brief account of the synthesis routes of the most important classes of silicon-based pre-ceramic polymers is given in the following subsections.

### *6.5.1 Polysiloxanes*

Polysiloxanes or silicones are the most important class of polymers amongst preceramic polymers. Their low cost, easy synthesis and processing, and good thermal and chemical stability make them eligible for widespread applications (Abe and Gunji 2004; Reidel et al. 2006; Kendrick et al. 1995). The most important synthesis route for polysiloxanes is the reaction of chloro(organo)silanes with water. Other synthesis routes have also been proposed, such as polycondensation reactions of suitably substituted linear silanes, and ring-opening polymerization of cyclic ethers having silicon in the ring. On the other hand, pyrolysis of polysiloxanes in non-oxidizing environments gives silicon oxycarbide amorphous ceramic-containing networks of both Si–O and Si–C bonds at relatively low temperatures: such networks cannot be obtained by conventional processing methods.

### *6.5.2 Polysilazanes*

Polysilazanes are polymers in which the main backbone consists of alternate silicon and nitrogen atoms. They generally offer enhanced performances, both in the polymeric and in the pyrolysed state, as compared to the most widely used polysiloxanes. As polymers, polysilazanes are very versatile materials, and represent an excellent choice when high-temperature stability, corrosion resistance and durability are critical factors (Kroke et al. 2000).

Presently, polysilazanes are commercially available and mainly used for the preparation of high-temperature SiN or SiCN ceramic coatings/components, especially as coatings in the polymeric state. Polysilazanes have demonstrated to be excellent materials for protective coatings, owing to their excellent oxidation and corrosion resistance, UV radiation stability and high hardness. They have been successfully applied as clear, corrosion-protective coatings for many ferrous and non-ferrous metal alloys in marine, aerospace and automobile components, and for electrical insulation, waterproof surfaces and mechanical protection of optical surfaces and indoor flooring (Colombo et al. 2010). The general synthetic procedure for polysilazanes

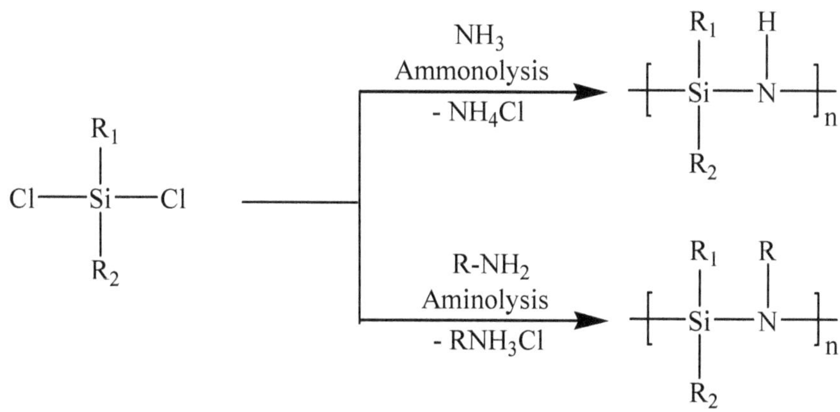


**Fig. 6.3** Synthesis of poly(organo)silazanes starting from chlorosilanes (Kroke et al. 2000)

involves ammonolysis or aminolysis of chlorosilanes with ammonia or other amines, respectively, as shown in Fig. 6.3.

### 6.5.3 *Polysilanes (PSs)*

PSs are a class of Si-based polymers in which a one-dimensional backbone is composed only of Si atoms, and with organic side chain groups (Miller and Michl 1989). The first PS derivatives were reported by Kipping and Sands (1921) in 1921 by reacting dichlorodiphenylsilane with sodium, and even today, his method of sodium mediated dechlorination (Wurtz type coupling) remains one of the most common methods for PSs synthesis. Other important methods include silylene polymerization and dehydrogenative coupling. The Wurtz type coupling is schematically shown in Fig. 6.4.

Wurtz coupling was used by Burkhard to synthesize polydimethylsilane (Burkhard 1949). It is highly crystalline, practically insoluble in any organic solvent, and on

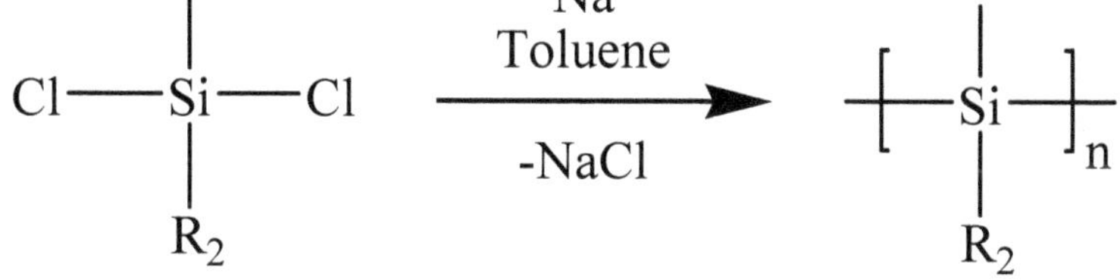


**Fig. 6.4** Synthesis of PSs by Wurtz type coupling

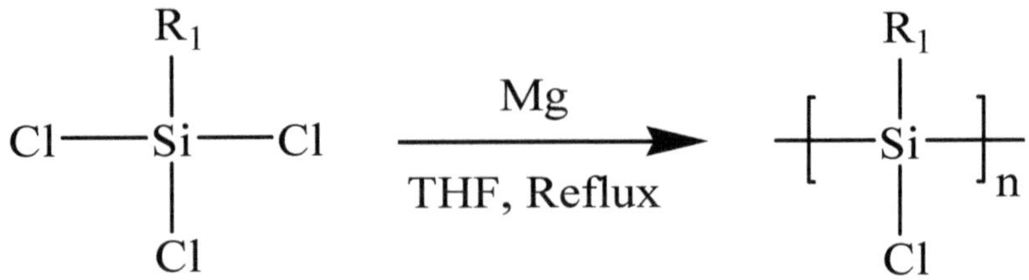


**Fig. 6.5** Grignard reaction for synthesis of PSs

heating decomposes above 250 °C. Burkhard's description suggested that polydimethylsilane (PDMS) is quite intractable. Eugene G. Rochow, the father of the silicone industry, also dismissed PSs for any use. According to him, PSs would be prone to oxidation, thermally unstable and easily hydrolysable by alkalis. These were possibly the reasons nobody considered PSs for the next three decades until the classical work of Yajima on PCSs in 1975. Yajima established that permethyl polymer $(Me_2Si)_n$ or its cyclic oligomers $(Me_2Si)_6$, can be converted into PCS that gives SiC on pyrolysis. Soon after, polydimethyl-co-polymethylphenylsilane was synthesized. It is soluble and can be melted (West et al. 1981).

Silicon polymers synthesized using Wurtz type coupling (Koe 2009) had complicated, highly cross-linked structures and were insoluble. Linear and soluble PSs were synthesized by the Grignard condensation reaction (Birot et al. 1995), see Fig. 6.5. Usually, functionalization of PSs after polymerization process was employed to introduce the required functional groups (Horiguchi et al. 1988).

As discussed above, earlier PSs were thought to be either intractable or unstable. However, we now know that the reality is different. The main weakness, as a material, is none of the above: it is susceptibility to UV light. PSs are potential UV acting photo resists for microelectronics, photoconductors in electro-photography. They also find used in organic reactions as free radical photo initiators. They possess good nonlinear optical properties which make them very useful for laser and other optical technologies. Today, they are even commercially available as precursors of SiC based materials.

### *6.5.4 Polycarbosilanes (PCSs)*

Polymers which have both Si and C as part of their backbone structure are called PCSs. A simplified general structure is shown in Fig. 6.6. But in reality PCSs have very complex structures comprised of rings and cages. Besides Si–C bonds, Si–Si bonds may be present. The structure may also contain rings, cages and complex hyper-branching, thus modifying the overall Si:C ratio and final ceramic yield.

PCSs represent one of the most interesting and technologically important classes of pre-ceramic polymers. The main interest that has led to the synthesis and continued

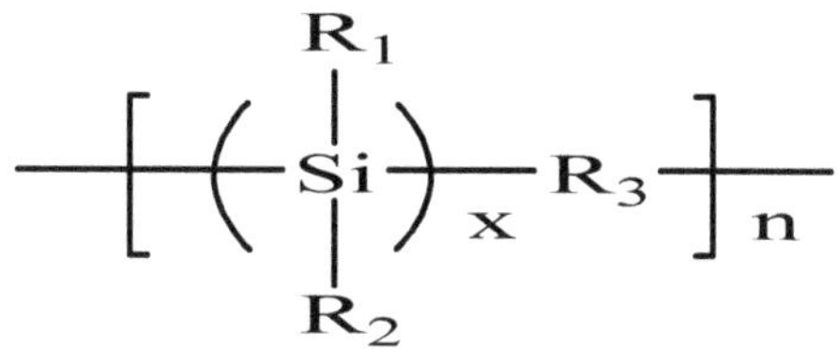

**Fig. 6.6** Simplified general structure of PCS

improvement of PCSs is in the manufacture of high-performance SiC fibres (e.g., Nicalon™ by Nippon Carbon Co., and Tyranno™ by UBE Industries).

Besides the use of PCSs as SiC precursors (Cao et al. 2002; Yu et al. 2011; Zeigler and Fearson 1989), they can also find applications as functional materials, such as photoresists and semiconductors. Because of their electric conductivity, unsaturated PCSs could be used in electronic applications such as compact capacitors, electromagnetic shielding of computers, piezoelectric sensors, smart windows (with tunable colour and transparency), antistatic coatings, transistors, LEDs, lasers and solar cells (Zeldin et al. 1988). Other newer but important applications are in nanotechnology (Tsirlin et al. 2002), bio-medicine, surface science, electrochemistry, catalysis, liquid crystals (Allcock et al. 1992), and nanocomposites.

Several synthetic routes have been developed for PCSs. The most important ones are discussed in Sect. 6.5.4.1–6.5.4.2.

#### 6.5.4.1 Yajima Strategy

The first and most famous approach for synthesis of PCSs was proposed by Yajima (Yajima et al. 1975a, 1978). Dichlorosilanes were used to get PSs that were subjected to thermally induced backbone rearrangement, i.e. Kumada rearrangement (Birot et al. 1995; Shiina and Kumada 1958) to get PCSs. Kumada rearrangement involves insertion of carbon between Si–Si bonds of PSs as shown in Fig. 6.7 for PDMS.

The actual structure of the Yajima PCS (as obtained by Kumada rearrangement) is complicated, which accounts for its largely insoluble character and easy cross-linking in air. The most acknowledged structure of the Yajima PCS is shown in Fig. 6.8.

The PCS was prepared from PDMS under pressure in an autoclave (50–60% yield), and was called Mark I PCS. The Mark II PCS was obtained by adding a minor fraction of $(Ph_2Si)_n$ to PDMS or $Ph_2SiCl_2$ to $Me_2SiCl_2$, using the same processing

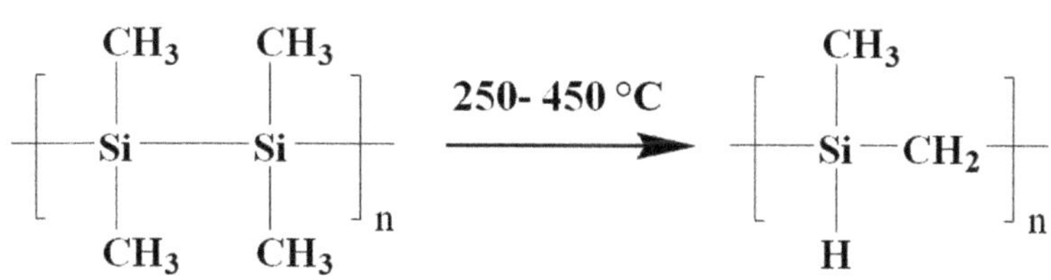


**Fig. 6.7** Kumada rearrangement of PDMS

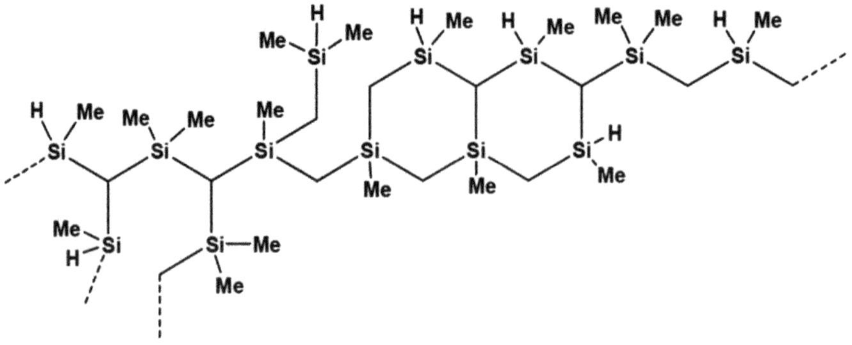


**Fig. 6.8** Model structure of Yajima PCS

method. It was reported that Mark II was easier to spin into fibres as compared to Mark I. In both cases an autoclave was necessary (Birot et al. 1995). The Mark III PCS was obtained, in an alternative way, under atmospheric pressure conditions and avoided the use of an expensive autoclave (Bacque et al. 1988). Here the PDMS was heated at 350–400 °C under atmospheric pressure with, as 'catalyst', poly-borodiphenylsiloxane (PBDPSO) to give a high ceramic-yielding-PCS (Ishikawa et al. 1990).

#### 6.5.4.2 PCSs Other Than Yajima-Method PCS

The effect of Yajima's work was numerous attempts to synthesize SiC precursors suitable for fibres without using PDMS. Several alternative synthetic routes have been proposed to prepare other types of PCSs. A brief description of these is presented here.

PCSs from Ring Opening Polymerization (ROP) Reactions

Ring opening polymerization of1,3-disilacyclobutane polymerization in the presence of a Pt-group metal catalyst (see Fig. 6.9) gives a PCS. Poly(silylmethylene) was prepared from 1,3- disilacyclobutane using this procedure. This polymer resulted in an 85% yield of SiC ceramic after 1 h in argon at 900 °C. Fourier transform infrared (FTIR) spectroscopy showed that cross-linking began after 500 °C and involved dehydrogenative coupling between Si–H and C–H. Many researchers have synthesized PCSs by ring opening of four- or five-membered cyclic silanes (Bacque et al. 1988; Zhou et al. 1990; Charles and Weber 1992).

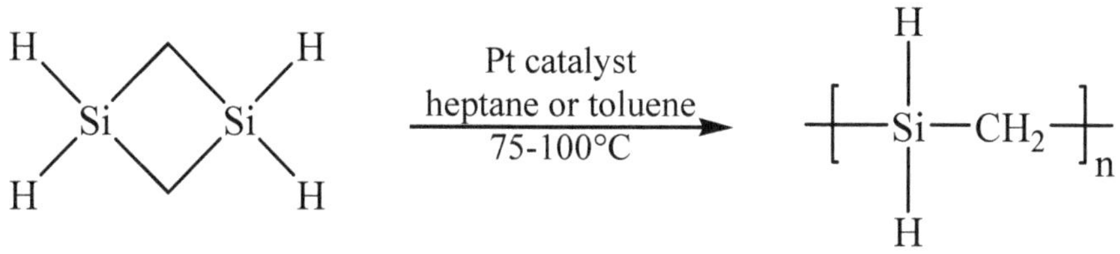


**Fig. 6.9** Synthesis of PCS from 1,3-disilacyclobutane

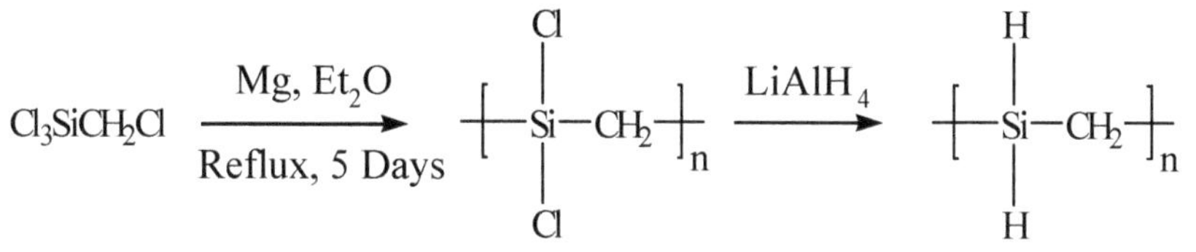


**Fig. 6.10** Synthesis of PCS by polycondensation of chloromethylchlorosilane

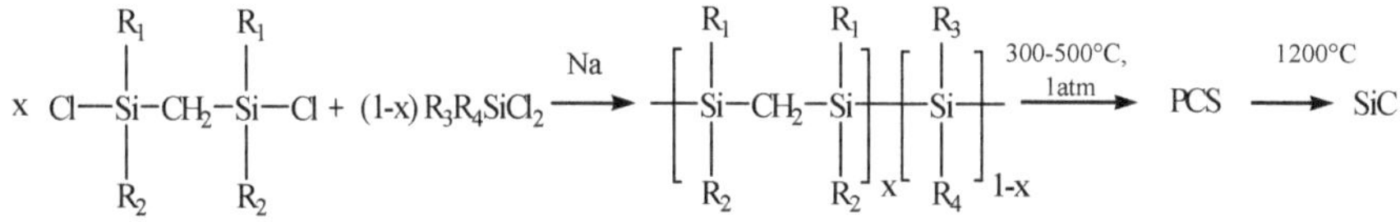


**Fig. 6.11** Conversion of 1,3-dichloro-1,3-disilylmethanes into a PSCS and PCS and then SiC

## PCSs from Polycondensation Reaction of Chloromethylchlorosilane

Whitemarsh et al. (Whitmarsh and Interrante 1991) used Mg mediated Grignard coupling of (chloromethyl)trichlorosilne followed by reduction with $LiAlH_4$ (see Fig. 6.10) to prepare a highly branched PCS. Many other reports are also available describing polycondensation of chlorosilanes to give PCSs (Dvornic and Owen 2009).

## PCSs from Copolymers with 1,3-Dichloro-1,3-Disilylmethanes

PDMS is a practically insoluble, highly crystalline solid (Wesson and Williams 1979). Several researchers have tried to prepare co-PSs $(Me_2Si)_n(RR'Si)_n$. The solubility can be improved by decreasing the crystallinity; and the C/Si ratio in the polymer can also be changed by manipulating R and R'. This approach helps to synthesize polysilacarbosilanes (PSCSs) that at atmospheric pressure are easily thermally convertible into PCS and finally, by pyrolysis, into a SiC ceramic, see Fig. 6.11. Alkali metal mediated condensation of 1,3-dichloro-1,3-disilylmethanes and dichlorosilanes resulted in a soluble PSCS greasy material.

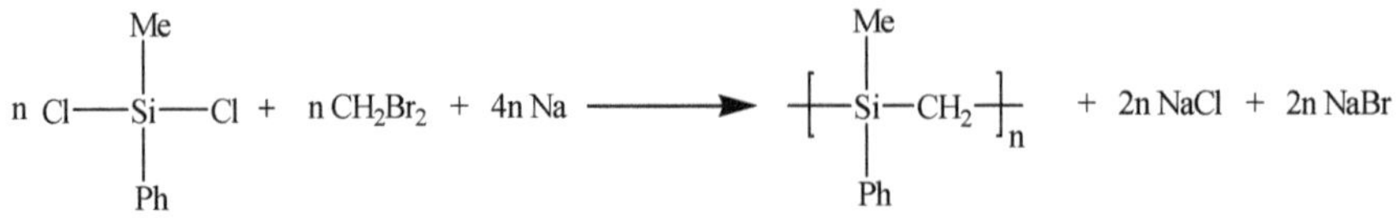

**Fig. 6.12** Sodium mediated synthesis of PCS from chlorosilane and methylene bromide

Soluble PSCSs give low ceramic yields (<10%) due to a lack of cross-linking. A heat treatment under inert gas atmospheric pressure at 450–500 °C, gave PCSs yields in the range of 20–50%. These PCSs exhibited ceramic yields up to 80%. This strategy has been used to prepare spinnable precursors with tunable C/Si ratios (Birot et al. 1995).

PCSs from ethylene Halides and Chlorosilanes

A series of dichlorosilanes and methylene bromide were reacted with sodium by Sartori to prepare PCSs which converted into ceramics, see Fig. 6.12 (Habel et al. 2008).

PCSs from Chlorosilanes and Acetylides

The use of polysilylacetylenes as precursors to SiC fibres was almost simultaneously investigated by two groups, namely those of Barton and Corriu. Co-condensation reaction of dichlorosilanes with dilithioacetylenes gives soluble polymers. Barton (Maghsoodi et al. 1990; Meyer et al. 1991) obtained polymers with an average molecular weight range of 20–30,000 and 20–80% thermogravimetric analysis (TGA) residues at 1100 °C. He could make SiC fibres from these PCSs after spinning and curing thermally or photochemically, followed by pyrolysis.

PCSs Using hydrosilylation Reactions

Corriu (Boury et al. 1991) used the hydrosilylation of vinylhydrogenosilanes, first described by Curry (Curry 1956), for the preparation of linear polysilethylenes. They used Pt- or Rh-based catalysts for this reaction. Nuclear magnetic resonance (NMR) spectral data revealed that the product of β-hydrosilylation gave 70–85% linear –Si–$CH_2$–$CH_2$–Si– rather than branched –SiCHMe- units.

N.B: Several methods other than those described in the foregoing subsections, have also been employed for obtaining polymeric precursors of SiC (Habel et al. 2008; Bernardo et al. 2014). This is an ongoing research topic.

## 6.6 Processing of PCSs

### 6.6.1 *Shaping*

Easy shaping of pre-ceramic polymers is one of the most advantageous aspects of the precursor polymer route. Compared with common powder processing routes, the polymeric system has the potential advantage of easy fabrication of green compacts with no need for additional binders. Moreover, the possibility of obtaining a machinable component prior to the polymer-to-ceramic conversion is a remarkable advantage, since it permits a more precise shape control without all the problems related to ceramic brittleness and tool wear (Rocha et al. 2005). The common polymer shaping methods such as pressing (Galusek et al. 2007), extrusion (Perale et al. 2008), injection moulding (Zhang et al. 1995), spray coating, dip coating, spin coating, chemical vapour deposition (CVD) (Colombo et al. 1997; Smirnova et al. 2003) can in principle be used for processing pre-ceramic polymers. Special polymer forming methods, when used with pre-ceramic polymers, make it possible to realize some special shapes and components generally extremely difficult to obtain via the powder processing route. Some typical examples are fibre drawing (Bunsell and Piant 2006), infiltration of pre-formed scaffolds to get ceramic matrix composites (CMCs) (Erdal et al. 1999), ceramic joining (Colombo et al. 1988), and preparation of highly porous ceramics components by direct foaming or other techniques such as sacrificial templates (Vakifahmetoglu et al. 2009).

It is important to mention here that although much progress has been made in the development of high-ceramic-yielding precursors, only low-dimensional products like fibres and coatings have become successful applications. The main problem hindering the production of large-size, thicker components lies in the pyrolysis step, which is accompanied with release of gases, introduction of porosity, and shrinkage due to the density change from polymer to ceramic. These problems need to be addressed with priority.

### 6.6.2 *Cross-Linking*

After getting the required shape using some shaping process, the cross-linking of pre-ceramic polymers is a must to induce non-meltability during the pyrolysis step transforming the polymer into the ceramic.

Providing that suitable functional groups are available in the polymer structure (such as –OH, –H or –vinyl groups that could undergo condensation or addition reactions) then the polymer can generally be cured thermally by heating up to 150–250 °C. The thermal curing temperatures can be lowered if an appropriate catalyst or radical initiator is used, with the further advantage of an increase in the ceramic yield due to a decreased evaporation of oligomers (Janakiraman and Aldinger 2009).

Other cross-linking methods could involve the use of UV-radiation, γ-rays, electron-beams, and the use of reactive substances, e.g. plasmas (Liew et al. 2002; Idesaki et al. 2004, 2001; Schulz et al. 2004). Precursor polymer cross-linking is also of fundamental importance in order to obtain high ceramic yields upon pyrolysis. An effective cross-linking process results in a highly branched network of polymeric molecules, that in turn gives lower oligomer content and fewer low-molecular-weight chains that could volatilize during pyrolysis: this means an increase of the final ceramic yield (Colombo et al. 2010).

### 6.6.3 Conversion of Polymer into Ceramic (Ceramization)

The polymer-to-ceramic conversion (ceramization) probably represents the most important step of the whole pre-ceramic polymers processing route of ceramics, and is the result of a thermal decomposition process above ~400 °C, where a totally inorganic, non-volatile residue is formed.

The mechanism involved in the pyrolysis process is generally quite complex. This involves structural rearrangements and radical reactions that result in the dissociation of bonds (e.g. Si–H, Si–C and C–H), the release of organic functional groups (e.g. $CH_4$, $C_6H_6$,) and the formation of an inorganic network (Reidel et al. 2006). The whole ceramization process comprises of a sequence of different distinct rearrangement reactions, each occurring in its specific temperature range. In the case of PCSs, for example, it has been shown that in the 550–750 °C range methane is the main volatile species released, while at higher temperatures (700–1100 °C) the release of hydrogen is observed (Hurwitz et al. 1993). At temperatures of 400–600 °C a porous network with open channels is generated that can diminish (transient porosity) at temperatures > 800 °C. A simplified representation of the weight loss (TG curve) and structural rearrangements of a PCS pre-ceramic polymer during heat treatment in inert (argon) atmosphere is shown in Fig. 6.13 (Greil 2000b).

Ceramic yield is a parameter of utmost importance in the pre-ceramic method. The yield is defined as the percentage obtained when the mass of the residue obtained after ceramization is divided by the mass of the starting pre-ceramic polymer. In most cases the yield lies in the range 60–90 wt%, but higher values of even up to ~105 wt% have been obtained when reactive atmospheres are used (Isoda et al. 1992). Since the beginning of the concept of pre-ceramic polymers, extreme efforts have been made to develop useful precursors that give high ceramic yields.

The ceramics obtained after completion of the polymer-to-ceramic transformation show an amorphous structure. The amorphous network is made up of a mixture of covalent bonds. The most important bonds are Si–C, Si–O, Si–N and C–C. Other bonds with other atoms like B, Ti, Al or Zr may also be present, depending on the type of precursor polymer and possible modifications of the polymer made before the ceramization step. This typical amorphous structure of ceramics is a special characteristic of pre-ceramic polymer-derived ceramics, which so far has been unobtainable through traditional processing routes (e.g. reactive sintering). The amorphous

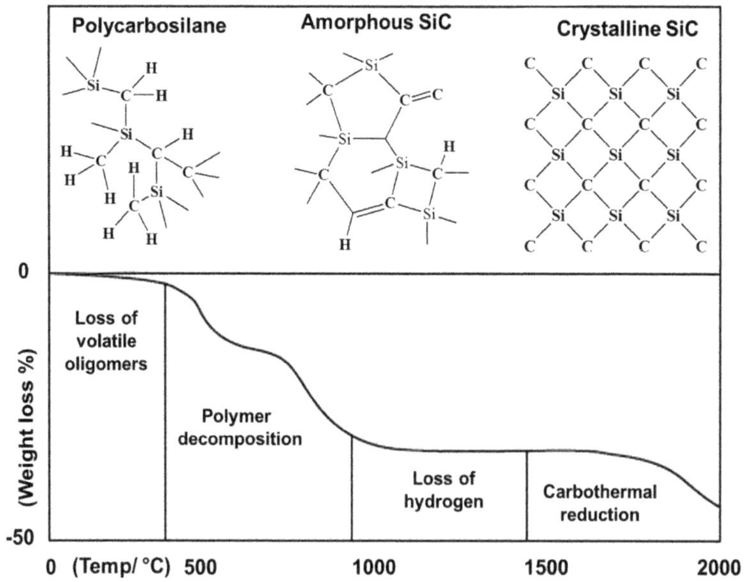


**Fig. 6.13** Typical structural transformations during an inert atmosphere pyrolysis of pre-ceramic polymers (PCSs)

structure is responsible for high thermomechanical properties of precursor-derived ceramics.

SiOC glasses are stable up to 1200 °C, and creep, oxidation and chemical resistance properties are maintained. Above this temperature the properties degrade owing to phase separation/crystallization that occurs around 1300 °C (Soraru et al. 2002). SiCN amorphous ceramics maintain their properties up to ~ 1400 °C: the presence of nitrogen atoms has beneficial effects on the thermo-mechanical properties (Loehman et al. 1985). The presence of additional atoms, such as boron in SiBCN ceramics, increases the crystallization temperature upto 1600 °C. For this reason, SiCN and SiBCN ceramics show some outstanding properties even at very high temperatures of 1800–2000 °C (Zhang et al. 2012).

## 6.7 Current Scenario in India

Researchers from the Indian sub-continent have made important contributions in the area of precursor-derived ceramics. Significant work has been carried out by a group at Kanpur (India) and their collaborators at different laboratories in India. In the following text a brief account of the important contributions is given.

### 6.7.1 Precursor Synthesis Methods at the DMSRDE

In order to replace the metal alloy components used in the high temperature zones of aeroengines with SiC ceramic composites, DRDO had initiated a research programme "Technology Development of Ceramic Matrix Composite (CMC) for Hot-end Components of Aero Engines" in the early years of this current century. In this programme the research group at DMSRDE Kanpur was entrusted with the responsibility to develop the 'Synthesis of Organometallic Precursors for Ceramic Materials'. These precursors were aimed to be used as polymer precursors for SiC matrix and fibre products to help realize lightweight high performance SiC–SiC composite components. The developed PCS was converted into fibres by the melt spinning method. As-spun green fibres were stabilized by oxidative curing and pyrolysed at elevated temperatures (1100–1250 °C) under an inert atmosphere to obtain SiC fibres.

In the above mentioned programme, researchers synthesized a variety of homo- and co-polysilanes using modified Wurtz coupling reactions, and attained yields > 85% (Shukla et al. 2004). They employed thermal studies, TGA and differential scanning calorimetry (DSC), to identify suitable polysilanes for synthesizing PCS and concluded that PDMS is a suitable precursor for PCS because of the lower reaction temperature and high yield of reaction. They established a 2.0 kg batch process for PDMS synthesis with ~ 90% yield, and converted it into PCS under ambient pressure conditions by thermal backbone rearrangement at 250–450 °C in an inert atmosphere in the presence of a polyborodiphenylsiloxane (Python) catalyst, thereby obtaining a high yield of ~ 66%. Gel permeation chromatography (GPC) analysis of the synthesized PCS revealed it is poly-dispersed in nature. The conversion process of PDMS into PCS and pyrolysis behavior of PCS was also studied in detail.

### 6.7.2 Synthesis of Polyborodiphenylsiloxane (Python) Catalyst

The so-called Python catalyst is used in smaller amounts, usually 1–5 wt%, with PDMS to enhance the spinning properties of PCS. It is synthesized by reacting boric acid (1 mol) with dichlorodiphenylsilane (1.5 mol) in dry n-dibutyl ether solvent under reflux conditions, followed by solvent removal and heating at 300 °C under reduced pressure. The catalyst is obtained as a glassy solid product.

## 6.7.3 *Synthesis of PDMS*

PDMS is the polysilane of choice for PCS synthesis. The reaction is shown in Fig. 6.14. The laboratory conditions for its synthesis are as follows: a 20 L three-neck round bottom flask is charged with 10 L of dry binary solvent (toluene and 1,4-dioxane), sodium pieces and 15-Crown-5 catalyst. This mixture is refluxed under inert atmosphere with constant mechanical stirring to ensure melting and dispersion of the sodium metal. Dichlorodimethylsilane (4.383 L, 36 mol) diluted with toluene is added drop-wise in 5-6 h to the refluxing solution, and the reaction mixture is further refluxed for 6 h. Sodium is taken in 1.1 molar ratio to the chlorine of chlorosilane. Approximately 50% solvent is distilled off and isopropyl alcohol is added drop-wise to quench the unreacted sodium, followed by addition of excess water. PDMS, in the form of white material, comes to the surface of the reaction mixture. The PDMS is filtered and washed several times with water to remove any traces of salt and alkali. Final washing with dry methanol, and drying at 250 °C under inert atmosphere, results in a non-meltable, non-soluble, free-flowing white powder of PDMS with ~86% yield.

### 6.7.3.1 Synthesis of PCS

PCS is synthesized from PDMS using a thermal backbone rearrangement. The typical laboratory synthesis process, according to the reaction shown in Fig. 6.7, is as follows: 8.0 kg pre-dried PDMS is thoroughly mixed with 1 wt% of Python catalyst and charged in a 20 L three-necked round bottom flask fitted with a double surface Davies condenser, digital temperature indicator and inert gas inlet tube. The reaction mixture is heated at 250–425 °C for a period of ~ 125 h using an electrically heated mantle. This heat treatment turns white PDMS into transparent golden yellow material. The reaction mixture is cooled to room temperature and the residue is dissolved in petroleum ether. Crude PCS solution is filtered and solvent-distilled under reduced pressure to result in 3.2 kg of PCS as a golden yellow solid with a yield of ~ 40% and melting range of 110–120 °C. Liquid PCS is also obtained as a by-product with

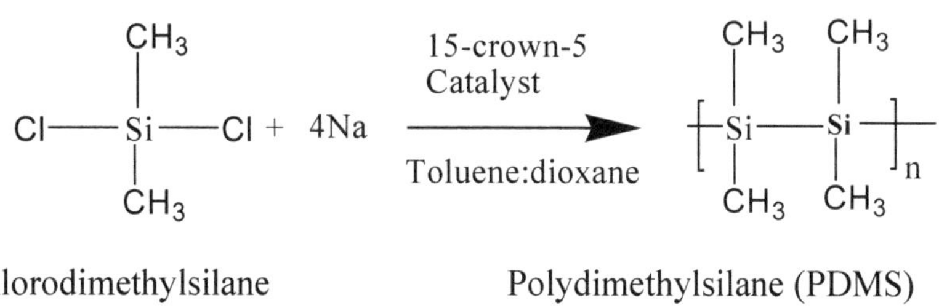


**Fig. 6.14** Synthesis of polydimethylsilane

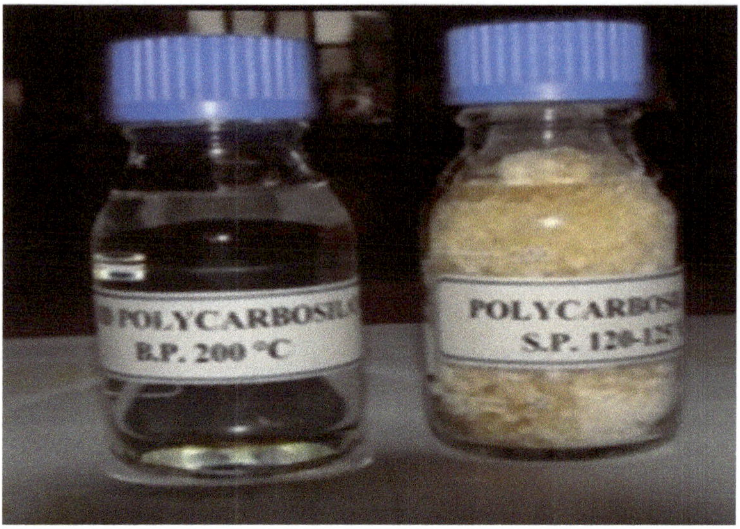

**Fig. 6.15** Polycarbosilane (liquid and solid)

approximately 26% yield: this has found very useful application as a non-corrosive CVD precursor of SiC. PCS in solid and liquid forms is shown in Fig. 6.15.

### *6.7.4 Rheological Behaviour of PCS and M-PCS*

SiC ceramic fibres are most commonly obtained by melt spinning of ceramic material precursors like PCS and the hetero-metal substituted derivatives (M-PCS), followed by controlled stabilization and pyrolysis. R. K. Gupta et al. have studied the rheological behaviour of PCS melts with respect to time, temperature, atmosphere, molecular weight and polydispersity because these are very important parameters for successful processing and for preparing PCS green fibres (Gupta et al. 2011). In this study PCS samples having different softening points were synthesized from PDMS and the variation in melt viscosity and thermosetting behaviour was studied. Temperature, time, atmosphere (inert and air) and shear rate were chosen as variables for measurement of melt behaviour.

Under inert atmosphere the melt viscosity of all the PCS samples decreased with increasing temperature; then under isothermal conditions it remained almost constant for the low softening point samples (high polydispersity), whereas it increased for the high softening point samples (low polydispersity). It was observed that in air all of the PCSs crosslinked rapidly as compared to an inert atmosphere.

Temperature was found to have a prominent effect on the increase in SiH/$SiCH_3$ ratio; and the processing time was reduced with an increase in processing temperature. PCSs having higher molecular weight easily became thermosets due to enhanced crosslinking of the polymer as compared to low molecular weight PCSs, irrespective of the atmosphere. The authors concluded that for processing of a PCS it is essential to have a polydispersed sample and to maintain an inert atmosphere to prevent undesired crosslinking.

In a related study R. K. Gupta et al. synthesized a few aluminium-containing PCSs, named AlOR-PCS and Alac-PCS, by the reaction of aluminium isopropoxide (AlOR) and aluminium acetylacetonate (Alac) with PCS. These compounds were studied for rheological properties with respect to time, temperature and atmosphere (inert and air) (Gupta and Saxena 2014). They observed that the increase in metal content enhances the cross-linking of the PCS chains, and under similar conditions the Alac-PCS derivatives cross-linked faster than AlOR-PCS. The molecular weight and ceramics yield of all these alumino-PCSs increased with increasing metal content as compared to virgin PCS. Alac-PCS had higher molecular weight and ceramic yield than AlOR-PCS for similar molar ratios of metal complexes. The authors also reported making aluminium-incorporated SiC fibres by drawing thin Al-PCS green fibres from an Al-PCS melt, followed by curing and pyrolysis.

## 6.8 Applications of Polymer Precursors

Polymer precursors are used in many high technology areas including SiC matrix composites, SiC foams in space mirrors, molten metal filtration, catalyst support, thermal insulation, SiC thin films, membranes, coatings, and high temperature electronics. Some of the contributions made by Indian authors are discussed in this section.

### *6.8.1 Polycarbosilane as Source of SiC Coating on Oxidation-Prone Substances*

Carbon-based materials like carbon fibres, C–C and C–SiC composites, graphite or carbon nanotubes (CNTs) possess high thermal stability but only in non-oxidative environments. For example, graphite can withstand a temperature of ~3000 °C in inert environments or vacuum, but is not suitable for continuous use at a temperature beyond ~500 °C in air/oxygen atmospheres owing to its poor oxidation resistance. Many methods have been studied in order to improve the oxidation resistance of graphite and similar materials: fabricating oxidation-resistant protective SiC coatings is one of the widely studied techniques.

CNTs are coming into use for different applications like conductive and high strength composites, chemical sensors, energy storage and energy conversion devices, nanometer sized semiconductor devices, probes and interconnects. However, CNTs have poor thermo-oxidative stability. A simple and convenient process to

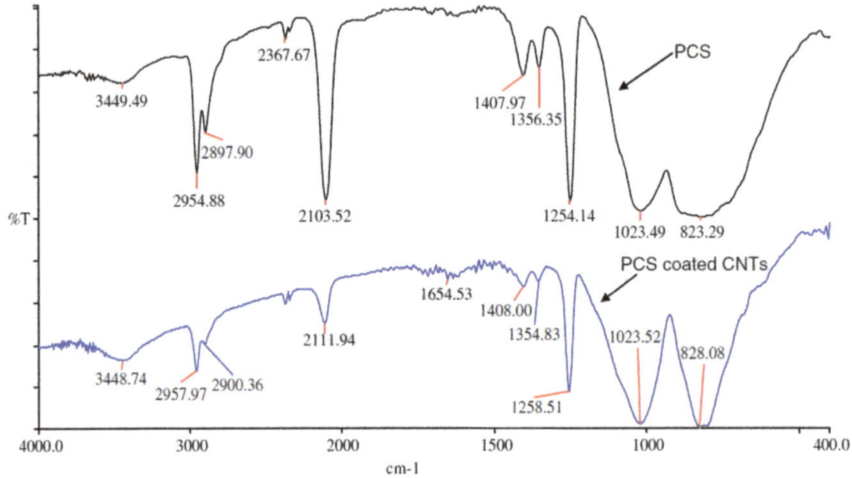


**Fig. 6.6. 6.16** FTIR spectra of PCS and PCS coated CNTs (https://s100.copyright.com/AppDispatchServlet)

generate a SiC protective coating on multi-wall carbon nanotubes (MWCNTs), its processing and spectral studies (XRD, TGA) has been reported by R K Gupta et al. (Gupta et al. 2009).

Figure 6.16 compares the FTIR spectra of coated and uncoated MWCNTs. This figure shows a reduced Si–H peak for the coated MWCNTs at ~ 2100 $cm^{-1}$ owing to hydosilylation of the polycarbosilane and the CNTs. The PCS coated MWCNTs give SiC coated MWCNTs after pyrolysis, as shown in Fig. 6.17.

### *6.8.2 Polycarbosilane Based C/SiC Composites and Nano Fibrous Webs*

Suresh Kumar et al. have fabricated unidirectional (UD) C/SiC composites using PCS (Kumar et al. 2015). Figure 6.18 gives the schematic for the entire process. Moulding of the composites was carried out at 300 °C, and pyrolysis was done at 1400 °C or 1600 °C. The composites were repeatedly infiltrated and pyrolysed to achieve densities in the range of 1.8–1.85 g/$cm^3$. The composites fabricated at 1600 °C showed better mechanical properties than composites fabricated at 1400 °C. The mechanical properties enhancements at 1600 °C are due to formation of silicon carbide nano-pins, see Fig. 6.19.

M. K. Sinha et al. have prepared SiC nano-fibrous webs from polycarbosilane by electro-spinning, see Fig. 6.20. The nano-fibrous webs of polycarbosilane were cured at 180 °C. The cured polycarbosilane webs were pyrolysed at 1000 °C to obtain

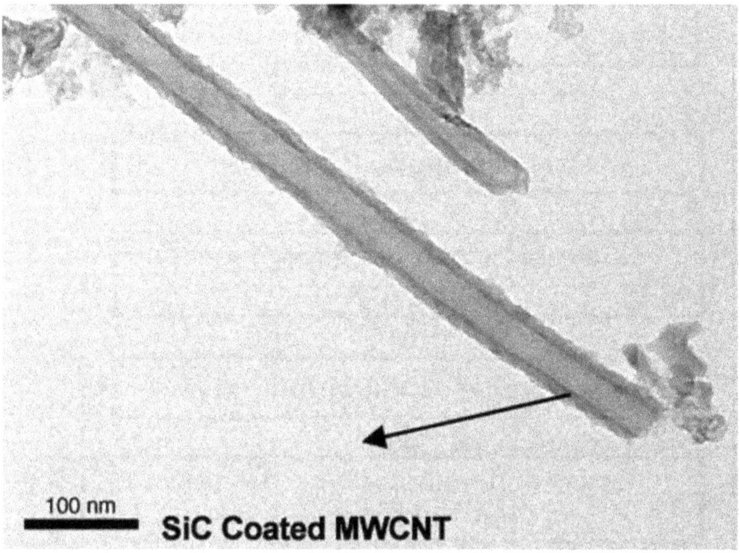


**Fig. 6.17** HRTEM Micrograph of SiC coated MWCNTs

silicon carbide nano-fibrous webs. For example, FESEM micrographs in Fig. 6.21 show the interconnected fibres in SiC nano-fibrous webs (Sinha et al. 2014).

### *6.8.3 Processing, Properties and Microstructure of SiC Foam Derived from Epoxy-Modified PCS*

Manoj Kr. Misra et al. fabricated SiC foams (Fig. 6.22) having controlled porosity using epoxy-modified polycarbosilane (EMPCS), and studied their properties (Mishra et al. 2017). The EMPCS was synthesized by refluxing epoxy resin and PCS in a THF solution at 150 °C, see Fig. 6.23. There were three EMPCS types, with epoxy contents of 0, 10 and 20% by weight: these were designated as PCS, 10EMPCS and 20EMPCS. Thermal foaming of these EMPCSs was carried out at 1000 °C under inert atmosphere followed by ceramization at 1200 °C, 1400 °C and 1600 °C under vacuum. The cell size of the ceramized SiC foam was found to vary between 100 and 700 μm, see Fig. 6.24, with increasing cell sizes for higher contents of epoxy in the EMPCSs. The crushing strength was inversely proportional to the wt% of epoxy resin in the PCS.

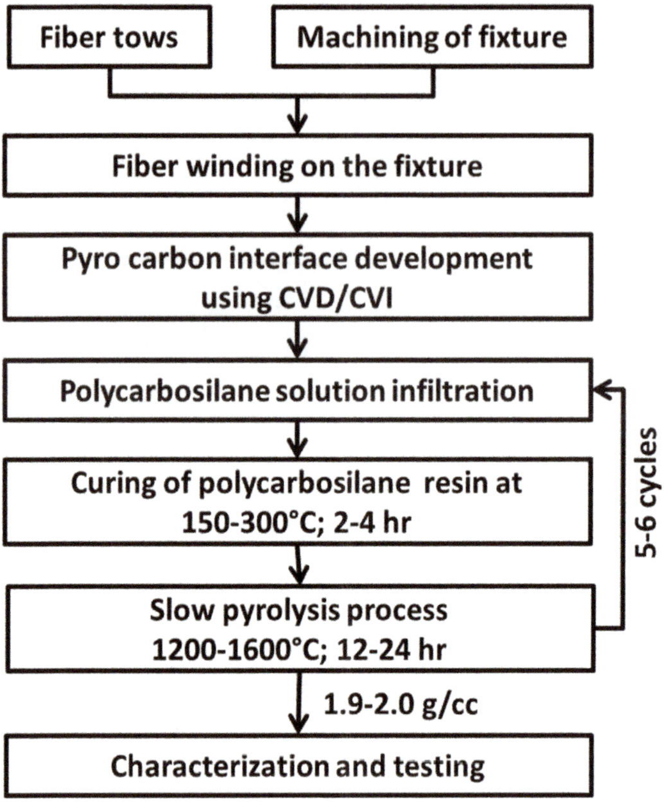


**Fig. 6.18** Schematic layout for fabricating a unidirectional (UD) C/SiC composite (https://s100.copyright.com/AppDispatchServlet)

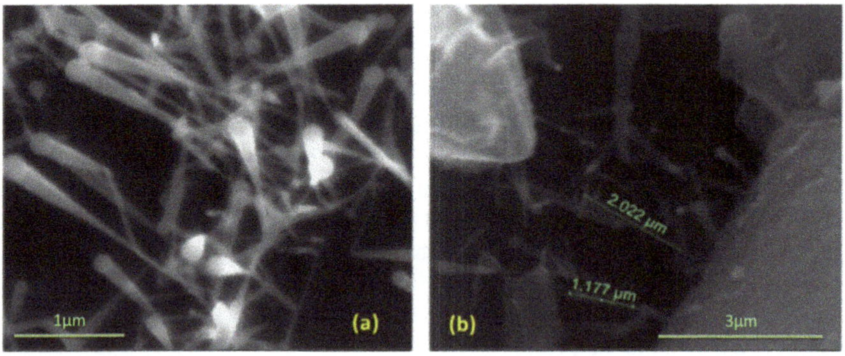


**Fig. 6.19** SiC nano-pins in UD C/SiC composites pyrolysed at 1600 °C (**a**) in the matrix (**b**) near a fibre surface

**Fig. 6.20** Electro-spinning set up fornano-fibrous webs

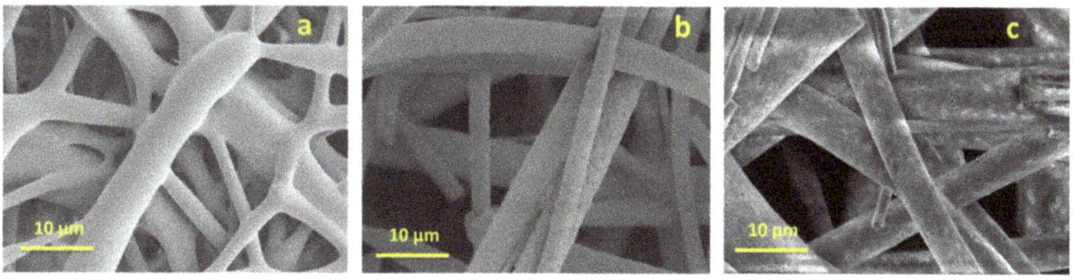


**Fig. 6.21** FESEM images of nano webs: **a** PCS as-spun **b** PCS cured **c** PCS pyrolysed

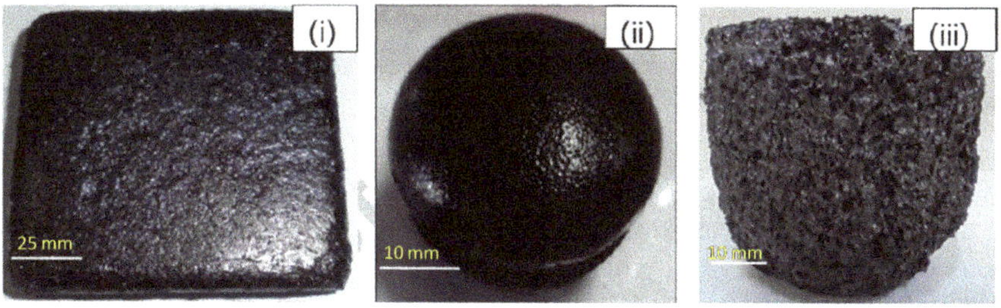


**Fig. 6.22** Typical shapes and sizes of the SiC foams: (i) 100 × 100 mm; (ii) Dia = 35 mm; (iii) Dia = 50 mm

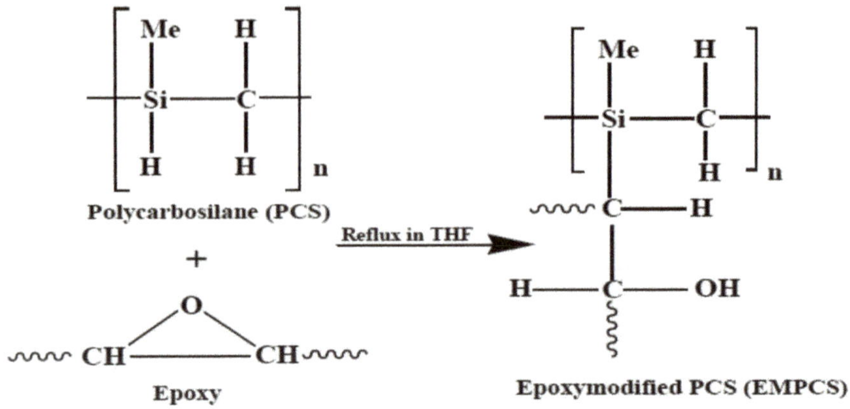


**Fig. 6.24** Synthesis of EMPCS (https://s100.copyright.com/AppDispatchServlet)

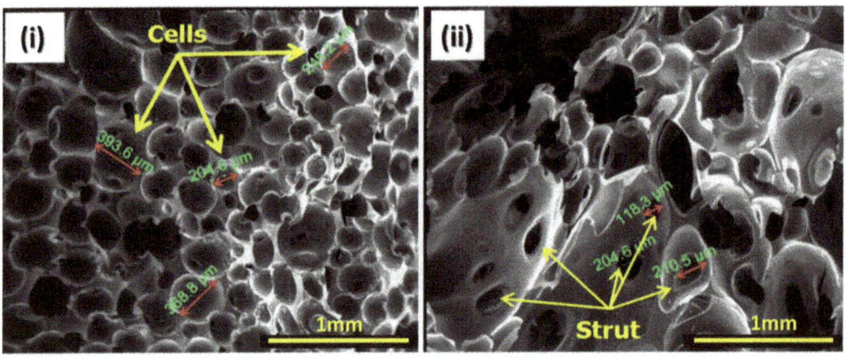


**Fig. 6.24** SEM image of SiC foam; Effect of epoxy content in EMPCS powder (i): 10EMPCS, (ii): 20EMPCS

### *6.8.4 Enhanced Nano-mechanical and Wear Properties of PCS Derived SiC Coating on Silicon*

Jonaki Mukherjee et al. (Mukherjee et al. 2015) have used liquid PCS synthesized by DMSRDE to deposit β-SiC and α-SiC on silicon wafers by the CVD method at three moderate high temperatures (800, 900 and 1000 °C), followed by evaluation of the nano-mechanical properties of the SiC coatings. They observed that the nano-mechanical properties of the coatings improve with an increase in deposition temperature. The reason for this enhancement is the increase in crystallinity along with good adherence of SiC to the silicon wafer at higher temperatures.

The tribological properties of the coated samples showed similar trends with increase in deposition temperature, although the trend was much stronger for the β-SiC coated samples than for those coated with α-SiC. All the coated samples showed a brittle mode of failure during tribological testing. It was found that the 900 °C deposited film had better nano-mechanical as well as good tribological properties compared with the other deposition temperatures. This favourable combination was due to better crystallinity and adherence. The authors concluded that the enhanced mechanical and tribological properties of SiC coatings make them useful as protective films on silicon for application in micrometers and other microdevices.

### *6.8.5 PCS as SiC Source in Ultra High Temperature Ceramics (UHTCs)*

Indigenous PCS has been used in several studies for in-situ generation of SiC during the fabrication of UHTC ceramics. Saxena et al. synthesized zirconium diboride–zirconium carbide binary ceramic powder and used PCS as the in-situ source of SiC to fabricate a SiC–$ZrB_2$–ZrC ternary CMC by vacuum hot pressing (Saxena et al. 2014). Debnath et al. studied mechanical, tribological and thermal properties of three variants of a hot-pressed $ZrB_2$–SiC composite, using SiC with different morphologies. Different SiC powders were used for two of the composites, whereas indigenous PCS was used as the SiC source for the third composite (Debnath et al. 2015). In similar work, microscopic, mechanical and thermal properties of a $ZrB_2$–SiC based composite were studied by S. Chakraborty et al. (Chakraborty et al. 2015). In this case the authors used the spark plasma sintering method for fabrication of the composite, and SiC was derived from indigenous PCS.

### *6.8.6 Preparation of SiC Ceramic Fibres from PCS*

The research group at DMSRDE prepared PCS green fibres by melt spinning PCS using a piston-cylinder set-up. PCS fibres were collected on a rotating drum and infusibility was obtained by oxidative curing in air. The stabilized fibres were subjected to controlled pyrolysis in an inert atmosphere, resulting in SiC fibres with diameters in the 20–50 $\mu$m range. The fibres had circular cross-sections and no defects were visible from scanning electron microscopy (Saxena et al. 2016).

## 6.9 Summary and Conclusions

Polymeric precursors for ceramics have been investigated for more than four decades. Many researchers have contributed to understanding the chemistry of precursors, their transformation into ceramics, and various factors that determine the suitability of any polymer to be an acceptable precursor material. The knowledge base that has evolved has enabled researchers to develop complicated ceramic structures and shapes that are extremely difficult to be obtained by the conventional powder processing method.

There are several advantages of the precursor-derived ceramic route, such as: (i) control of the polymer structure giving ceramics controllable composition; (ii) formation of complicated shapes using conventional polymer forming methods; (iii) obtaining ceramics at comparatively low temperatures; (iv) formation of near-net shape structures; (v) achieving ceramic microstructures that are not obtainable by the powder route. A few examples of complex structures derived from PCS are: fine diameter SiC fibres, thin SiC films, nano-structures, and complex-shaped SiC matrix composites.

In India the previous two decades have witnessed a significant amount of work on precursor-derived ceramics. Multi-kilogram synthesis capability for PDMS and PCS has been established, and indigenous PCS has been used in developing the PIP technology for fabricating UD-, 2D-, and 3D-stitched C/SiC composites (including shaped composite articles). SiC fibres, SiC foams, magnetic SiC and nano-SiC structures have also been realised. Furthermore, liquid PCS has been shown to be a potential substitute for methyltrichlorosilane (hazardous) used to CVD-coat SiC ceramic onto various substrates.

Worldwide efforts are continuing in order to improve the thermo-oxidative stability and mechanical properties of ceramics by finding newer precursors or by modifying the existing ones with various additives, like metal salts. Although there has been tremendous advancement in the precursor-derived ceramic technology, much more needs to be done. Thinner structures like SiC fibres have been commercialized successfully, but thicker structures have not seen commercial success owing to underlying challenges of the precursor route in obtaining a defect-free and dense structure. There is thus an enormous scope to develop cost-effective precursor polymers that (i) are stable under normal process conditions; (ii) can be easily processed into the desired shapes; (iii) can be made infusible before pyrolysis; (iv) can be pyrolyzed into ceramics with retention of shape; (v) give high ceramic yield and a pure form of the desired ceramic.

We also need to develop more understanding of the rheological behaviour of precursor polymers, since this behaviour is extremely important for processing into fibres or other shapes. In India we need to develop industrial synthesis facilities for organochlorosilanes, siloxanes, silanes, polycarbosilanes and other silicon-based polymers for self-reliance in this area. Efforts are being made to develop modern facilities for processing precursors into usable ceramic products, and non-destructive testing of these products. With concerted efforts this country will soon become self-reliant in critical precursor-derived ceramic technology for various advanced high temperature applications in the aerospace, nuclear and defence sectors.

**Acknowledgements** The authors are grateful for the funding from DRDO to conduct these studies via various DRDO projects and the DMSRDE task activities. They would like to profoundly thank the former and present Chairmen of DRDO, Dr. S Christopher, Dr. G Satheesh Reddy and Dr. Samir V. Kamat for their exemplary support and constant encouragement. They also thank profoundly Dr.

Dipankar Banerjee of IISc., Bangalore (Former CC R&D, DRDO) and Professor LM Manocha, Former DRDO's Raja Ramanna Fellow at DMSRDE for their keen interest and many an advice.

## References

Abe, Y., Gunji, T.: Oligo- and polysiloxanes. Prog. Polym. Sci. **29**(3), 149–182 (2004)

Ainger, F.W., Herbert, J.M.: The preparation of phosphorus-nitrogen compounds as non-porous solids. In: Popper, P. (ed.) Special Ceramics, pp. 168–182. Academic Press, New York (1960)

Allcock, H.R.: Chemistry and Applications of Polyphosphazenes. Wiley, USA (2002)

Allcock, H.R., Mark, J.E., West, R.: Inorganic polymers. Prentice Hall, USA (1992)

Babonneau, F., Soraru, G.D.: Synthesis and characterization of Si-Zr-C-O ceramics from polymer precursors. J. Eur. Ceram. Soc. **8**, 29–34 (1991)

Bacque, E., Pillot, J.P., Birot, M., Dunogues, J.: New polycarbosilane models. 2. First synthesis of poly(silapropylene). Macromolecules **21**, 34–38 (1988)

Bernardo, E., Ponsot, I., Colombo, P., Grasso, S., Porwal, H., Reece, M.J.: Polymer-derived SiC ceramics from polycarbosilane/boron mixtures densified by SPS. Ceram. Int. **40**(9) Part A, 14493–14500 (2014)

Birot, M., Pillot, J., Dunogues, J.: Comprehensive chemistry of polycarbosilanes, polysilazanes, and polycarbosilazanes as precursors of ceramics. Chem. Rev. **95**, 1443–1477 (1995)

Boury, B., Corriu, R., Leclercq, D., Mutin, P., Planeix, J.M., Vioux, A.: A poly (vinylsilane): a precursor to silicon carbide. 1 Preparation and Characterization. Organometallics **10**, 1457–1461 (1991)

Brook, M.A.: Silicon in Organic, Organometallic, and Ppolymer Chemistry. Wiley and Sons, New York (2000)

Bunsell, A.R., Piant, A.: A review of the development of three generations of small diameter silicon carbide fibers. J. Mater. Sci. **41**, 823–839 (2006)

Burkhard, C.: Polydimethylsilanes. J. Am. Chem. Soc. **71**, 963–964 (1949)

Cao, F., Kim, D., Li, X.: Preparation of hybrid polymer as near-stochiometric SiC precursor by polycarbosilane and polymethylsilane. J. Mater. Chem. **12**, 1213–1217 (2002)

Chakraborty, S., Debnath, D., Mallick, A.R., Gupta, R.K., Ranjan, A., Das, P.K., Ghosh, D.: Microscopic, mechanical and thermal properties of spark plasma sinteredZr$B_2$ based composite containing polycarbosilane derived SiC. Int. J. Refract. Met. H **52**, 176–182 (2015)

ChantrellPG, P.P.: Inorganic polymers and ceramics. In: Popper, P. (ed.) Special ceramics, pp. 87–103. Academic Press, New York (1965)

Charles, X.L., Weber, W.P.: Synthesis and characterization of poly(1-methyl-1-vinyl-1-silabutane), poly(1-phenyl-1-vinyl-1-silabutane), and poly(1,1-divinyl-1-silabutane). Macromolecules **25**, 1639–1641 (1992)

Chojnowski, J., Cypryk, M.: Synthesis of linear polysiloxanes. In: Jones, R.G., Ando, W., Chojnowski, J. (eds.) Silicon-Containing Polymers, pp. 3–42. Kluwer Academic Publishers, Dordrecht, Netherlands (2000)

Choyke, W.J., Matsunami, H., Pensl, G. (eds.): Silicon Carbide: Recent Major Advances in 2 Vol., Springer, USA (2004)

Colombo, P., Sglavo, V., Pippel, E., Woltersdorf, J.: Joining of reaction-bonded silicon carbide using a preceramic polymer. J. Mater. Sci. **33**, 2409–2416 (1988)

Colombo, P., Mera, G., Riedel, R., Soraru, G.D.: Polymer-derived ceramics: 40 years of research and innovation in advanced ceramics. J. Am. Ceram. Soc. **93**(7), 1805–1837 (2010)

Colombo, P., Paulson, T.E., Pantano, C.G.: Synthesis of silicon carbide thin films with polycarbosilane (PCS). J. Am. Ceram. Soc. **80**, 2333–2340 (1997)

Curry, J.W.: The synthesis and polymerization of organosilanes containing vinyl and hydrogen joined to the same silicon atom. J. Am. Chem. Soc. **78**, 1686–1689 (1956)

Curtis, L., Schilling, J., Williams, T.C.: Hydrosilyl-modified polycarbosilane precursors for silicon carbide. US Patent no. 4,608,242 (1986)
Debnath, D., Chakraborty, S., Malick, A.R., Gupta, R.K., Ranjan, A., Das, P.K.: Mechanical, tribological and thermal properties of hot pressed $ZrB_2$-SiC composite with SiC of different morphology. Adv. Appl. Ceram **114**(1), 45–54 (2015)
Dvornic, P.R., Owen, M.J.: Silicon-containing dendritic polymers. Adv. Silicon Sci. **2**, Springer, USA (2009)
Erdal, M., Guceri, S.I., Danforth, S.C.: Impregnation molding of particle-filled preceramic polymers: process modeling. J. Am. Ceram. Soc. **82**, 2017–2028 (1999)
Feinroth, H.: Silicon carbide TRIPLEXTM fuel clad and SiC channel boxes for accident resistance and durability. E-J. Adv. Maint. **5**(2), 128–131 (2013)
Friedel, C., Ladenburg, A.: Uebersilicopropionsäureundihrenaether. justusliebigs. Ann. Chem. **159**, 259–274 (1871)
Fritz, G., Raabe, B.: Bildung siliciumorganischer verbindungen. V. Die thermische zersetzung von $Si(CH_3)$ und $Si(C_2H_5)_4$. Anorg. Allg. Chem. **286**, 149–167 (1956)
Fujiki, T., Kawasaki, S., Murase, H., Nishida, R., Sakai, S., Zaima, H.: Epoxy resin composition and moldings. European Patent no. EP0970981A1 (2000)
Galusek, D., Sedlacek, J., Riedel, R.: $Al_2O_3$-SiC composites by warm pressing and sintering of an organosilicon polymer-coated alumina powder. Jeur. Ceram. Soc. **27**, 2385–2392 (2007)
Ginzburg, M., MacLachlan, M.J., Yang, S.M., Coombs, N., Coyle, T.W., Raju, N.P., Greedan, J.E., Herber, R.H., Ozin, G.A., Manners, I.: Genesis of nanostructured, magnetically tunable ceramics from the pyrolysis of cross-linked polyferrocenylsilane networks and formation of shaped macroscopic objects and micron scale patterns by micro molding inside silicon wafers. J. Am. Chem. Soc. **124**(11), 2625–2639 (2002)
Greil, P.: Polymer derived engineering ceramics. Adv. Eng. Mater. **2**(6), 339–348 (2000)
Greil, P.: Polymer derived engineering ceramics. Adveng. Mater. **2**(6), 339–348 (2000)
Gupta, R.K., Saxena, A.K.: Studies on the rheological behavior of polycarbosilane Part II: Effect of heterometal (Al) content and nature of bonding with Si of polycarbosilane. SILICON **6**, 233–246 (2014)
Gupta, R.K., Mishra, R., Mukhopadhyay, K., Tiwari, R.K., Ranjan, A., Saxena, A.K.: A new technique for coating silicon carbide onto carbon nanotubes using a polycarbosilane precursor. SILICON **1**(2), 125–129 (2009)
Gupta, R.K., Mishra, R., Tiwari, R.K., Ranjan, A., Saxena, A.K.: Studies on the rheological behavior of polycarbosilane Part I: Effect of time, temperature & atmosphere. SILICON **3**(1), 27–35 (2011)
Habel, W., Haeusler, W., Oelschläger, A., Sartori, P.: New Modified Polycarbosilanes. In: Auner, N., Weis, J. (eds.) Organosilicon Chemistry II: From Molecules to Materials. Wiley-VCH, Chapter 89 (2008)
Hench, L.L., Ulrich, D.R.: Ultrastructure Processing of Ceramics, Glasses and Composites. John Wiley and Sons, New York, USA (1984)
Horiguchi, R., Onishi, Y., Hayase, S.: High molecular weight polysilanes with phenol moieties. Macromolecules **21**, 304–309 (1988)
Hurwitz, F.I., Heimann, P., Farmer, S.C., Hembree, D.M.: Characterization of the pyrolytic conversion of polysilsesquioxanes to silicon oxycarbides. J. Mater. Sci. **28**(24), 6622–6630 (1993)
Idesaki, A., Narisawa, M., Okamura, K., Sugimoto, M., Morita, Y., Seguchi, T., Itoh, M.: Application of electron beam curing for silicon carbide fiber synthesis from blend polymer of polycarbosilane and polyvinylsilane. Radiat. Phys. Chem. **60**, 483–487 (2001)
Idesaki, A., Sugimoto, M., Tanaka, S., Narisawa, M., Okamura, K., Itoh, M.: Synthesis of a minute SiC product from polyvinylsilane with radiation curing Part I Radiation curing of polyvinylsilane. J. Mater. Sci. **39**, 5689–5694 (2004)

Ishikawa, T., Shibuya, M., Yamamura, T.: The conversion process from polydimethylsilane to polycarbosilane in the presence of polyborodiphenylsiloxane. J. Mater. Sci. **25**, 2809–2814 (1990)

Isoda, T., Kaya, H., Nishii, H., Funayama, O., Suzuki, T., Tashiro, Y.: Perhydropolysilazane precursors to silicon nitride ceramics. J. Inorg. Organomet. Polym. **2**(1), 151–160 (1992)

Janakiraman, N., Aldinger, F.: Fabrication and characterization of fully dense Si-C-N ceramics from a poly(ureamethylvinyl)silazane precursor. J. Eur. Ceram. Soc. **29**, 163–173 (2009)

Kendrick, T.C., Parbhoo, B., White, J.W.: Siloxane polymers and copolymers. In: Patai, S., Rappoport, Z. (eds.) The Chemistry of Organic Silicon Compounds, pp 1289–1361. John Wiley & Sons, Chichester (1995)

Kimoto, T., Cooper, J.A.: Fundamentals of Silicon Carbide Technology: Growth, Characterization. Wiley-IEEE Press, Devices and Applications (2014)

Kipping, F.S.: Organic derivative of silicon. Preparation of alkylsilicon chlorides. Pr. Chem. Soc. **20**, 15–16 (1904)

Kipping, F.S., Sands, J.E.: Organic derivatives of silicon. Part XXV. Saturated and unsaturated silicohydrocarbons, $Si_4Ph_8$. J. Chem. Soc. **119**, 830–847 (1921)

Koe, J: Contemporary polysilane synthesis and functionalization. Polym. Int. **58**, 255–260 (2009)

Kohyama, A., Kotani, M., Katoh, Y., Nakayasu, T., Sato, M., Yamamura, T., Okamura, K.: High-performance SiC/SiC composites by improved PIP processing with new precursor polymers. J. Nucl. Mater. **283**, 565–569 (2000)

Krenkel, W.: Carbon Fibre Reinforced Silicon Carbide Composites (C/SiC, C/C-SiC). In: Bansal, N.P. (ed.) Handbook of Ceramic Composites, pp. 117–148. Springer, Boston, MA (2005)

Kroke, E., Li, Y.L., Konetschny, C., Lecomte, E., Fasel, C., Riedel, R.: Silazane derived ceramics and related material. Mater. Sci. Eng. R **26**(4–6), 197–199 (2000)

Kumar, S., Kumar, A., Sampath, V., Prasad, V.V.B., Chaudhary, J.C., Gupta, A.K., Devi, G.R.: Fabrication and erosion studies of C-SiC composite Jet Vanes in solid rocket motor exhaust. Jeur. Ceram. Soc. **31**, 2425–2431 (2011a)

Kumar, S., Kumar, A., Devi, G.R., Gupta, A.K.: Preparation of 3D orthogonal woven C-SiC composite and its characterization for thermo-mechanical properties. Mater. Sci. Eng. A **528**, 6210–6216 (2011b)

Kumar, S., Misra, M.K., Mondal, S., Gupta, R.K., Mishra, R., Ranjan, A., Saxena, A.K.: Polycarbosilane based UD C/SiC composites: effect of in-situ grown SiCnano-pins on mechanical properties. Ceram. Int. **41**(10), 12849–12860 (2015)

Lee, Y.J., Lee, J.H., Kim, S.R., Kwon, W.T., Klepeis, J.P., Teat, S.J., Kim, Y.H.: Synthesis and characterization of novel preceramic polymer for SiC. J. Mater. Sci. **45**, 1025–1031 (2010)

Liew, L.A., Liu, Y., Luo, R., Cross, T., An, L., Bright, V.M., Dunn, M., Daily, J.W., Raj, R.: Fabrication of SiCN MEMS by photopolymerization of pre-ceramic polymer. Sens. Act. A **95**, 143–151 (2002)

Loehman, R.E.: Oxynitride glasses. In: Glass, I.V., Tomozawa, M., Doremus, R.H. (eds.) Treatise on Materials Science and Technology, pp. 119–149. Academic Press, Orlando, USA (1985)

Loh, P.P., Bao, X., Nangrejo, M.R., Edirisinghe, M.J.: Preparation of silicon carbide foams using polymeric precursor solutions. J. Mater. Sci. Lett. **35**, 4365–4372 (2000)

Ly, H.Q., Taylor, R., Day, R., Heatley, F.: Conversion of polycarbosilane (PCS) to SiC-based ceramic Part 1. Characterisation of PCS and curing products. J. Mater. Sci. **36**, 4037–4043 (2001)

Maghsoodi, S.I., Pang, Y., Barton, T.J.: Efficient, "one-pot" synthesis of silylene–acetylene and disilylene–acetylene preceramic polymers from trichloroethylene. J. Polym. Sci. Part A Polym. Chem. **28**, 955–965 (1990)

Mera, G., Riedel, R.: Organosilicon-based polymers as precursors for ceramics. In: Colombo, P., Riedel, R., Soraru, G.D., Kleebe, H.J. (eds.) Polymer Derived Ceramics: From Nanostructure to Applications. DE Stech Publications Inc., Lancaster, PA, USA, pp 51–89 (2009)

Meyer, M.K., Akinc, M., Maghsoodi, S.I., Zhang, X., Barton, T.J.: Silylene-acetylene polymers as precursors to SiCfibers. Ceram. Eng. Sci. Proc. **12**, 1019–1031 (1991)

Miller, R.D., Michl, J.: Polysilane high polymers. Chem. Rev. **89**, 1359–1410 (1989)

Mishra, R., Tiwari, R.K., Saxena, A.K.: Synthesis of Fe–SiC nanowires via precursor route. J. Inorg. Organomet. Polym. **19**, 223–227 (2009)
Mishra, M.K., Kumar, S., Ranjan, A., Prasad, N.E.: Processing, properties and microstructure of SiC foam derived from epoxy-modified polycarbosilane. Ceram. Int. **44**(2), 1859–1867 (2017)
Mukherjee, J., Ghosh, S., Ghosh, A., Ranjan, A., Saxena, A.K., Das, P.K., Banerjee, R.: Enhanced nano-mechanical and wear properties of polycarbosilanederived SiC coating on silicon. Appl. Surf. Sci. **325**, 39–44 (2015)
Okamura, K., Sato, M., Matsuzawa, T., Seguchi, T., Kawanishi, S.: Siliconbased ceramic fibers. Ceram. Eng. Sci. Proc. **9**, 909–918 (1988)
Pawlenko, S.: Organosilicon Chemistry. Walter de Gruyter, New York (1986)
Perale, G., Giordano, C., Daniele, F., Masi, M., Colombo, P., Gottardo, L., Maccagnan, S.: A novel process for the manufacture of ceramic microelectrodes for biomedical applications. Int. J. Appl. Ceram. Technol. **5**, 37–43 (2008)
Price, F.P.: Mechanism of alkaline cleavage of silicon-hydrogen bonds: temperature coefficient of the rate of cleavage of several trialkylsilanes. J. Am. Chem. Soc. **69**(11), 2600–2604 (1947)
Reed, R.C.: The Super Alloys: Fundamentals and Applications. Cambridge Univ. Press, Cambridge, UK (2006)
Reidel, R., Mera, G., Hauser, R., Klonczynski, A.: Silicon-based polymer-derived ceramics: synthesis, properties and application- a review. J. Ceram. Soc. Jpn **114**(6), 425–444 (2006)
Rocha, R.M.D., Greil, P., Bressiani, J.C., Bressiani, A.H.D.A.: Complex- shaped ceramic composites obtained by machining compact polymer-filler mixtures. Mater. Res. **8**(2), 191–196 (2005)
Saxena, A.K., Ranjan, A., Kumar, S., Mishra, R., Gupta, R.K.: Process for synthesis of zirconium diboride zirconium carbide binary ceramic powder and using it to prepare SiC:$ZrB_2$:ZrC CMC. Indian Patent Application no. 301/DEL/2014 (2014)
Saxena, A.K., Ranjan, A., Tiwari, R.K., Gupta, R.K., Mishra, R.: A process for the preparation of silicon carbide ceramic fibers. Indian Patent Application no. 201611011921 (2016)
Schulz, M., Borner, M., Gottert, J., Hanemann, T., Hausselt, J., Motz, G.: Cross linking behavior of preceramic polymers effected by UV- and synchrotron radiation. Adv. Eng. Mater. **6**, 676–680 (2004)
Seyferth, D.: Polycarbosilanes: An overview. Inorganic and organometallic polymers. ACS Symp. Ser. **360**, 21–42 (1988)
Shiina, K., Kumada, M.: Thermal rearrangement of hexamethyldisilaneto trimethyl(dimethylsilylmethyl)silane. J. Org. Chem. **23**, 139–139 (1958)
Shukla, S.K., Tiwari, R.K., Ranjan, A., Saxena, A.K., Mathur, G.N.: Some thermal studies of polysilanes and polycarbosilanes. Thermochim Acta **424**(1–2), 209–217 (2004)
Sinha, M.K., Das, B.R., Mishra, R., Ranjan, A., Srivastava, A., Saxena, A.K.: Study of electrospunpolycarbosilane (PCS) nanofibrous web by needle-less technique. Fash. Text. **1**, 2–15 (2014)
Smirnova, T.P., Badalian, A.M., Yakovkina, L.V., Kaichev, V.V., Bukhtiyarov, V.I., Shmakov, A.N., Asanov, I.P., Rachlin, V.I., Fomina, A.N.: SiCN alloys obtained by remote plasma chemical vapour deposition from novel precursors. Thin Solid Films **429**, 144–151 (2003)
Soraru, G.D., Modena, S., Guadagnino, E., Colombo, P., Egan, J., Pantano, C.: Chemical durability of silicon oxycarbide glasses. J. Am. Ceram. Soc. **85**(6), 1529–1536 (2002)
Stark, F.O., Falender, J.R., Wright, A.P.: Silicones. In: Wilkinson, G. (ed.) Comprehensive Organometallic Chemistry, vol. 2. Pergamon Press, Oxford (1982)
Tsirlin, A.M., Shcherbakova, G.I., Florina, E.K., Popova, N.A., Gubin, S.P., Moroz, E.M., Riedel, R., Kroke, E., Steen, M.: Nano-structured metal containing polymer precursors for high temperature non-oxide ceramics and ceramic fibers—syntheses, pyrolysis and properties. J. Eur. Ceram. Soc. **22**, 2577–2585 (2002)
Vakifahmetoglu, C., Menapace, I., Hirsch, A., Biasetto, L., Hauser, R., Riedel, R., Colombo, P.: Highly porous macro-and micro-cellular ceramics from a polysilazane precursor. Ceram. Int. **35**(8), 3281–3290 (2009)

Wesson, J.P., Williams, T.C.: Organosilane polymers. I. Poly(dimethylsilylene). J. Polym. Sci. Polym. Ed. **17**, 2833–2843 (1979)
West, R., David, L.D., Djurovich, P.I., Stearley, K.L., Srinivasan, K.S.V., Yu. H.: Phenylmethylpolysilanes: formable silane copolymers with potential semiconducting properties. J. Am. Chem. Soc. **103**, 7352–7354 (1981)
Whitmarsh, C.K., Interrante, L.V.: Synthesis and structure of a highly branched polycarbosilane derived from (chloromethyl)trichlorosilane. Organometallics **10**, 1336–1344 (1991)
Winter, G., Verbeek, W., Mansmann, M.: Formkoerper aus homogenen mischungen von siliciumcarbid und siliciumnitrid und verfahren zu ihrer herstellung. Ger. Offen., 2243527 (1974)
Wynne, K.J., Rice, R.W.: Ceramics via polymer pyrolysis. Ann. Rev. Mater. Sci. **14**, 297–334 (1984)
Yajima, S., Hasegawa, Y., Okamura, K., Matsuzawa, I.: Development of high tensile strength silicon carbide fibre using an organosilicon polymer precursor. Nature **273**, 525–527 (1978)
Yajima, S., Hayashi, J., Omori, M.: Continuous silicon carbide fiber of high tensile strength. Chem. Lett. **4**(9), 931–934 (1975)
Yajima, S., Okamura, K., Hayashi, J.: Structural analysis in continuous SiCfibre of high tensile strength. Chem. Lett., 1209–1212 (1975)
Yajima, S., Okamura, K., Hasegawa, Y.: Method of producing silicon carbide fibers. US Patent no. 4,283,376 (1981)
Yang, H.S., Kwon, O.H., Lee, J.D., Kang, P.H.: Preparation and characteristics of titanium-modified silicon carbide fiber. J. Ind. Eng. Chem. **2**(2), 106–115 (1996)
Yu, Z., Fang, Y., Huang, M., Li, R., Zhan, J., Zhou, J., Zhou, C., He, G., Xia, H.: Preparation of a liquid boron-modified polycarbosilane and its ceramic conversion to dense SiC ceramics. Polym. Adv. Technol. **22**(12), 2409–2414 (2011)
Yu, Y., Tai, J., Tang, X., Guo, Y., Tang, M., Li, X.: Continuous Si–C–O–Al fiber derived from aluminum-containing polycarbosilane precursor. Compos.: Part A **39**, 1101–1105 (2008)
Zeigler, J.M., Fearson, F.W.G.: Silicon-based polymer science. Advances in Chemistry Series 224, ACS, Washington DC (1989)
Zeldin, M., Wynne, K.J., Allcock, H.R.: Inorganic and organometallic polymers. In: ACS Symposium Series, vol. 360, Am. Chem. Soc., Washington, DC (1988)
Zhang, T., Evans, J.R.G., Woodthorpe, J.: Injection moulding of silicon carbide using an organic vehicle based on a preceramic polymer. J. Eur. Ceram. Soc. **15**, 729–734 (1995)
Zhang, P., Jia, D., Yang, Z., Duan, X., Zhou, Y.: Progress of a novel non-oxide Si-B-C-N ceramic and its matrix composites. J. Adv. Ceram. **1**(3), 157–178 (2012)
Zhou, S.Q., Manuel, G., Weber, W.P.: Copolymerization of 1,1-dimethyl-1-silacyclopent-3-ene and 1,1-Diphenyl-1-silacyclopent-3-ene. Characterization of copolymer microstructures by 1H, 13C and 29Si NMR Spectroscopy. Macromolecules **23**, 1583–1586 (1990)

# Chapter 7
# Precursor Materials for Semiconductor Thin Films

**Santosh Kr. Tripathi, Himanshi Chaurasia, Kingsuk Mukhopadhyay, and Eswara Prasad Namburi**

**Abstract** The present chapter emphasizes detailed discussion of recent approaches in which (i) the desired elements for compound semiconductors are obtained from a single molecule (the 'single-source' precursor route) and (ii) the development of suitable precursors available for deposition of semiconducting thin films or nanoparticles. Specifically, this chapter concentrates on more recent studies and approaches using single-molecular precursors to develop advanced materials such as metal chalcogenides and group III metal chalcogenides, nitrides, and a novel approach for the fabrication of photovolataic cells using semiconductor nanoinks.

## 7.1 Introduction

Thin films technology plays a crucial role in many present and future electronic applications. The semiconductor materials based on combinations of elements from Groups II and VI, III and V, III and VI, and IV and VI, such as cadmium tellurides (CdTe), cadmium selenide (CdSe), zinc selenide (ZnSe), cadmium mercury telluride (CdHgTe), gallium nitride (GaN), gallium arsenide (GaAs) and copper indium gallium selenide (CuInGaSe, usually written CIGS), are used in a wide range of applications in solar cells, night vision devices, detectors, window materials, satellite TV

S. Kr. Tripathi · K. Mukhopadhyay
Defence Materials and Stores Research and Development Establishment (DMSRDE), Defence Research and Development Organisation (DRDO), Kanpur, Uttar Pradesh, India

H. Chaurasia
Formerly With Defence Materials and Stores Research and Development Establishment (DMSRDE), Defence Research and Development Organisation (DRDO), Kanpur, Uttar Pradesh, India

E. P. Namburi (✉)
Former Outstanding Scientist and Ex-Director, Defence Materials and Stores Research and Development Establishment (DMSRDE), Defence Research and Development Organisation (DRDO), Kanpur, Uttar Pradesh, India
e-mail: neswarap@gmail.com

 
E. P. Namburi et al. (eds.), *Novel Defence Functional and Engineering Materials (NDFEM) Volume 1*, Indian Institute of Metals Series,
https://doi.org/10.1007/978-981-99-9791-6_7

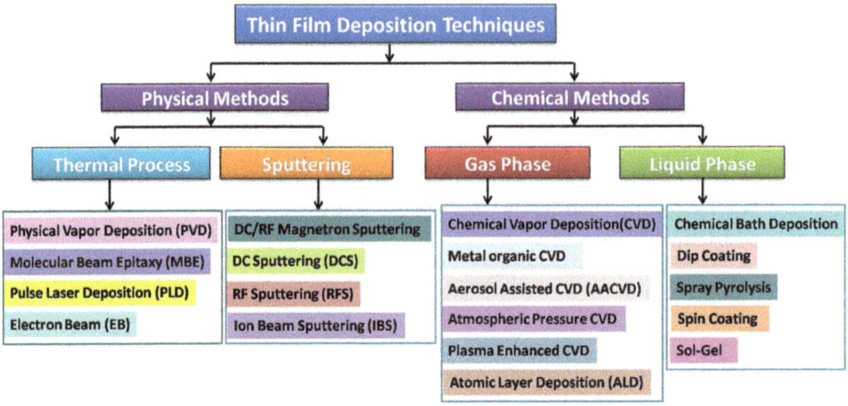


**Fig. 7.1** Various physical and chemical deposition processes

receivers, optical fibre communications and full colour advertising displays. There is an increasing demand for new materials and cost-effective methods of depositing these materials, while imposing minimum environmental hazards. The chemical methods are more cost effective than physical ones. Many techniques have been developed for the preparation of thin films. Based on the precursors, the deposition process can be separated into two categories, i.e. physical and chemical methods, Fig. 7.1.

## 7.2 Semiconductor Thin Films: A Brief Overview

The group II–VI, III–V, III–VI and IV–VI compound semiconductors are important for modern technologies. Materials like GaAs, GaN, AlGaAs, InP, ZnSe, CdTe, and CIGS are potentially useful in electronics and optoelectronic devices such as communications, compact disk players, bar code readers, full colour displays, and high efficiency photovoltaics. Manasevit was the first to grow thin films of GaAs in 1968 using high-purity metal–organic compounds (Manasevit 1981). This is known as the Metalorganic Chemical Vapour Deposition (MOCVD) method. MOCVD has been classified according to the use of precursor compounds containing (i) metal atoms bonded to organic radicals, "Metalorganics"; and (ii) compounds having one or more direct metal–carbon covalent bonds,"Organometallics'. The use of organometallic precursors has led to renaming the deposition process as organometallic CVD (OMCVD) to reflect the choice of the precursor.

MOCVD is commonly used for deposition of thin films for semiconductors by using conventional (dual-source) or single-source precursor systems. Progress of the techniques has been the result of developments in precursor chemistry with improved synthesis and purification. One of the driving forces in recent years in the organometallic chemistry of Group 12 (Zn, Hg, Cd) and Group 13 (Al, Ga, In) elements has been increased due to increasing demands from the electronics industry. Examples are the semiconducting III–V materials such as GaAs, GaN and InP, and II–VI materials such as CdTe, ZnSe, HgTe and CdHgTe. Also in demand are some semiconductors obtained from organometallic metal chalcogenides such as binary (SnS, $In_2S_3$), and ternary phases ($CuInSe_2$). Several reviews have been published on the growth techniques and structural, electrical and optical properties, as well as the applications, which include solar cells, light emitting diodes, photocathodes, lasers and thermal imaging devices (Malik et al. 2010; Ambacher 1998; Marchand et al. 2013; Kurek et al. 2014; Strite et al. 1993; Jones and O'Brien 1997; Knapp and Carmalt 2016). MOCVD (or OMCVD) can be used to deposit a wide range of materials in the form of amorphous, epitaxial and polycrystalline films. Table 7.1 shows examples that are commonly used to grow III–V, II–VI and IV–VI semiconducting materials, metallic films and dielectric films, together with their applications (Manasevit 1981; Choy 2003).

The purity of precursors is the most important factor that affects the composition and structural quality of the deposited material. In the case of CVD, the precursors should decompose by thermal treatment to produce thin films. Ideally, the organic moities associated with the precursors should be lost into the gas phase. However, when complex ligands (heteroleptic compounds) are involved, more complex fragmentations take place. Hence it is essential to study the precursor decomposition mechanisms in the gas phase by investigating the surface reactions of these precursors as well as their fragmentation.

### *7.2.1 Precursor Requirements*

As mentioned above, the purity is a prime requirement for a precursor. Thus development of precursors requires investigation into purification techniques and potential alternative synthetic routes to produce high-purity compounds. The precursor must be suitable for preparation and isolation in large quantities, with minimal environmental or cost implications for scale-up and industrial use. The precursor should have stability and low toxicity for ease of use and storage.

Dual precursors are usually used in CVD processes for the deposition of thin films, but efforts to lower the growth temperature have resulted in the development of single-source precursors, which already contain M–E bonds (M = group 12, 13, 16 elements; E = N, P, As, Sb, S, Se, Te, etc.) at the centre of the molecule, along with other organic moieties attached to each of the elements. The desired reaction pathway involves adsorption of the precursor without breaking the M–E bond but discarding the peripheral ligands. These single-source precursors have several potential advantages over conventional CVD precursors. These advantages include limited pre-reaction since there is only one precursor; reduced toxicity (e.g. for III–V materials the highly toxic compounds $AsH_3$ and $PH_3$ are not involved in the CVD process); possible growth at lower temperatures; and stability with respect to air and moisture. However, commercialization of single-source precursors is still limited, owing

**Table 7.1** Precursors used togrow III–V, II–VI and IV–VI semiconducting and metallic thin films

| Materials | Precursors* | Applications |
|---|---|---|
| **III–V** | | |
| GaAs or GaAs on GaP substrates | TMGa, TEGa, $AsH_3$ | Light emitting diode (LEDs) injection laser |
| AlGaAs/GaAs/ AlGaAs | TMAl, TMGa, TEGa, $AsH_3$ | Double heterojunction and quantum well lasers |
| InGaAsP | TMInTMGa, TEGa, $AsH_3$, $PH_3$ | Fibre optics, communications |
| InP/InGaAs | | Double heterojunction bipolar transistor |
| InGaAsP on GaAs substrates | | Al-free injection lasers ($\lambda$ = ~0.98 μm) |
| AlGaInP | | |
| InAlGaP | TMAl, TMIn, TMGa, $AsH_3$, $PH_3$ | Red and yellow LED Injection laser emitting ($\lambda$ = ~0.98 μm) |
| InGaAs on InP substrates | TMIn, TMGa, TEGa $AsH_3$ | High speed optoelectronic devices (e.g. high electron mobility and heterojunction bipolar transistors) |
| **III–V nitride** | | |
| InAlGaN | TMIn, TMAl, TMGa, $NH_3$ | Blue LEDs |
| **II–VI** | | |
| Wide band gap | | |
| ZnCdSe | DEZn, DMCd, DTBSe | Blue-green LEDs |
| ZnCdSSe | DEZn, DMCd, DTBSe, DES, $H_2S$(MeCd)2Mg, DMSe | LEDs, laser diodes |
| ZnMgSSe | | |
| Narrow band-gap | | |
| HgCdTe | DMHg, DMCd, DETe, DATe, DIPTe | IR detectors |
| HgCdZnTe | DMCd, DMZn | |
| **IV–VI** | | |
| PbSnTe | DEPb, DESn, $H_2Te$ | IR detectors |
| **Metals** | | |
| Cu | $(CF_3COCH_2COCF_3)_2Cu$ | Al, Cu and CuAl TFs used for high density interconnect Si integrated circuits |
| Al | TMAl | |

**Note* DM, Dimethyl; DE, Diethyl; TM Trimethyl; DA, DiAllyl; MA, Methylallyl; DTB, Ditertiarybutyl; DIP, DiIsopropyl to their disadvantages. These include difficulties in controlling the stoichiometry of ternary or quaternary materials, for example $In_xGa_{1-x}As_x$; low volatility, making the use of established MOCVD equipment difficult; and also the hindering of epitaxial growth owing to low surface mobility of polynuclear decomposition fragments.

During gas phase CVD the precursor has to decompose on the surface of a substrate at elevated temperatures in order to deposit the required thin film. Hence volatility is very important to enable the precursor to volatilise below its decomposition temperature and yield high quality thin films. This important problem arises because most of the precursors have low intrinsic volatility. To solve this problem liquid phase (solution-based) CVD routes have been developed, since for these techniques the precursor solubility is more important than its volatility. This approach greatly extends the range of potential precursors. For example, aerosol- assisted Chemical Vapour Deposition (AACVD) is a solution-based technique that has generated much interest for deposition of a wide range of films. In the AACVD process aerosol droplets of the precursor solution are obtained via a nebulizer, thereby allowing a carrier gas to transport the precursor to the reaction chamber, see Fig. 7.2. On reaching the substrate the solvent evaporates from the gaseous precursor owing to the resulting increase in temperature. The vaporised precursor is then deposited onto the substrate, yielding the desired film material.

Advantages of AACVD are: (i) it is not always necessary to isolate and purify a precursor (needed in conventional CVD processes) because the precursor can be synthesised in situ in the solution used in the AACVD process; (ii) deposition of ternary materials may be obtained by simply mixing different precursors in the solution. Hence AACVD has eliminated many of the disadvantages associated with single-source precursors, e.g. volatility and control of stoichiometry.

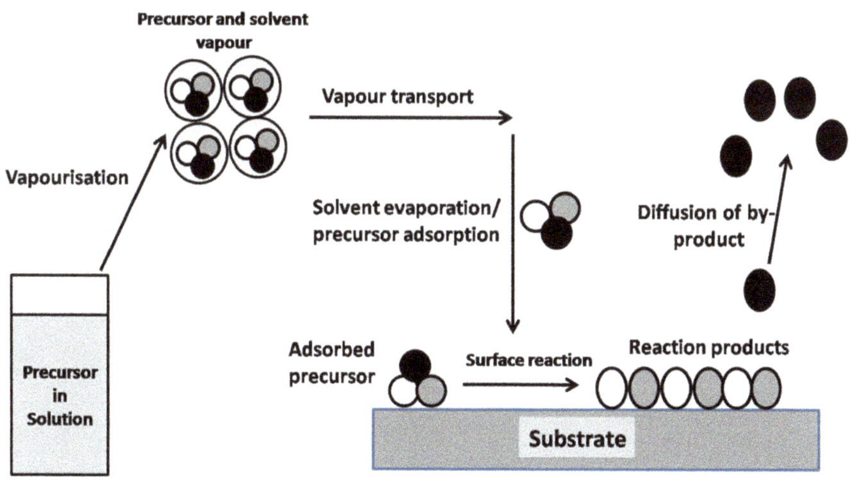


**Fig. 7.2** Schematic illustration of the process involved in AACVD (https://s100.copyright.com/AppDispatchServlet#formTop)

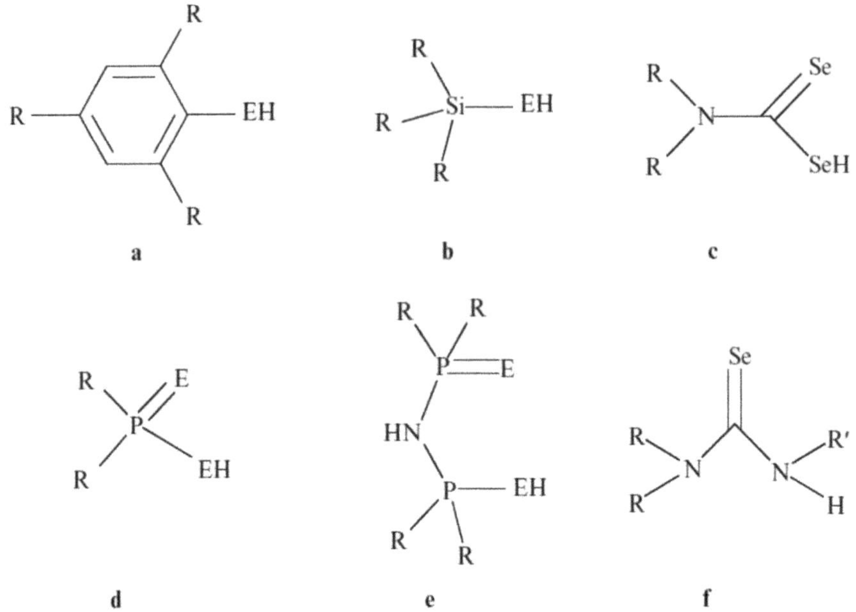


**Fig. 7.3** Organic moities used to prepare single-source precursors: **a)** 2,4,6 substituted aryl chalcogenide, R = $^{t}$Bu; **b)** trimethylsilylchalcogenol, R = [Si(Me)$_3$]; **c)** diselenocarbamate, R and R′ are alkyls; **d)** diselenocarbamic, R is an alkyl, E = E′ = Se, E = S and E′ = Se; **e)** dichalcogenophosphinicacid, R is an alkyl, E = E′ = Se or Te, E = S E′ = Se or Te, E = Se E′ = Te; **f)** imidochalcogenophosphinicacid, (E = O, S, Se; R = Me, $^{i}$Pr, $^{n}$But, Ph)triorganylselenourea

The main disadvantages of AACD are the deposition parameters, which can greatly affect the microstructures and morphologies of the films. These parameters include the compositions/formulations of the precursor, the choice of solvent and substrate, and the temperature. Also, particulate materials can sometimes form (prereaction) before the aerosols reach the deposition zone. Therefore new precursor systems will be required to overcome this problem, and also other issues such as controlled volatility, reduced toxicity, the quality of the grown layers, and reduction of the deposition temperature. The details of preparation, including organometallic and organometallic dual or conventional precursors for deposition of thin films, and their potential applications are summarised in Table 7.1.

## 7.3 Precursors for Group II–VI (Metal Chalcogenide) Materials

Two types of precursors are used for the deposition of II–VI materials (thin films or nanoparticles). These are (i) dual (traditional) precursors; (ii) single source molecular precursors.

### *7.3.1 Dual Precursors*

The use of dual precursors was first described by Manasevit for the deposition of II–VI metal chalcogenide thin films using volatile metal alkyls such ($Me_2Cd$) or ($Me_2Zn$) with $H_2S$, $H_2Se$, or $Me_2Te$ (Manasevit 1981). The MOCVD process was used with hydrogen as the carrier gas. Various II–VI thin films were grown by this process, including ZnS, ZnSe, CdS, CdSe and CdTe (Manasevit 1981). The major drawbacks with these precursors are high toxicity, flammability, sensitivity to air and moisture, and also the need for a high temperature to deposit thin films of the metal chalcogenides. In other words, this is a very problematic process.

### *7.3.2 Single-Source Molecule Precursors (SSMPs)*

In recent decades chemists have developed a variety of single-source molecular precursors (SSMPs) for thin films and nanocrystals. These SSMPs have advantages over other routes such as stabilized bonds (M–E) that can lead to materials with fewer defects and/or improved stoichiometry. For CVD thin film deposition the flow and temperature requirements become simpler, enabling depositions in simple installations. Most of the SSMPs) are stable in air, hence easier to handle and characterize. The metal chalcogenides (group II–VI materials) are obtained by thermal decomposition of chalcogen-containing metal complexes. However, owing to high molecularity the cadmium or zinc complexes with some ligands (such as thiolates) generally form involatile polymeric structures with a tetrahedral metal centre. Hence these complexes are not useful as MOCVD precursors even when trying to improve volatility by low pressure processing.

Fortunately, discovery of the AACVD approach makes it possible to deposit high quality thin films from non-volatile or less volatile complexes. In AACVD the most important condition is that the precursors should be soluble in organic solvents. Various types of precursors, developed by a number of research groups, are used for depositing semiconductor thin films and are summarized in Table 7.2. Some ligand moities used for synthesis of SSMPs are shown in Fig. 7.3; and a few crystal structures of SSMPs are shown in Figs. 7.4, 7.5 and 7.6.

### *7.3.3 Precursors for III–V Materials*

***(A) Conventional or Single Precursors for Group V Materials***: The group III–V thin films are grown from metal–organic single (conventional) precursors such as trialkyls of group III, trimethylgallium ($Me_3Ga$), trimethylaluminium ($Me_3A1$), and trimethylindium ($Me_3In$), with arsine ($AsH_3$) and phosphine ($PH_3$) gases of group V hydrides. During the pyrolysis of $AsH_3$ or $PH_3$, large quantities of active atomic

**Table 7.2** Precursors for Group II–VI thin films

| Types of precursors | Group II–VI precursors | Methods of preparation | Operating temperature and pressure | Targeted materials | Refs. |
|---|---|---|---|---|---|
| A. **Conventional precursors** | | | | | |
| Chalcogen-containing single precursors | Metal Alkyls and hydrides<br>$Me_2Cd$, $Me_2Zn$, $H_2S$, $H_2Se$ or $Me_2Te$<br>AlkylChalcogenides<br>MeSeH, $^tBuSeH$, $Et_2Se$, $^tBu_2Se$, $Me_2Te_2$, $MeTe^iPr$ | MOCVD, OMVPE | ~750 °C 600 °C, 500 °C | ZnS, ZnSe, CdS, CdSe, and CdTe | Manasevit (1981), Malik et al. (2010), Singh (1996), Kuhn et al. (1992) |
| **Chalcogen-containing single source molecular precursors (SSMPs)** | | | | | |
| Metal alkyl adducts | $MeZn(S^tBu)]_5$, $[MeZn\text{-}(S^iPr)]_8$, $[EtZn(SEt)_{10}]$, $[MeZn(S^tBu)_5]$, (Fig. 7.4a), $MeZn(S^tBu)(py)]_2$(Fig. 7.3b) | MOCVD, AACVD | 450 °C | ZnS, | Hursthouse et al. (1991) |
| Precursors containing bulky ligands | 1,2-bis-(diethylphosphino) ethane (depe), $M(ER)_2$,<br>$[Cd_2(SeC_6H_5)_4(depe)]n$ and $[Hg(SeC_6H_5)_2(depe)]_2$,<br>$[Cd_2(SeC_6H_5)_4(depe)]n$,<br>$[Zn(SEt)Et]_{10}$<br>$[Cd(SC_6H_2tBu_3\text{-}2,4,6)]_2$ and $[Hg(SC_6H_2tBu_3\text{-}2,4,6)]_2$ (Jun et al. 2001)<br>RMSeR′ ($R' = C_6H_2{}^iPr_3$-2,4,6; M = Zn R = Me, Et, Pr, *i*Pr; M = Cd, R = Me)[($N_3E_2$, NEM and $N_4O_2\Xi_2$ cyclic and Acyclic, [Zn(TePh)2][TMEDA] | AACVD Solvothermal | 250 °C 450 °C | CdTe, CdSe, ZnS, CdS, HgSe, HgS ZnTe | Brennan et al. (1989) Hursthouse et al. 1991; Brennan et al. 1989; Bochmann and Webb 1991; Bochmann et al. 1992; Tripathi et al. 2002; Santosh et al. 2005) Jun et al. 2001) |

(continued)

**Table 7.2** (continued)

| Types of precursors | Group II–VI precursors | Methods of preparation | Operating temperature and pressure | Targeted materials | Refs. |
|---|---|---|---|---|---|
| Dichalcogenophosphinatocomplexes | $[Cd(S_2P^iBu_2)_2]_2$ and $[Zn(S_2P^iBu_2)_2]_2$, $Cd_2[(i\text{-}C_3H_7O)_2PS_2]_4$, $[(R_2PSe_2)Se]$ (R = Ph, $^i$Pr) $[M(R_2PSe_2)n]$, (M = Zn, Cd, Pb, In, Ga, Cu, Bi, Ni; R = $^i$Pr, Ph) and $[Mo_2O_2Se_2(Se_2P^iPr_2)_2]$, $[In(^iPr_2PSe_2)_3]$ $[Ga(^iPr_2PSe_2)_3]$, $[Ga(^iBut_2PS_2)_3]$ | CW LPMOCVD or AACVD | | MS,Se type M = Zn,Cd,Pb, In,Ga,Cu,Bi,Ni | Adeogun et al. (2006), Park et al. (2001) |
| Bis(dialkydichalcogeno-carbamato) metal species | $[(M(E_2CNR_2)_2]$, $[M(E_2CNRR')_2]$, $Zn(S_2CNEt_2)_2]$, $[Cd(S_2CNEt_2)_2]$ (Jones 1997) $[Zn(S_2CNEt_2)_2]$, $[Cd(S_2CNEt_2)_2]$, $[Zn(S_2CNMeEt)]_2$, $[Zn(S_2CNMe^nPr)]_2$, $[Zn(S_2CNMe^nBu)]_2$, $[Zn(S_2CNMe^nBu)]_2$ $[Cd(S_2CNMe^nBu)_2]$, $[Zn\ (Se_2CNMe^nHex)_2]$, $[Cd\ (Se_2CNMe^nHex)_2]$ | LP-MOCVD, PEMOCVD, CW-LP-MOCVD | 450 °C | ZnS, CdS, ZnSe, CdSe and $Zn_xCd_{1\text{-}x}S$ GaAs or InP substrates | Malik et al. (2010) |
| N-Alkyldithiocarbamato complexes | $[M(S_2CNHR)_2]$, (M = Zn, Cd; R = *n*Hex,*n*dodecyl) $[Zn(S_2CNHHex)_2$ | | | CdS, ZnS | Malik et al. (2010), Knapp and Carmalt (2016) |
| Alkyl/Dithio or diselenocarbamato complexes | $Me_2Zn$ with $[Cd(Se_2CNEt_2)_2]$, $[Me_2MM,(E_2CNEt_2)_2]$\{MM, = Cd, Zn; E = S, Se] $[Zn(SOCCH_3)_2(tmeda)]$, $[Cd(SOCCH_3)_2\ (tmeda)]$ $[Zn(thq\text{-}dtc)_2]$ (Fig. 7.5) | LP-MOCVD AACVD Solvothermal | $10^{-2}$–$10^{-3}$ T Torr | CdS, ZnS, and CdxZn1-xS ZnS NPs | Malik et al. (2010), Knapp and Carmalt (2016), Linda et al. (2014) |
| Xanthate complexes | $[M(S_2COC_3H_7)_2]$ (M = Zn, Cd), $[M(S_2COCHR)_2]$ (M = Cd; R = Et or $^i$Pr) | CVD | | CdS, ZnS | |

(continued)

**Table 7.2** (continued)

| Types of precursors | Group II–VI precursors | Methods of preparation | Operating temperature and pressure | Targeted materials | Refs. |
|---|---|---|---|---|---|
| Dichalcogenoimid-diphozsphinato complexes | $Cd[(SeP^iPr_2)_2N]_2$,$[M[(EP^iPr_2)_2N]_2]$, $[M[(SePPh_2)_2N]_2]$, $[MeCd[(SeP^iPr_2)_2N]]_2]$ , $[M[(TeP^iPr_2)_2N]_2]$ (M = Cd, Hg) $[Cd[(TeP^iPr_2)_2N]_2]$, $[(TeP^iPr_2N^iPr_2PTe_2)_2]$ $[Cd\{(SP^iPr_2)N]$ | MOCVD AACVD | 400- 500 °C 375–475 °C 450–550 °C | CdSe, CdTe, HgTe | |
| Thiobiurets and dithiobiurets | $[Zn(S_2N_3C_{10}H_{20})_2]$, $[M(N(SCNR_2)_2)_2]$ [M = Zn, Cd; R = Me, Et] $[M(SON(CNR_2)_2)_2]$ [M = Zn, R = isopropyl and M = Cd, R = isopropyl] (Fig. 7.6) | AACVD | >350- > 400 oC | ZnS, CdS, $Zn_xCd_{1-x}S$ | Memon et al. (2015), Ramasamy et al. (2011) |

hydrogen are released, resulting in clean removal of carbon-containing fragments from the growth surface.

The hydrides of group V, such as $AsH_3$ and $PH_3$, are extremely toxic gases, requiring much care in handling and storage: high-pressure cylinders are usually required for storing them. Therefore much work has been carried out (Jones 1997) to develop safer liquid alternatives, which disperse more slowly in the atmosphere and minimize accidental risk. A series of trialkylarsine compounds, such as trimethylarsine ($Me_3As$) and triethylarsine ($Et_3As$), have been synthesized but these give heavily carbon-contaminated GaAs layers due to the absence of active AsH species, which are essential for removal of carbonaceous species. Hence one or more hydrogen atoms containing arsenic precursors, such as the ethylarsines ($Et_2AsH$, $EtAsH_2$) or *tert*-butylarsine ($^tBuAsH_2$) serve as better sources. $^tBuAsH_2$ is the most efficient liquid arsenic source: this has a suitable vapour pressure and pyrolyzes at a lower temperature than $AsH_3$ (50% pyrolyzed at 425 °C compared with 575 °C for $AsH_3$). The more efficient pyrolysis of $^tBuAsH_2$ relative to $AsH_3$ allows the growth of GaAs at lowerV/III ratios, and the increased concentration of active AsH species on the substrate reduces carbon contamination (Larsen et al. 1988).

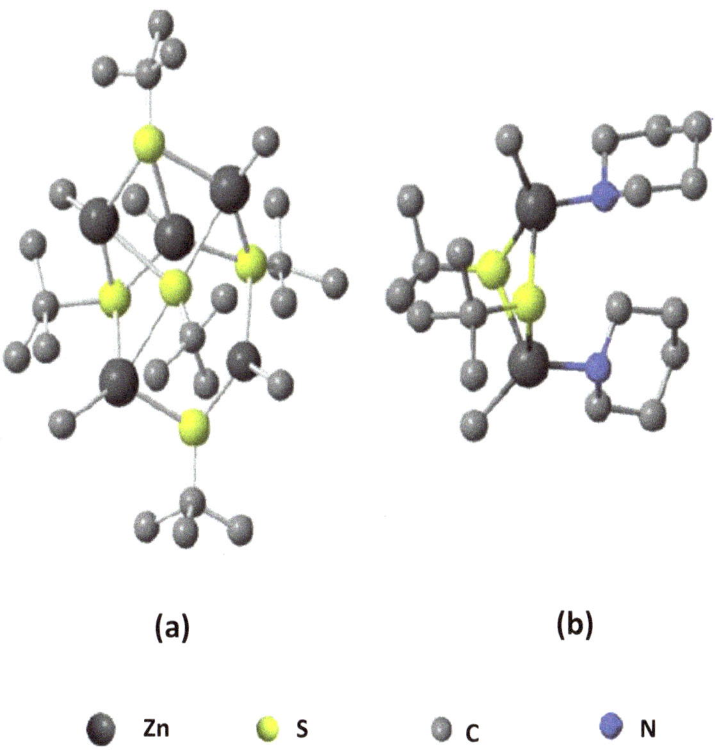


**Fig. 7.4** Structures of **a)** $[MeZn(S^tBu)]_5$ and **b)** $[MeZn(S^{t-}Bu)(Py)]_2$

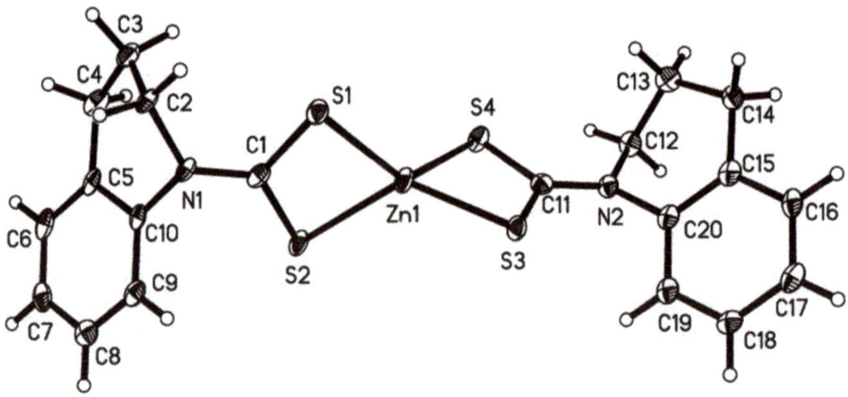


**Fig. 7.5** X-ray structure of $[Zn(thq\text{-}dtc)_2]$

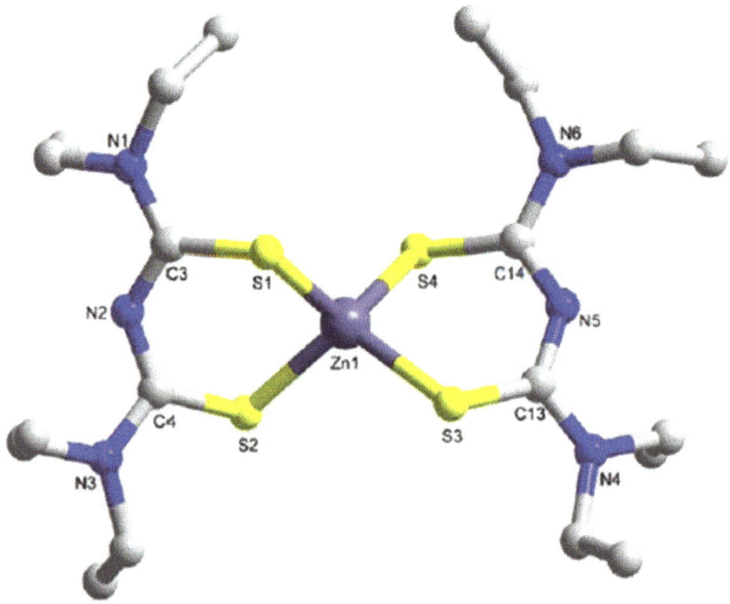


**Fig. 7.6** Crystal structure of 1,1,5,5-tetraalkyl-2-thiobiurets $[M(SON(CNR_2)_2)_2]$ [M = Zn, R = isopropyl]

The *tert*-butylphosphine ($^tBuPH_2$) precursor is currently the most successful alternative phosphorus source. It is a liquid with a vapour pressure (184.9 mbar at 10 °C) highly suitable for a wide range of MOVPE applications. Metalorganic Vapor Phase Epitaxy (MOVPE) is one of the key epitaxial growth technologies, that produces single or polycrystalline thin films for electronics well as optoelectronic applications such as mobile phone components (GaAs), semiconductor lasers and LEDs (III–Vs, nitrides), optical communications (oxides), infrared detectors, photovoltaics (II–IV materials), etc. The intrinsic toxicity of $^{t}BuPH_2$ is also lower than that of $PH_3$, and it pyrolyzes at a significantly lower temperature (50% pyrolyzed at 450 °C), probably via homolytic phosphorus–carbon bond fission. The reduced thermal stability of $^{t}BuPH_2$ relative to PH3 allows the growth of InP at lower V–III ratios and leads to substantial improvements in the uniformity of the deposited InGaAsP. Moreover, $^{t}BuPH_2$ has more favourable gas-phase chemistry than other $RPH_2$ precursors. Details of precursors used for deposition of group III–V thin films and nanoparticles are summarised in Table 7.3.

***(B) Single-Source Molecular Precursors (SSMPs) for Group III Chalcogenides***: The group III chalcogenides have potential use in optoelectronic and photovoltaic devices since they possess direct, mid- or wide-band gaps. Gallium sulphide has been used as a passivating layer for III–V thin films. Recent research has found that $Ga_2Te_3$ could be a potential phase-change memory material (Zhu et al. 2010). Complexes of gallium bound to chalcogen elements have become important in synthesis and as SSMPs for preparing thin films of gallium chalcogenides.

There are very few reports on SSMPs for the CVD of GaTe or $Ga_2Te_3$. Cubane precursors $[RGa(\mu_3\text{-}Te)]_4$ (R = $^{t}Bu$, $CEtMe_2$, $CEt_2Me$) have been used to deposit GaTe. A mixture of $Ga_2Te_3$, GaTe and Te is deposited from the ditelluroimidodiphosphinato complex $\{Ga(\mu\text{-}Te)[N(iPr_2PTe)_2]\}_3$ (Garje et al. 2006). Recently, George et al. (2013) used telluroether-based SSMPs for deposition of metal tellurides such as $GaTe_3$ and GaTe. The precursors $[GaCl_3(^{n}Bu_2E)]$ (E = Se, Te) and $[(GaCl_3)_2\{^{n}BuE(CH_2)_nE^{n}Bu\}]$ (E = Se, n = 2; E = Te, n = 3) have been used for deposition of $Ga_2E_3$ thin films by LPCVD. All these precursors are air-sensitive (E = Se) or when E = Te, oils-sensitive, except$[(GaCl_3)_2\{^{n}BuSeCH_2CH_2Se^{n}Bu\}]$.

***(C) Single-Source Molecular Precursors (SSMPs) for Group III Nitrides***: The most desirable requirement for deposition of metal nitride thin films is that the precursors should be nitrogen-rich to deliver sufficient active nitrogen to the surface to produce stoichiometric growth. AlN films have been deposited by CVD methods using a number of organoaluminum precursors such as $[Al(NR_2)_3]_2$, $[HAl(NR_2)_2]_2$ (R = Me, Et), $(Et_2AlN_3)_3$, $(Me_2AlNH_2)_3$, and $(Me_2AlNiPr_2)_2$. In general, $NH_3$ has been used as a nitrogen source, but due to high thermal stability a high substrate temperature (>900 °C) is required to produce AlN thin films. This leads to evaporation of nitrogen from the deposited AlN thin films. Also, the films contain large amounts of residual impurities due to carbon, and this directly affects the thermal conductivity and electrical resistivity of the films (Fischer et al. 1997). For example, the trimethylaluminum–ammonia adduct ($Me_3AlNH_3$) has been used to deposite stoichiometric AlN thin films, but they were contaminated by carbonaceous substances. Some specific points about various SSMPs follow.

- The monomeric $Me_3AlNH_3$species can be expected to have a higher volatility compared to trimeric amide $(Me_2AlNH_2)_3$ and to have a more suitable vapour pressure for MOCVD applications. Upon pyrolysis at 400–600°C $Me_3AlNH_3$ decomposes with the evolution of methane to form alkyl aluminum amides or

**Table 7.3** Details of group III–V precursors and their deposition processes

| Types of precursors | Group III–V precursors | Methods of preparation | Operating temperature | Targeted materials | Refs |
|---|---|---|---|---|---|
| **Conventional precursors** | | | | | |
| Dual precursors | $Me_3Ga$, $Me_3Al$, $Me_3In$, $AsH_3$, $PH_3$<br>$Et_2AsH$, $EtAsH_2$, $^tButAsH_2$ | MOVPE | >500 °C<br>425 °C | GaAs, InGaAsP | Jones (1997), Larsen et al. (1988) |
| **Single-source molecular precursors** | | | | | |
| Chalcogen (Se. Te) containing single source molecular precursors (SSMPs) | $[RGa(\mu 3\text{-}Te)]_4$ (R = $^t$Bu, $CEtMe_2$, $CEt_2Me$)<br>$\{M(\mu\text{-}Te)[^iPr_2PTe)_2]\}_3$ (M = Ga, In)<br>$[GaCl_3(nBu_2E)]$ (E = Se, Te)<br>$[GaCl_3\{^nBu_2E(CH)_nE^nBu\}]]$ (E = Se, N = 2, E = Te, n = 3) | LPCVD<br>AACVD | 500–600 °C<br>325–475 °C | $Ga_2Te_3$<br>$In_2Te_3$<br>$Ga_2Se_3$ | Zhu et al. (2010), Garje et al. (2006), George et al. (2013) |
| Nitrogen-containing single source precursors | $[Al(NR_2)_3]_2$, $[HAl(NR_2)_2]_2$ (R = Me, Et), $(Et_2AlN_3)_3$, $(Me_2AlNH_2)_3$, and $(Me_2AlNiPr_2)_2$ and $NH_3$<br>$Al_2(NMe_2)_6$ and hydrazidoalane dimers, $[R_2Al\text{-}\mu\text{-}N(H)NMe_2]_2$ (R = Me, Et) | CVD | >900 °C<br>250 °C | AlN | Fischer et al. (1997), Interrante et al. (1989) |
| | $[R_2Ga(N_3)]_3$ (R) Me,Et),<br>$[(\mu\text{-}NMe_2)Ga(N_3)(NMe_2)]_2$, and $[HClGaN_3]_4$, and $NH_3$<br>$(H_2GaN_3)$ and $(N_3)_3$ Ga(R) (R = Me, Et)<br>$[(N_3)_2\ Ga(CH_2)_3NMe_2]$<br>$Et_2Ga(N_3)\cdot MeHNNH_2]$ of $(H_2GaN_3)$ | CVD | | GaN | Sung et al. (2002) |
| | $[(N_3)In[(CH_2)_3NMe_2]_2$ | MOCVD | 200 °C | InN | Schaefer et al. (1999) |
| | $[M(H_2NCONH_2)_6]Cl_3$(M = Al,Ga)(Fig. 7.9a) and $In(H_2NCONH_2)_3Cl_3$(Fig. 7.8b)in *n*-trioctylamine | Solvothermal in Octylamine solvent | 200 °C | AlN, GaN and InN | Sardar et al. (2005) |

(continued)

**Table 7.3** (continued)

| Types of precursors | Group III–V precursors | Methods of preparation | Operating temperature | Targeted materials | Refs |
|---|---|---|---|---|---|
| Arsenic-containing SSMPs | $[Me_2Ga(\mu\text{-}As^tBu_2)]_2$,$[Et_2Ga(\mu\text{-}As^tBu_2)]_2$, and$[Ga(As^tBu_2)_3]$<br>$[Me_2Ga(\mu\text{-}AstBu_2)]_2$by using $H_2$ or He<br>$[Me_2Ga(\mu\text{-}As^tBu_2)]_2$ or $[Et_2Ga(\mu\text{-}As^{t-}Bu_2)]_2$ | LP MOCVD | 450–700 °C<br>400–500 °C | GaAs | Malik et al. (2010) |
| | $[Me_2Ga(\mu\text{-}AsMe_2)]_3$ and $[Me_2Ga(\mu\text{-}As^iPr_2)]_3$, | UHVMOCVD | | No GaAs | Malik et al. (2010) |
| | $[nBu_2Ga(\mu\text{-}As^tBu_2)]_2$,<br>$[Me_2In\text{-}P^tBu_2]_2$, | | | GaAs<br>InP | |
| Antimony-containing SSMPs | $[Me_2MSb^tBu_2]_3$(M = Ga, In) | HWMOCVD | ($10^{-3}$ mbar) | GaSb, InSb | Malik et al. (2010), Schulz et al. (1982) |
| | $[R_2MSbR'_2]_x$(M = Ga, In; $x = 2, 3$)<br>(R = Me, Et, $^i$Pr, $^n$Pr, tBu), $[^tBuGaSbEt_2]$ | CW MOCVD | 400–500 °C | GaSb, InSb | |
| | $[Et_2AlSb(SiMe_3)_2]_2$ or$[^iBu_2AlSb(SiMe_3)_2]_2$ | CW MOCVD | 300 and550°C | AlSb | |

imides that contain a strong (Al-N)bond, and the following reaction sequence has been proposed:

$$[((CH_3)_2AlNH_2)]_3 \rightarrow 3AlN + 6CH_4 \tag{7.1}$$

- The precursors $Al_2(NMe_2)_6$ and hydrazidoalane dimers, $[R_2Al\text{-}\mu\text{-}N(H)NMe_2]_2$ (R = Me, Et) have been used in CVD to produce carbon-free AlN at even lower temperatures (200–250 °C) (Interrante et al. 1989).
- All of the above-mentioned compounds require the use of $NH_3$ and $Me_3Al$. A low carbon contamination and good N/Al atomic ratio of AlN films are obtained by the MOCVD method when using an $AlCl_3$:$^t BuNH_2$ (1:1 adduct of $AlCl_3$ and $^t BuNH_2$) air-stable precursor. However, Cl contamination from $AlCl_3$ is then another concern. Urea complexes of Al, In and Ga are also used to prepare good quality metal nitrides. Wurtzite (hexagonal) nanocrystals of AlN, GaN, and InN are obtained by refluxing metal chloride and urea adducts, e.g. $[M(H_2NCONH_2)_6]Cl_3$(M = Al, Ga) and $In(H_2NCONH_2)_3Cl_3$ in *n*-trioctylamine (Sardar et al. 2005). However, the nitrogen-containing Ga and In species are generally thermally unstable, resulting in nonstoichiometric and poor quality films. This is because the M–N bond strengths decrease in the order AlN > GaN > InN.
- The precursor $[Ga(NMe_2)_3]_2$ in the presence of ammonia results in amorphous GaN films at 200 °C substrate temperatures. This clearly indicates that the activation energy of deposition may be lowered significantly when amide precursors are used.
- In general, organometallic and inorganic azide compounds, such as $[R_2Ga(N_3)]_3$ (R = Me, Et), $[(\mu\text{-}NMe_2)Ga(N_3)(NMe_2)]_2$, and $[HClGaN_3]_4$, have been shown to be useful precursors for deposition of Gallium nitride thin films. The introduction of azide units appears to be a good source of nitrogen. However, their exothermic decomposition is a major concern. The nitrogen-rich compounds such as $(H_2GaN_3)$ and $(N_3)_3Ga(R)$ (R = Me, Et) may even explode under certain conditions, because the M-$N_3$ bonds produce explosive hydrazoic acid upon hydrolysis. This is a major obstacle with respect to the technological use of azide-type compounds.

  The intramolecularly Lewis-base stabilized compound $[(N_3)_2Ga(CH_2)_3NMe_2]$ has been synthesized to overcome the above problem. Transparent GaN films have been grown from this compound in a cold-wall reactor with or without the use of ammonia. However, films deposited without using ammonia contained a small amount of carbon. It is interesting to note that the composition of the gas phase in the boundary layer above the substrate was monitored in order to obtain information on the decomposition pathway of $[(N_3)_2Ga(CH_2)_3NMe_2]$ (Malik et al. 2010) and some species such as $HGaN_6$, $GaN_6$, $HGaN_2$, and $GaN_2$ were identified. The data analysis gave clear evidence for the production of nitrogen-rich species in the boundary layer.
- The Lewis acid–base adduct$[Et_2Ga(N_3)\cdot MeHNNH_2]$ of $(H_2GaN_3)$ has been used to deposit epitaxial *h*-GaN films on Si(111) substrates by MOCVD (Sung et al.

2002). This precursor undergoes protonation of the ethyl ligands by methylhydrazine, which occurs via loss of the stable $N_2$, $C_2H_6$, and MeNNH components, thereby yielding good quality GaN films with low contamination.
- Unlike AlN and GaN, the growth of InN is most difficult because of its low bond energy. Thus a low decomposition temperature is required to grow InN thin films. Deposition of InN can be achieved at 300–400 °C with $[(N_3)In[(CH_2)_3NMe_2]_2$ (Schaefer et al. 1999). This molecule is stabilized by the intermolecular Lewis-base adduct, which is air-stable (meltingpoint 67 °C) and neither pyrophoric nor explosive. The gas phase is dominated by Indium atoms at higher temperatures.

## 7.4 Recent Scientific Findings of DMSRDE/DRDO, India

### *7.4.1 Metalchalcogenides from Alkyl Chalcogenolate Metal Complexes*

Alkyl chalcogenolate complexes are potential SSMPs for the preparation of metal chalcogenides materials. Tripathi et al. prepareda series of complexes of type $(H_2N(H_2)_3Te)_2MX_2$ (M = Zn, Cd, Hg; X = Cl, Br, I), by in situ reduction of bis(3-aminopropyl)dichalcogenide to 3-aminopropylsodiumchalcogenolate, followed by metal halides. CdTe, ZnTe and HgTe nanoparticles have been prepared by solvothermal decomposition of these complexes in quinolone, a high boiling point solvent, at 200 °C. The prepared nanoparticles have been characterized by XRD, UV–Visible and Photoluminescence (PL) spectrometry, SEM, and TEM.

The TEM images showed that the nanoparticles were in the range of 10–20 nm (Tripathi and Nasim 2008). Our group has also reported cadmium telluride nanoparticles by refluxing Bis(isopropyletelluro)methane Cd(II) Cl precursor in quinolene solvent. The precursor has been synthesized as shown in Fig. 7.7 (Selvakumar et al. 2008). The X-ray diffraction pattern confirmed a face-centered-cubic structure, and the average diameter of the nanocrystallites was found to be 29.5 nm. SEM micrographs clearly showed the needle-like structure, which is further supported by TEM imaging where pearl necklace-type CdTe nanoparticles were clearly seen (Fig. 7.8).

Tripathi et al. also prepared a Bis(Isopropyletelluro)methane Cd(II) acetate SSMP. The CVD was done at temperatures in the range 300–500 °C under reduced pressure (1 torr) to optimise the deposition parameters for high quality CdTe thin films. At 450 °C the precursor gave cubic CdTe thin films on an Si(100) substrate. SEM micrographs confirmed the homogeneous deposition of CdTe on the substrate. EDAX analysis showed a 1/1.1 ratio of Cd/Te. The band gap of the deposited films was about 1.45 eV, which is close to that of bulk CdTe materials (Tripathi et al. 2020).

This research group also reported new Lewis-base (LB) stabilized gallium-telluride precursors. Gallium tellurolate complexes with the formula $[LB][Ga(TePh)_3]x$, [x = 1, LB = 4-dimethylaminopyridine; x = 2, LB = 4,4-methylene bis(N,N-dimethylaniline] were prepared by reacting the corresponding

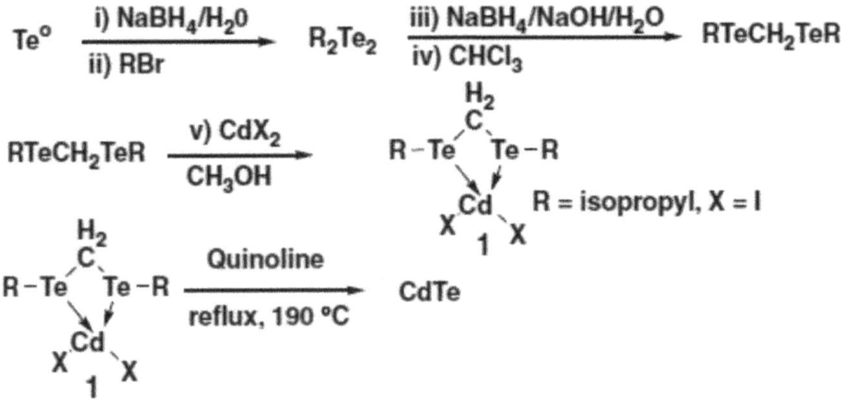


**Fig. 7.7** Synthesis of precursor Bis($^i$prTe)$_2$CH$_2$, CdCl$_2$, and CdTe nanoparticles

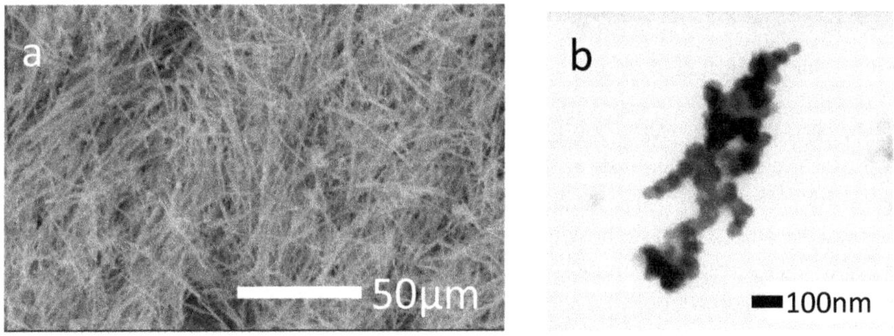


**Fig. 7.8** **a)** SEM and **b)** TEM images of cadmium telluride nanoparticles

Lewis-base adduct of gallium (III) iodide and phenyl lithium tellurolate. These complexes have been characterized by element analyses, ICP-MS, multinuclear NMR, and thermal and mass spectrometry. These complexes are potential SSMPs for III–VI semiconductor materials (Selvakumar et al. 2005).

### *7.4.2 Precursors for III–V Nitrides*

Most of the precursors used for the deposition of group III–V nitrides are highly air sensitive and pyrophoric in nature, hence difficult to handle. Because of this DMSRDE recently prepared the single source molecular precursors (SMMPs) such as $Al(H_2NCONH_2)_6{\cdot}Cl_3$, $Al(H_2NCH_2CH_2NH_2)_3{\cdot}Cl_3$, $[Al(CH_3)_2NCH_2CH_2NH_2)_3].Cl_3$ and $[Al(CH_3)_2NCH_2CH_2N(CH_3)_2)_3]{\cdot}Cl_3$, see Fig. 7.9. These SSMPs are air stable, nonpyrophoric and easily used in CVD to produce carbon-free AlN thin films or bulk AlN powder at 1000 °C and 1 torr

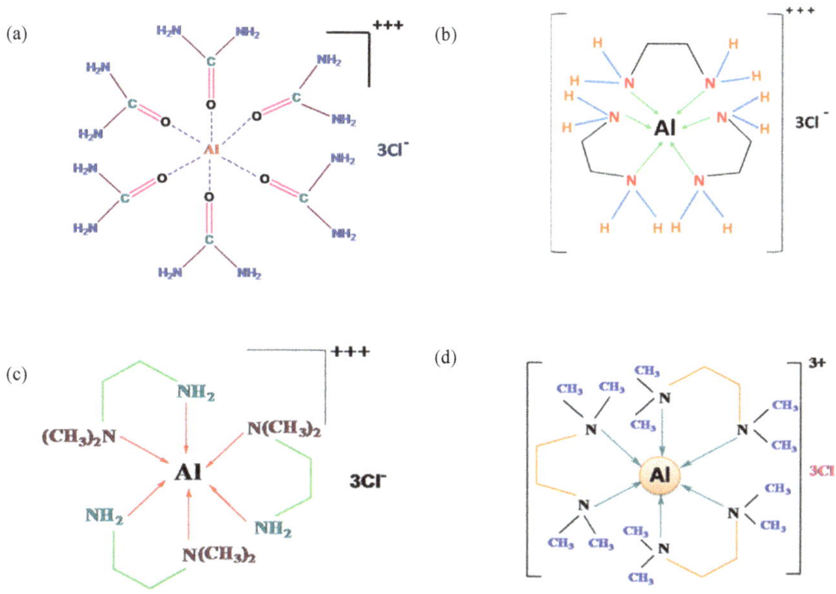


**Fig. 7.9** **a)** Hexaurea aluminate (III) chloride complex; **b)** Trisethylenediamine aluminate (III) chloride complex; **c)** N,N Dimethyl ethylenediamine aluminate (III) chloride complex; **d)** Tetramethylethylenediamine aluminate (III) chloride complex

in a nitrogen atmosphere (Chaurasia et al. 2019a, b, 2023, Unpublished). SEM images showed crack-free deposition of AlN films on Si(100) substrates. The atomic compositions were analysed by EDAX and found to have a 2/1 wt% ratio for Al/N atoms.

### *7.4.3 Copper Tin Selenium Nanoink and Copper Zinc Tin Selenium for Solar Cell Applications*

$Cu_2SnSe_3$ (CTSe) nanocrystals were prepared by the solvothermal route using a copper, tin and selenium mixture heated in an autoclave at 180 °C for 18 h. A black sediment contained the polycrystalline $Cu_2SnSe_3$ nanocrystals with average size of 14–19 nm, as confirmed by XRD (Fig. 7.10a) and TEM (Fig. 7.10b). This is the first time we have fabricated heterojunction hybrid solar cells using a blend of poly(3-hexyle thiophene)(P3HT): phenyl-C61-butyric-acid-methyl-ester (PCBM) and $Cu_2SnSe_3$ (CTSe) nanoink as an additional photoactive layer (Fig. 7.11).

Electrochemical impedance spectroscopy (EIS) of fabricated devices was carried out under dark conditions with an applied bias of 0.1 V and frequency ranging from 1 Hz to 1 MHz. It was observed that the charge storage capacity of the device based on CTSe nano-ink/P3HT:PCBM as a photo-absorber, was greater than that of a similar

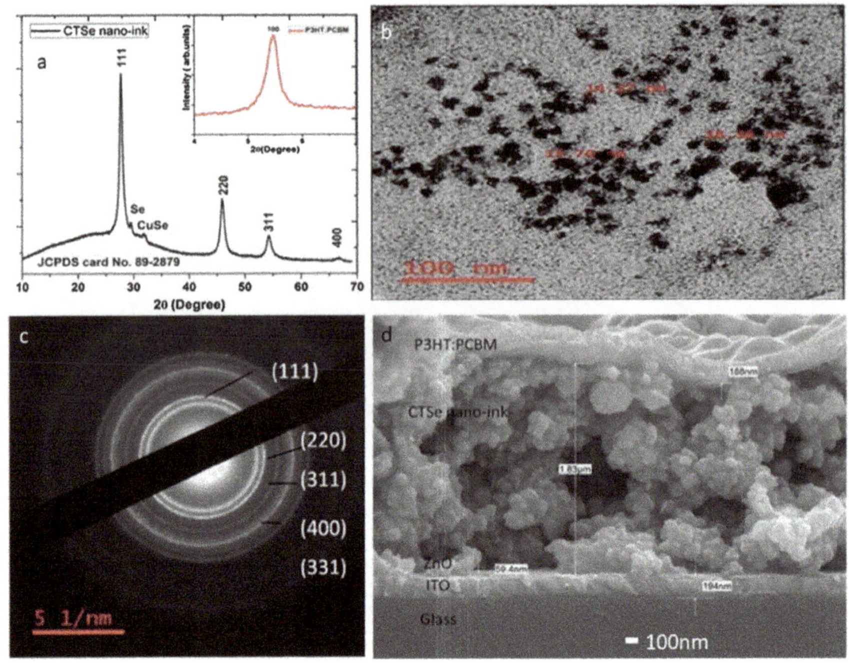


**Fig. 7.10** **a**) XRD trace of $Cu_2SnSe_3$nanoink-based thin film and P3HT:PCBM(inset) annealed at 120 °C; **b**) TEM micrograph of $Cu_2SnSe_3$ nanocrystals; **c**) selective area electron diffraction (SAED) of $Cu_2SnSe_3$ nanocrystals; and **d**) cross sectional view of a fabricated solar cell

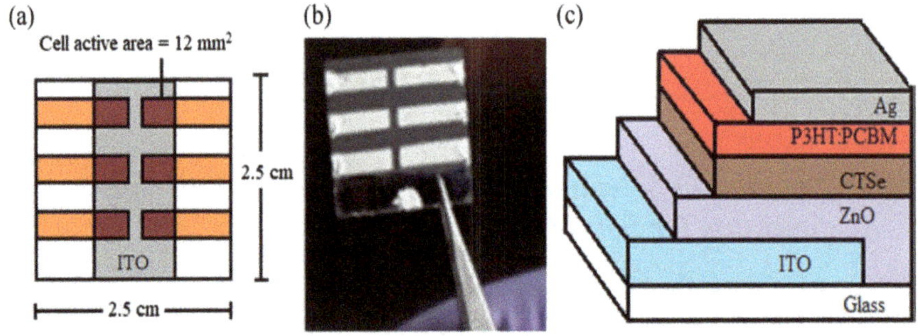


**Fig. 7.11** **a**) Top view of patterned indium tin oxide (ITO) solar cell with active area; **b**) photograph of fabricated hybrid solar cell; and **c**) structure of the fabricated heterojunction hybrid solar cell, showing the added CTSe layer

device without the P3HT:PCBM layer. We have obtained an open circuit voltage (Voc) of 565 mV, current density (Jsc) of 0.87 mA/cm$^2$ and fill factor (FF) of 0.48 from the hybrid solar cells (Fig. 7.12). In addition the nanoink preparation method is cost-effective for preparing stable nanoinks suitable for large-scale production of printable optoelectronics devices (Dwivedi et al. 2019a).

Other thin film solar cells with ITO/ZnO/P3HT:PCM:CTSe NCs(nanocrystals)/Ag structure were made using a blended solution of P3HT:PCBM:CTSe NCs deposited by spin casting, followed by thermal annealing steps. Inverted architectures of devices with structure ITO/ZnO/P3HT: PCBM:CTSe NCs/Ag have been fabricated with different concentrations of CTSe NCs in a poly(3-hexyle thiophene) (P3HT): [6,6]phenyl-C61-butyric-acid-methyl-ester (PCBM) matrix to optimize the cell performance. The effect of CTSe NCs on the performance of hybrid solar cells with an optimized blend ratio of P3HT:PCBM and CTSe NCs has been investigated. The charge carrier extraction and recombination at the interfaces of donor–acceptor materials were studied using EIS under dark conditions. It was found that the charge

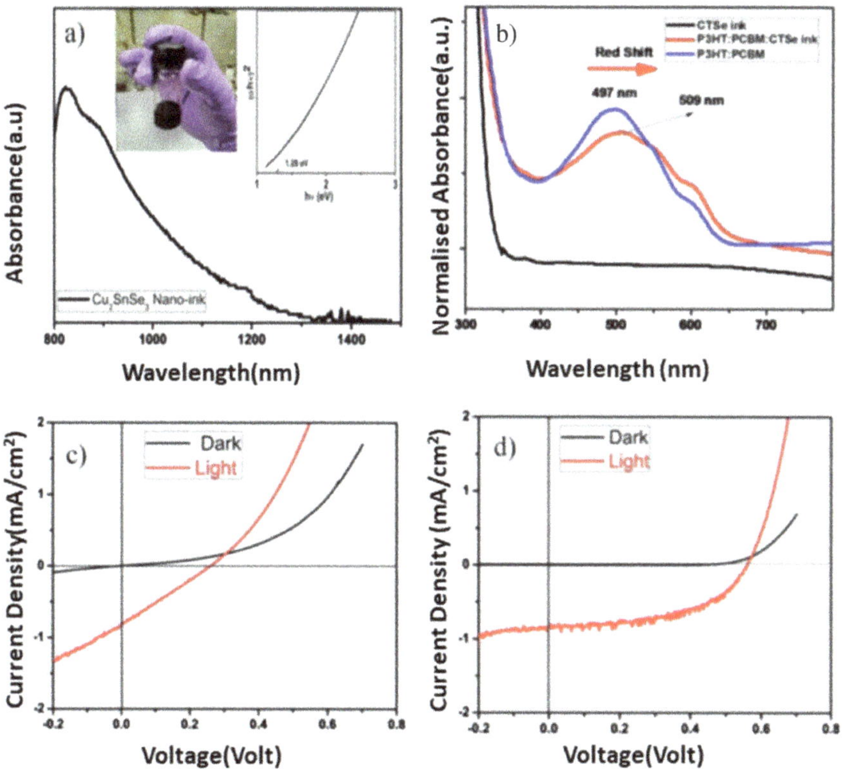


**Fig. 7.12** **a)** UV–Vis–NIR absorption spectrum of CTSe nanoink (inset diagram is a Tauc bandwidth plot between (αhυ)2 as a function of photon energy of $Cu_2SnSe_3$ nanocrystals; insert photograph shows an ink solution containing stably suspended $Cu_2SnSe_3$ nanocrystals); **b)** absorption spectra of P3HT:PCBM/CTSe layer, P3HT:PCBM layer, and CTSe nanoink in the visible region; **c)** current density vs voltage graph of a device based on CTSe nanoink as an absorber layer; and **d)** current density vs voltage graph of a device based on CTSe nano-ink/P3HT:PCBM as an absorber layer under 100 mW/cm$^2$ illumination (AM 1.5G sun)

transfer rate was higher for the device having an optimized wt% of CTSe NCs (10 wt%) in the P3HT:PCBM active layer, and also that significant improvements in device performances were obtained by incorporating CTSe NCs. The device exhibited an open circuit voltage (Voc) of 0.475 V, short circuit current density (Jsc) of 6.95 $mA/cm^2$, fill factor (FF) of 0.41 and power conversion efficiency (PCE) of 1.35% (Dwivedi et al. 2019b).

Very recently, this DRDO laboratory has fabricated crystalline CZTS thin films by the sol–gel method followed by sulphurization using 1,2-Ethanedithiol (EDT) at 400 °C in a three-zone horizontal tubular reactor equipped with a computer controlled vacuum system. These films were characterized by XRD, Raman spectroscopy, Hall measurements and UV–Vis–NIR spectroscopy. The XRD and Raman studies confirmed highly crystalline and single phase CZTS. The Hall measurements revealed p-type semiconducting CZTS films having a charge carrier density of 1.70 $\times$ 1018 $cm^{-3}$ and resistivity ($\rho$) of the order of $5.00 \times 10^{-1}$ $\Omega$-cm.

It was found from the optical study that the CZTS films possess a direct band gap of 1.44 eV. Furthermore, double metal oxide layers having slightly different band gaps were used to form a p–n junction: cadmium-free solar cells with structure ITO/ZnO/CZTS/Ag (cell-1) and ITO/ZnO/com-$TiO_2$/CZTS/Ag (cell-2) were fabricated. EIS measurements were performed in the low frequency range with an applied bias of 0.1 V for the devices. The EIS studies clearly showed that a solar cell with an ultrathin compact $TiO_2$ layer provides a favourable path for efficient charge transport at the interface. The external quantum efficiency of the fabricated cells was measured at 520 nm and found to be 62.5%. Cell-2 exhibited improved photovoltaic parameters such as a short circuit current density (Jsc) of 4.721 $mA/cm^2$, fill factor of 0.46 and open circuit voltage (Voc) of 400 mV. The short circuit current density and fill factor of double metal oxide layers (cell-2) were increased by 78% and 46% as compared to a single metal oxide layer (cell-1) (Dwivedi et al. 2021).

## 7.5 Important Technological Applications

The group II–VI materials have various technological applications in the fabrication of high efficiency detectors, sensors, LEDs and solar cells:

(1) **CdTe** is a well-known light absorber with outstanding properties. It has a band gap of 1.45 eV, which is well-suited to absorbing solar radiation: light is fully captured by about two microns of material owing to its high optical absorption coefficient.

CdTe or cadmium zinc telluride (CZT) detectors have much better detection efficiencies than silicon-based detectors. More specifically, in clinical practice the photon energies lie in the range of 100–140 kVp, and silicon-based detectors have the disadvantage that a large part of the energy information for spectral imaging is lost due to Compton effects. This disadvantage limits the interest for using silicon detectors in spectral CT applications. Consequently CdTe/CZT detectors are taking a major role in photon counting CT systems with spectral imaging capabilities.

(2) **CdZnTe (CZT)** is a ternary semiconductor material with numerous applications in medical imaging, e.g. Computed Tomography (CT), Single Photon Emission Computed Tomography (SPECT), Integration-Mode X-Ray Radiography, Photon-Counting, Surgical-Oncology, Positron Emission Tomography (PET) (Marchand et al. 2013), and Dedicated Emission Mammotomography. CZT is also important for X-ray and gamma ray detectors, nuclear radiation detectors, substrates for epitaxial growth of the IR-detector material HgCdTe, electro-optical modulators, photo-conductors, light emitting diodes and solar-cells for defence applications. The tunable band-gap of CZT lies between 1.45 and 2.25 eV, whereby it can be altered by changing either the Cd or Zn contents. An additional advantage of CZT detectors is that in high energy (X-ray and gamma ray) applications these detectors can operate at room temperature.

(3) **Group III–V materials** also have potential uses in the fabrication of high efficiency devices and detectors, owing to their outstanding electronic and optical properties. These materials and related alloys/compounds, e.g. GaAs, AlGaAs, InP, are widely used in the microelectronic and optoelectronic fields. Practically all kinds of passive and active devices working in the communication spectral windows around 850 (GaAs), 1300 or 1550 (InP) nm can be fabricated with III–V semiconductors: waveguides, switches, modulators, lasers and photodetectors.

Gallium nitride and its alloys with InN and AIN have recently emerged as important semiconductor materials with emitter and detector applications in the yellow, green, blue and ultraviolet portions of the electromagnetic spectrum; and also for use in high power and high temperature electronics. These latter applications promise substantial weight savings for aircraft and spacecraft.

Blue and green nitride LEDs exhibit brightness levels and longevity well in excess of those required for outdoor applications. Combined with the available red LEDs, true full colour semiconductor flat panel displays can now be obtained. If used for traffic lights and illumination (pending further improvements in blue in some cases) these devices can outlast and outperform incandescent light bulbs while saving precious energy.

This material system is also intrinsically suitable for short wavelength semiconductor lasers to enable increased data storage. Very recently, pulsed room temperature operation of 410 nm semiconductor lasers, the shortest wavelength ever from a semiconductor, have been reported.

In view of all these possibilities, it is no surprise that much research and development (R&D) effort is being expended on the synthesis and characterization of these materials, and also the processes necessary to develop devices based on them.

## 7.6 Scope for Future Research

Organometallic Vapour Phase Epitaxy (OMVPE) or Metalorganic Chemical Vapour Deposition (MOCVD) is a clearly maturing technology that has found acceptance in many laboratories throughout the world. The technology in some cases has moved from the R&D phase into the manufacturing phase. This phase will likely become the technology of choice for fabrication of many Group II–VI and III–V devices, particularly when large volumes are required.

Many of the areas for further technological advancement have only been sketched out. However these techniques mostly used conventional reactants which require stringent safety measures owing to the toxic and pyrophoric natures of many of the precursors. Hence there is always a driving force for safer alternative sources. There is therefore a strong incentive to develop SSMPs, which overcome the above issues. However, because SSMPs are less volatile compared to dual precursors, a new variant of CVD—AACVD—has recently come to prominence. The AACVD approach requires soluble precursors that enter the reactor as aerosols, thereby avoiding the volatility problem when selecting precursors, but introducing a solubility issue. However, the benefits of the AACVD approach overcome its limitations, particulary because relying on solubility instead of volatility greatly extends the range of available SSMPs. Furthermore, the AACVD approach lends itself to better morphological control. Films can be deposited on glass, steel, alumina, silica and many other surfaces; and uniform films with excellent adhesion can be obtained by this technique.

Another related R&D area is the design of precursors and their ligands in order to improve the properties of the resultant films. Ligands have been designed to facilitate cleaner decomposition to reduce contamination, increase volatility and/or solubility, and lower the production costs. However, sensitivity to exposure to air and the shelf life are still significant issues. The expanding field of flexible electronics encourages seeking low temperature routes, and AACVD is an ideal technique to achieve low temperature depositions. Therefore the future of SSMPs, particularly combined with AACVD processing, appears very promising.

## 7.7 Summary

We have reviewed and also demonstrated, with examples, the role of precursors for the development of advanced functional materials. The use of group II–VI and III–V metal complexes have considerable scope for the preparation of potentially semiconducting materials as thin films or nanoparticles, as shown by several research groups: the single-source (SSMP) approach is to be seen as an alternative, perhaps a much better one, to more established methods. Most of the single-source molecules have low volatilities, but this disadvantage can be overcome by introducing various chemical modifications such as liquid delivery methods i. e. aerosols, spraying or injection.

The use of solvothermal methods are essential for the preparation of nanoparticles, since the precursors need to be pyrolysed in high boiling-point solvents. The important factors for designing precursors, such as purity, stability, toxicity, ease and costs of synthesis are also discussed.

This chapter also includes some information about the relatively new and advanced chalcogenides-based precursors, which have potential applications in hybrid solar cells and sensors. Lastly, efforts are ongoing to develop advanced precursor materials for important production methods such as low temperature processing and high reproducibility of electronic and optoelectronic devices.

## References


Adeogun, A., Nguyen, C.Q., Afzaal, M., Malik, M.A., O'Brien, P.: Facile and reproducible syntheses of bis(dialkylselenophosphenyl)-selenides and diselenides: X-ray structure of $(^iPr_2PSe)_2Se$, $(^iPr_2PSe)_2Se_2$ and $(Ph_2PSe_2)Se$. Chem. Commun. 2179 (2006)

Ambacher, O.: Growth and applications of Group III-nitrides. J. Phys. D Appl. Phys. **31**, 2653–2710 (1998)

Bochmann, M., Webb, K.J.: Sterically hindered thiolato, selenolato and tellurolato complexes of mercury (II). J. Chem. Soc., Dalton Trans. 2325 (1991)

Bochmann, M., Coleman, A.P., Powell, A.K.: Synthesis of some alkyl metal selenolato complexes of zinc, cadmium and mercury. X-ray crystal structure of Me, Hg, Se(2,4,6-Pr3i C62). Polyhedron **11**, 507 (1992)

Brennan, J.G., Segrist, T., Carroll, P.J., Stuczynski, S.M., Reynders, P., Brus, L.E., Steigerwald, M.L.: The preparation of large semiconductor clusters via the pyrolysis of a molecular precursor. J. Am. Chem. Soc. **111**, 4141 (1989)

Chaurasia, H., Tripathi, S.K., Bilgaiyan, K., Pandey, A., Mukhopadhyay, K., Kavita, A., Prasad, N.E.: Preparation and properties of AlN (Aluminum nitride) powder/thin films preparation and properties of AlN (Aluminum nitride) powder/thin films by single source precursor. New J. Chem. **43**, 1900–1909 (2019a)

Chaurasia, H., Tripathi, S.K., Bilgaiyan, K., Eswara Prasad, N.: Ethylendiamine aluminium (III) chloride: potential precursor for AlN materials/thin films. Ceram. Int. **45**, 23772–23780 (2019b)

Chaurasia, H., Tripathi, S.K., Bilgaiyan, K., Pandey, A., Eswara Prasad, N.: Structural and optical properties of aluminium nitride nanopowder prepared from tris (N,N dimethyl-ethylenediamine)$AlCl_3$ precursor. J. Coord. Chem. **76**, 134–154 (2023)

Chaurasia, H., Tripathi, S.K., Bilgaiyan, K., Eswara Prasad, N.: Microstructural charaterization of AlN thin films prepared from Tris(Tetramethylethylenediamine) Al(III) $Cl_3$ precursour, Unpublished work

Choy, K.L.: Chemical vapour deposition of coatings. Progress Mater. Sci. **48**, 57–170 (2003)

Dwivedi, S.K., Tripathi, S.K., Tiwari, D.C., Chauhan, A.S., Dwivedi, P.K., Eswara Prasad, N.: Low cost copper zinc tin sulphide (CZTS) solar cells fabricated by sulphurizing sol-gel deposited precursor using 1,2-ethanedithiol (EDT). Sol. Energy **224**, 210–217 (2021)

Dwivedi, S.K., Tiwari, D.C., Tripathi, S.K., Zaman, M.B., Dipak, P., Imamuddin, M., Poolla, R., Eswara Prasad, N.: P3HT:PCBM and Cu2SnSe3 nano-ink based hybrid solar cells. Solar Energy **177**, 382–386 (2019a)

Dwivedi, S.K., Tiwari, D.C., Tripathi, S.K., Dwivedi P.K., Dipak, P., Chandel, T., Eswara Prasad, N.: Fabrication and properties of P3HT: PCBM/Cu2SnSe3 (CTSe) nanocrystals based inverted hybrid solar cells. Solar Energy **187**, 167–174 (2019b)

Fischer, R.A., Miehr, A., Ambacher, O., Metzger, T., Born, E.: Novel single source precursors for mocvd of AlN, GaN, and InN. J. Cryst. Growth **170**, 139 (1997)

Garje, S.S., Copsey, M.C., Afzaal, M., O'Brien, P., Chivers, T.: Aerosol-assisted chemical vapour deposition of indium telluride thin films from $\{In(\mu\text{-}Te)[N(iPr_2PTe)_2]\}_3$. J. Mater. Chem. **16**, 4542 (2006)

George, K., (Kees) de Groot, C.H., Gurnani, C., Hector, A.L., Huang, R., Jura, M., Levason, W., Reid, G.: Telluroether and selenoether complexes as single source reagents for low pressure chemical vapor deposition of crystalline $Ga_2Te_3$ and $Ga_2Se_3$ thin film. Chem. Mater. **25**, 1829–1836 (2013)

Hursthouse, M.B., Motevalli, M., O'Brien, P., Walsh, J.R., Jones, A.C.: Triazineaducts of dimethylzinc and dimethylcadmium: X-ray crystal structure of $Me_2Zn[(CH_2NMe)_3]_2$. Organometallics **10**, 3196 (1991)

Interrante, L.V., Lee, W., McConnell, M., Lewis, N., Hall, E.: Preparation and properties of aluminum nitride films using an organometallic precursor. J. Electrochem. Soc. **136**, 472 (1989)

Jones, A.C.: Developments in metalorganic precursors for semiconductors growth from vapour phase. Chem. Soc. Rev. 101 (1997)

Jones, A.C., O'Brien, P.: CVD of Compound Semiconductors, pp. 319–320. Wiley-VCH Verlag GmbH, Weinheim, Germany (1997)

Jun, Y.W., Choi, C.S., Cheon, J.: Size and shape controlled ZnTe nanocrystals with quantum confinement effect. Chem. Commum. 101–102 (2001)

Knapp, C.E., Carmalt, C.J.: Solution based CVD of main group materials. Chem. Soc. Rev. **45**, 1036 (2016)

Kuhn, W., Naumov, A., Stanzl, H., Bauer, S., Wolf, K., Wagner, H.P., Gebhardt, W., Pohl, U.W., Krost, A., Richter, W., Dumichen, U., Thiele, K.H.: Low tempareture MOVPE growth of ZnSe by Ditertiarybutylselenide. J. Cryst. Growth **123**, 605 (1992)

Kurek, A., Gordon, P.G., Karle, S., Devi, A., Barry, S.T.: Recent advances using guanidinate ligands for chemical vapour deposition (CVD) and atomic layer deposition (ALD) applications. Aust. J. Chem. **67**, 989 (2014)

Larsen, C.A., Buchan, N.I., Li, S.H., Stringfellow, G.B.: GaAs growth using tertiarybutylarsine and trimethylgallium. J. Cryst. Growth **93**, 15 (1988)

Malik, M.A., Afzaal, M., O'Brien, P.: Precursor chemistry for main group elements in semiconducting materials. Chem. Rev. **110**, 4417–4446 (2010)

Manasevit, H.M.: Recollections and reflections of MOCVD. J. Cryst. Growth **55**, 1–9 (1981)

Marchand, P., Hassan, I.A., Parkin, I.P., Carmalt, C.J.: Aerosol-assisted delivery of precursors for chemical vapour deposition: expanding the scope of CVD for materials fabrication. Dalton Trans. **42**, 9406 (2013)

Memon, A.A., Dilshad, M., Revaprasadu, N., Malik, M.A., Raftery, J., Akhtar, J.: Deposition of cadmium sulfide and zinc sulfide thin films by aerosol-assisted chemical vapors from molecular precursors. Turk. J. Chem. **39**, 169–178 (2015)

Nyamen, L.D., Nejo, A.A., Pullabhotla, V.S.R., Ndifon, P.T., Malik, M.A., Akhtar, J., O'Brien, P., Revaprasadu, N.: The syntheses and structures of Zn(II) heterocyclic piperidine and tetrahydroquinolinedithiocarbamates and their use as single source precursors for ZnS nanoparticles. Polyhedron **67**, 129–135 (2014)

Park, J.-H., O'Brien, P., White, A.J.P., Williams, D.J.: Variable coordination modes in dialkyldithiophosphinato complexes of Group 13 metals: the X-ray single crystal structures of tris(diisobutyldithiophosphinato)gallium (II) and Indium (III). Inorg. Chem. **40**, 3629 (2001)

Ramasamy, K., Malik, M.A., Helliwell, M., Raftery, J., O'Brien, P.: Thio- and Dithio-Biuret precursors for zinc sulfide, cadmium sulfide, and zinc cadmium sulfide thin films. Chem. Mater. **23**, 1471–1481 (2011)

Sardar, K., Dan, M., Schwenzera, B., Rao, C.N.R.: A simple single-source precursor route to the nanostructures of AlN, GaN and InN. J. Mater. Chem. **15**, 2175–2177 (2005)

Schaefer, J., Wolfrum, J., Fischer, R.A., Sussek, H.: Detaction of $H_yInN_x$ species in InN deposition using organoindiumazides precursor $(N_3)In[(CH_2)_3NMe]_2$. Chem. Vap. Deposition **5**, 205 (1999)

Schulz, S., Fahrenholz, S., Kuczkowski, A., Assenmacher, W., Seemayer, A., Hommes, A., Wandelt, K.: Deposition of GaSb films from the single source orecursors [t-$Bu_2GaSbEt_2]_2$. Chem. Mater. **2005**, 17 (1982)
Selvakumar, D., Singh, R., Nasim, M., Mathur, G.N.: Synthesis and characterization of lewis base stabilized gallium-tellurium complexes. Phosphorus Sulfur Silicon **180**, 1011–1017 (2005)
Selvakumar, D., Tripathi, S.K., Singh, R., Nasim, M.: Solvothermal preparation of cadmium telluride nanoparticles from a novel single source molecular precursor. Chem. Lett. **37**, 34–35 (2008)
Singh, H.B., Sudha, N.: Organotellurium precursors for metal organic chemical vapour deposition (MOCVD) of mercury cadmium telluride (MCT). Polyhedron **15**, 745–763 (1996)
Strite, S., Lin, M.E., Morkoç, H.: Progress and prospectus for GaN and the III-V nitride semiconductors. Thin Solid Films **231**, 197–210 (1993)
Sung, M.M., Kim, C., Yoo, S.H., Kim, C.G., Kim, Y.: CVD of GaN films on si(111) chemically clean decomposition of $Et_2Ga(N_3)MeNNH_2$. Chem. Vap. Deposition **8**, 50 (2002)
Tripathi, S.K., Khandelwal, B.L., Gupta, S.K.: A new family of chalcogen bearing macrocycles: synthesis and characterization of $N_4O_2E_2$(E = Se, Te) type compounds. Phosphorus Sulfur Silicon Relat. Elem. Sulfur **177**, 2285–2293 (2002)
Tripathi, S.K., Mishra, S.B., Nasim, M., Khandelwal, B.L.: Organoselenium/Tellurium – bearing macoacyclic and cyclic ligand systems and their complaxation reactions. Phosphorus Sulfur Silicon **180**, 1019–1034 (2005)
Tripathi, S.K., Nasim, M.: Chemistry of aminopropylchalcogenolate complexes of zinc, cadmium, mercury and their nanoparticles. Phosphorus Sulfur Silicon Relat. Elem. Sulfur **183**, 1087–1097 (2008)
Tripathi, S.K., Verma, R.D., Mukhopadhyay, K., Prasad, N.E., Pandey, A.K., Kapoor, A.K.: Preparation of CdTe bulk/thin film by applying novel single source molecular precursor and process thereof. A Indian Patent No. 372542(2020)
Zhu, H., Yin, J., Xia, Y., Liu, Z.: $Ga_2Te_3$ phase change material for low-power phase change memory application. Appl. Phys. Lett. **97**, 083504 (2010)

# Chapter 8
# Functional Paints and Coatings

**Shilendra Kumar, Sunil Kumar, and Eswara Prasad Namburi**

**Abstract** This chapter discusses the importance of functional paints and coatings. The types of materials used, their processing, and different methods of testing of the final products are discussed. The range of applications is immense and especially important for strategic applications in defence and aerospace systems and platforms. Some specific functional applications include camouflage and stealth coatings; thermal imaging coatings; anti-static coatings; thermal insulating paints; high performance anti-corrosion coatings; anti-skid coatings; antifouling coatings; self-healing and self-cleaning coatings. Over the years DMSRDE, Kanpur, has made considerable efforts to develop specialized paints and coatings. In this chapter particular attention is paid to DMSRDE-developed camouflage and stealth coatings.

## 8.1 Introduction

A paint is a generally homogeneous mixture of major ingredients, namely a binder, pigment, volatile organic carriers and additives. When applied as a thin layer, the paint forms a solid dry adherent film. Sometimes the terms 'paint' and 'surface coating' are used interchangeably. Paints and coatings enable surface protection and provide decorative or camouflage functions. Innovation of new paints and coatings targets a range of industries including aircraft, railways, automobiles, ship building and wind turbine manufacturers.

S. Kumar · S. Kumar
Defence Materials and Stores Research and Development Establishment (DMSRDE), Defence Research and Development Organisation (DRDO), Kanpur, Uttar Pradesh, India

E. P. Namburi (✉)
Former Outstanding Scientist and Ex-Director, Defence Materials and Stores Research and Development Establishment (DMSRDE), Defence Research and Development Organisation (DRDO), Kanpur, Uttar Pradesh, India
e-mail: neswarap@gmail.com

 
E. P. Namburi et al. (eds.), *Novel Defence Functional and Engineering Materials (NDFEM) Volume 1*, Indian Institute of Metals Series,
https://doi.org/10.1007/978-981-99-9791-6_8

The inclusion of functional additives introduces unique properties to the paints and coatings. Functional paints or coatings serve specific purposes that are unavailable from traditional materials. However, the additives must (i) be compatible with the other coating ingredients, (ii) not affect general coating performance, (iii) be stable under typical storage conditions, and (iv) be capable of responding to the appropriate environmental stimuli quickly and reversibly. Also, with the current demand for green technology, the additives and manufacturing processes must be environmentally friendly.

## 8.2 Materials

Paints are often produced by mixing a binder (in a liquid medium) and solvents together with insoluble powders of pigments, extenders and specialty additives to form a premix. The pigments and extenders consist of primary particles in small agglomerations, and these are reduced, wetted (with a binder) and mechanically dispersed by grinding. After grinding, the paint can be diluted by using a solvent. There are many types of paints and coatings, e.g. organic solvent-based, waterborne, distempers, emulsions, and powder coatings. Each type is called a *formulation*. The nature and composition of the principal constituents decide the final properties of a paint or coating.

### *8.2.1 Binders*

Binders are either natural or synthetic organic resins that produce a continuous hard film adhering to the substrate. Synthetic resins are more generally used, since their properties can be 'tailored' by structural modification or control of the molecular weight. Some examples of natural resins are wood rosin and mastic-amber. Synthetic resin examples for paints and coatings are alkyds, phenolics, polyesters, epoxies, amino resins, acrylics and polyurethanes (Chen et al. 2011). Some commonly used synthetic resins are elaborated in the following subsections.

#### 8.2.1.1 Alkyd Resins

Among many, alkyd resins found some of the first applications as synthetic paints. These resins incorporated oil or oil-derived fatty acids into a polyester polymer structure. Major advantages of these resins include enhanced drying speed, durability, and superior properties. Many new and better resins emerged later and the paints based on the more sophisticated polymers served for specific or selective applications, for example, the long oil alkyd, optionally blended with other (modified) alkyds.

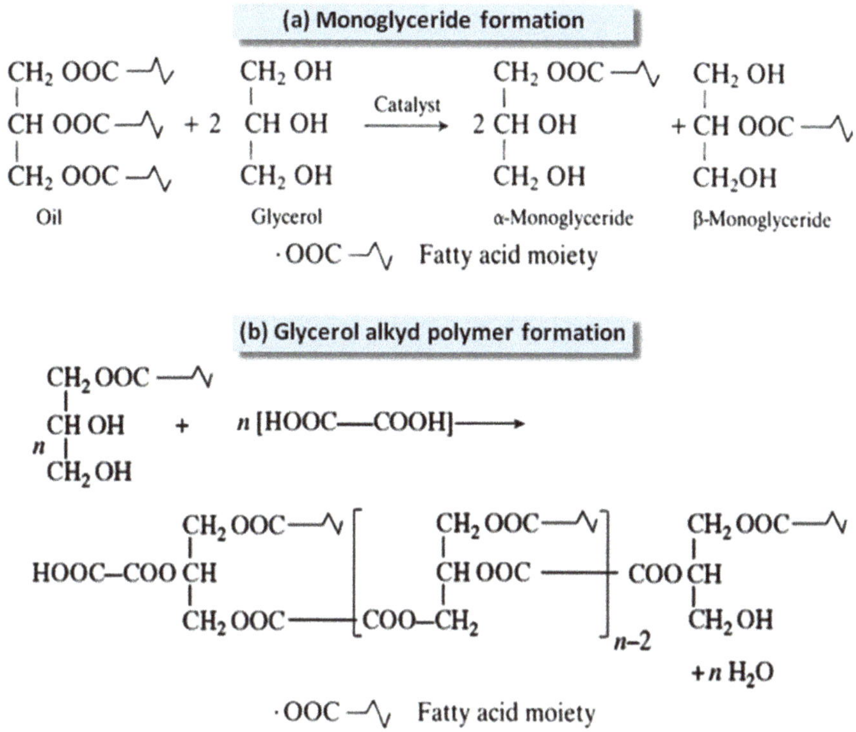


**Fig. 8.1** Schematic representations of **a** monoglyceride and **b** glycerol alkyd polymer formation

The salient aspects of synthesis of the alkyd resin based paints include: (i) Monoglyceride preparation, in which process, the glycerol (or another polyol) and oil are reacted together in the molar ratio of 2: 1 at around 240 °C in the presence of a basic catalyst (sodium hydroxide, lithium hydroxide, etc.) to form 'monoglyceride' (see Fig. 8.1a); (ii) Preparation of the final resin is carried out by further addition of polyol and a dibasic acid in the so-called "bodying stage". In this stage, multiple esterification reactions occur (Fig. 8.1b), and the water that forms as a by-product is removed from time to time. Yet another simpler process called 'fatty acid process' is also practiced, in which the fatty acids (not oil) are the major reactants. This process is widely used for shorter oil alkyds. These alkyd paints are the common printing inks, coatings for the decorative as well as industrial water-soluble and electro-deposition systems.

#### 8.2.1.2 Phenolic Resins

Phenolic resins are of two types, Novolac and Resol, see Fig. 8.2. Novolacs are produced when the molar ratio of formaldehyde to phenol is less than one, and an

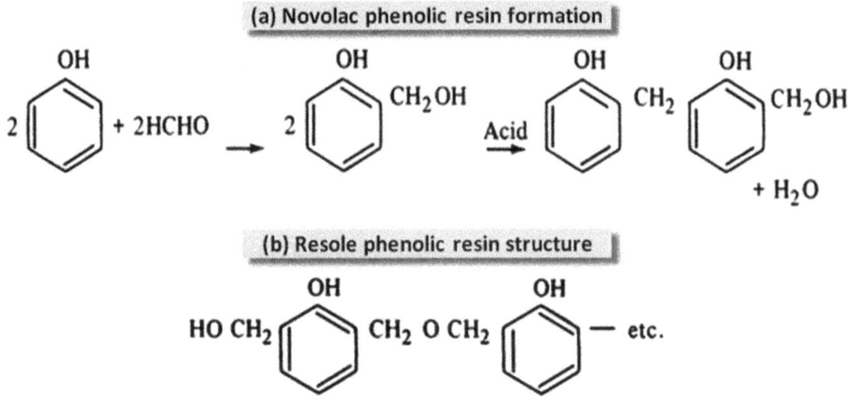


**Fig. 8.2** **a** Novolac phenolic resin formation and structure, and **b** Resole phenolic resin structure

acid catalyst is used in the synthesis. However, Novolacs are insoluble in oils and hydrocarbon solvents and are not used for surface coatings.

Resole types of phenolic resins are produced when the molar ratio of formaldehyde to phenol is greater than one, and an alkali catalyst like sodium hydroxide or lime is used. Resoles are thermo-setting resins and soluble in oils, and they are widely used in varnish making. Also, resoles react with rosin to give products that can be esterified with polyhydric alcohols: the resulting resins are soluble in oils. These are known as rosin-modified phenolics and find extensive applications in decorative undercoats, primers, marine paints and certain types of printings inks.

#### 8.2.1.3 Polyester Resins

Polyester resins typically consist mainly of co-reacted di- or polyhydric alcohols (polyols) and di- or tri-basic acids or anhydrides, and are thinned with normal solvents. Depending upon the raw materials used, polyesters can be either saturated or unsaturated (Zhan et al. 2013; Shi et al. 2005; Cherian and Thachil 2004; Cherian et al. 2006; Awasthi and Agarwal 2010). Unsaturated polyesters are cured by a free-radical-initiated addition reaction at elevated temperatures. Unsaturated polyesters are generally dissolved in a reactive solvent such as styrene prior to the production of the final resin. Saturated polyesters with high hydroxyl content are used in the production of polyurethanes.

In choosing polyols, two problems that affect durability must be considered:

- Steric factors and the 'neighboring group' or 'anchimeric' effects affect resistance to hydrolysis.
- The presence of hydrogen atoms on the carbon atom beta-to-hydroxyl group and subsequent ester group is detrimental to the resistance of the ester link to breakdown by heat or radiation.

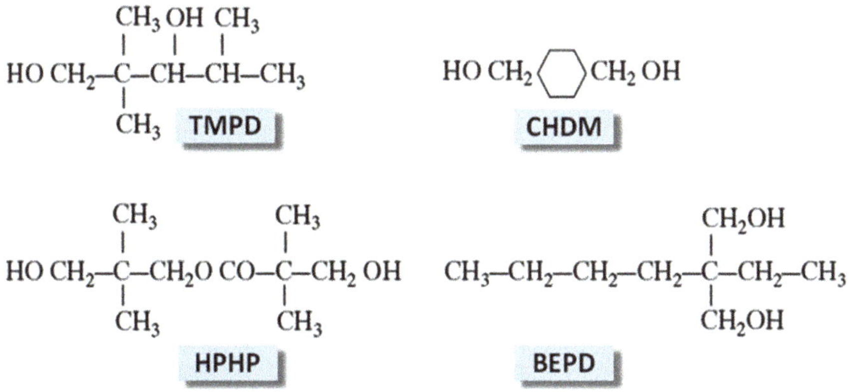


**Fig. 8.3** Chemical structures of some diols (with hindered beta positions relative to OH groups)

These problems are avoided by polyols in which the beta hydrogen content is reduced or hindered: examples are 1,4-cyclohexane dimethanol (CHDM); 2,2,4-trimethyl-1,3-pentane diol (TMPD); trimethylol propane (TMP); neopentyl glycol (NPG); hydroxypivalyl hydroxy pivalate (HPHP); and 2-butyl-2-ethyl propane diol (BEPD). Figure 8.3 shows the structures of some of these polyols, which in these cases are diols.

#### 8.2.1.4 Epoxy Resins

Epoxy resins are obtained by condensation of epichlorohydrin and 2, 2'-bis(4-hydroxyphenyl) propane in the presence of an alkali. The resulting product is known as the diglycidylether of bisphenol A (DGEBA), shown in the lower part of Fig. 8.4). Since epoxy resins contain both hydroxyl and epoxy groups, they can be cured either at room temperature or by heating with amino and phenolic resins, amines, anhydrides, polyamides and isocyanates. As coatings epoxy resins are resistant to chemicals and display high adhesion to metallic and other surfaces. They are widely used for coating metallic parts (Azcan et al. 2011; Qian et al. 2009; Bajpai and Bajpai 2010).

#### 8.2.1.5 Amino Resins

These are obtained by reaction between formaldehyde and either urea or melamine to produce urea–formaldehyde or melamine–formaldehyde resins, see Fig. 8.5. They are not used as such for coatings owing to brittleness. Instead they are used in combination with alkyd resins. Amino resins used in the surface coating industries are modified with butanol to enable solubility in normal coating solvents and better compatibility with other resins like saturated polyesters, epoxies and acrylics, hence participating

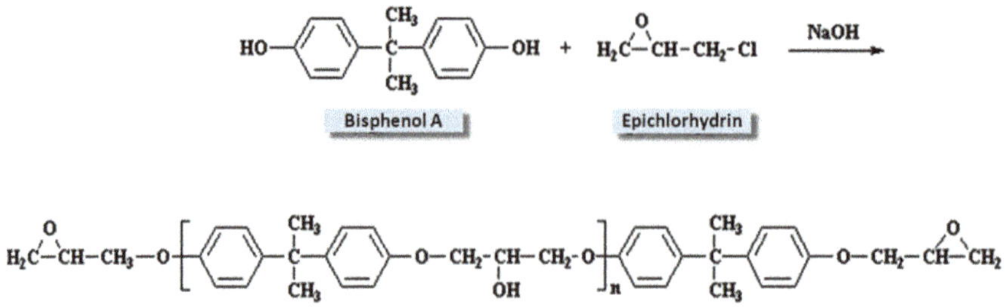


Fig. 8.4 Reaction mechanism of epoxy resin

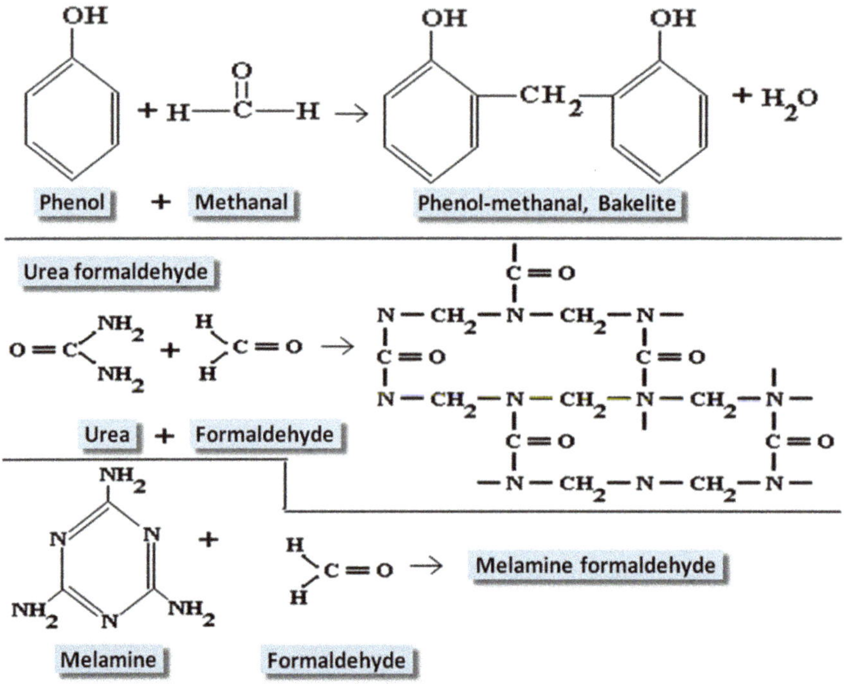


Fig. 8.5 Reaction mechanisms of amino resins

in co-curing systems. Melamine formaldehyde shows better properties than urea–formaldehyde in terms of colour retention and exterior durability. Also, amino alkyd systems are widely used in stoving finishes.

#### 8.2.1.6 Polyurethane Resins

Polyurethanes are based on reactions between a diisocyanate and compounds containing an active hydrogen atom to produce urethane linkages, as shown in

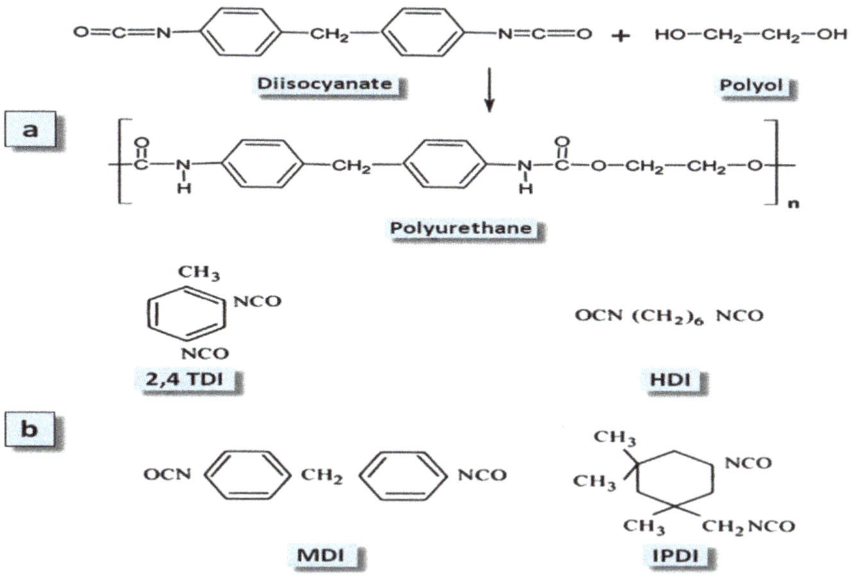


**Fig. 8.6** **a** Reaction mechanism for forming a polyurethane resin, and **b** some di-isocyanates used

Fig. 8.6. There are many compounds that contain active hydrogen, including water, alcohol, and amines containing hydroxyl groups. Thus a polyurethane coating may contain ester, ether, amide, urea, or other groups. The hydroxyl component may be polymeric polyols like polyesters and polyethers, and the di-isocyanates can be tolylene di-isocyanates (TDI), diphenyl diisocyanate (MDI), hexamethylene di-isocyanate (HDI) and isophorone di-isocyanate (IPDI).

Polyurethanes, either in a single pack or as a two-pack system, can undergo low temperature curing, and they provide a wide range of flexibility and hardness in the final products, e.g. good adhesion and excellent weathering resistance, besides resistance to moisture (Niihama and Yoshioka 2004), chemical attack and solvents (Popov et al. 2005; Crawford and Escarsega 2000).

#### 8.2.1.7 Acrylic Resins

Acrylic resins are yet another primary binder used in a wide variety of industrial coatings, the details of which are discussed at length by Oldring and Lam (Oldring and Lam 1996). The process involving vinyl/acrylic polymerization is particularly versatile compared to condensation polymerization and thus provides far more possibilities for controlling the polymer architecture and special features.

#### 8.2.1.8 Classification of Binders

Table 8.1 lists some important binders and their classification according to the chemical reactions involved (Wulf et al. 2002).

### 8.2.2 *Pigments*

Pigments are insoluble solids added to a resin formulation to provide a stable colour to coatings and resistance to wear and weathering. The pigments can be inorganic or organic. Inorganic pigments are usually metal oxides such as lead oxide, chromium oxide and cobalt blue. Figure 8.7 shows the general effects of the amount of pigment on gloss and moisture resistance. Clearly, there will often be compromises. Table 8.2 lists some important pigments, their classification, and functionality (Qian et al. 2009).

**Table 8.1** Classification of binders according to their chemical reactions

| Classification | Examples | Chemical reactions |
|---|---|---|
| Oxygen reactive binders | Alkyd | The binder molecules react with oxygen, and cross-linking of the resin molecules takes place |
| | Epoxy esters | |
| | Urethane alkyds | |
| Lacquers | Polyvinyl chloride polymers | Drying mechanism by solvent evaporation. The long-chain resins entangle with each other but no cross-linking occurs |
| | Chlorinated rubbers | |
| | Acrylics | |
| Heat conversion binders | Hot melts | Curing takes place upon heating as the components melt. Both cross-linked and non-cross-linked coatings are possible |
| | Organisols and plastisols | |
| | Powder coatings | |
| Co-reactive binders | Epoxies | The coating is formed by a polymerization between the resin and a curing agent. A 3D network is formed |
| | Polyurethanes | |
| Inorganic binders | Post-cured silicates | The binders are usually used in zinc dust pigmented primers, where a reaction between zinc and the binders takes place, resulting in a very hard coating |
| | Self-curing water silicates | |
| | Self-curing solvent based silicates | |
| Coalescent binders | Latex | Coatings form by coalescence of binder particles dispersed in water |

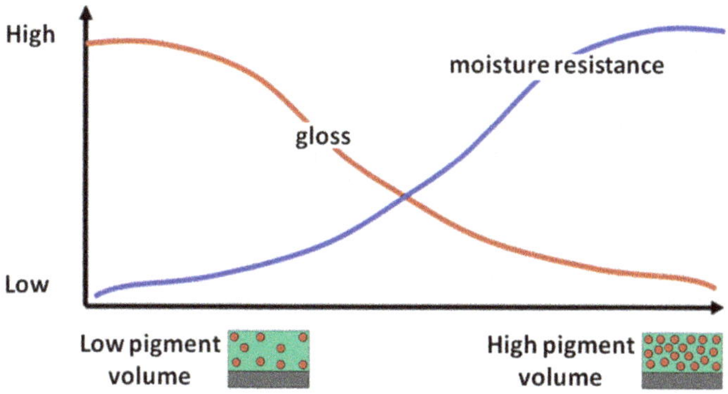


**Fig. 8.7** Effects of pigment on gloss and moisture resistance (Chen et al. 2011)

**Table 8.2** Classification of pigments according to their functionality

| Classification | Examples | Chemical reactions |
|---|---|---|
| Colour pigments | Titanium dioxide, iron oxides, organic azo pigments | Provide colour to the paint. Titanium dioxide is the most popular white pigment because of its high refractive index |
| Inhibitive pigments | Zinc phosphate, aluminium phosphate, zinc molybdate | Provide active corrosion inhibition to the metal substrate. The pigments are slightly water soluble. The dissolved ionic species react with the metal to form passivating reaction products |
| Barrier pigments | Aluminium flake, micaceous iron oxide | Increase the moisture permeation path length to the substrate |
| Sacrificial pigments | High purity zinc dust | Function as sacrificial anodes that provide cathodic protection of the substrate metal |
| Hiding pigments | Rutile titanium dioxide, zinc oxide | Pigments with high light-refractive index to provide good coverage |
| Extender pigments | Carbonates, silicates, sulphates, barites and mica | Act as reinforcement and flow control pigments. They are relatively inexpensive |

**Table 8.3** Classification of solvents with descriptions of their applications (Qian et al. 2009; Bajpai and Bajpai 2010)

| Classification | Example | Uses |
|---|---|---|
| Aliphatic hydrocarbons | Naphtha, mineral spirits, hexane, heptanes | Used with asphalt, oil and vinyl based coatings. Poor to moderate solvency and a wide range of evaporating rates. Least expensive of all solvents |
| Aromatic hydrocarbons | Toluene, xylene | Used with chlorinated rubbers, coal tars and certain alkyds. Greater solvent power than aliphatic hydrocarbons |
| Ketones | Acetone, methyl ethyl ketone, methyl isobutyl ketone | Used with vinyls and some epoxies. Varying evaporation rates and relatively strong solubility parameters. Strong hydrogen bonding and high polarity |
| Esters | Ethyl acetate, isobutyl acetate, ethylene glycol | Used as latent solvents with epoxy and polyurethane solvency power between those of aromatic hydrocarbons and ketones. Strong hydrogen bonding and a relatively high polarity |
| Alcohols | Ethanol, iso-propanol, n-butanol | Good solvents for highly polar binders such as phenolics. Alcohols are highly polar with a strong affinity for water |
| Ether and alcohol ethers | Ethyl ether | Excellent solvents for some natural resins, oils and fats |
| Water | Water | Used in latex paint |

### 8.2.3 *Volatile Components (Solvent)*

Solvents are used in paint to reduce its viscosity. After application the solvent should be fully evaporated from the coating. With the exception of water in waterborne paints, the emulsion types of paints use solvents that are volatile organic liquids, e.g. hydrocarbons, aromatic hydrocarbons, alcohol, esters, and ketones. Table 8.3 gives a list of solvents and their uses in paint formulations.

### 8.2.4 *Additives*

Additives are chemical compounds used to improve and modify the properties as well as the performance of the paints. For example, they are used to stabilize the colour on painted surfaces, and control viscosity, surface and interfacial stress, shine and drying properties of the coatings. The most common additives and their functions are listed in Table 8.4.

**Table 8.4** Classification of additives according to their functions in the coatings

| Classification | Examples | Functions |
|---|---|---|
| Antifoam additives | Mineral oil, silicone oil, wax dispersions | Prevent the formation of foam, e.g. by coalescing small bubbles into larger bubbles, which increases their buoyancy |
| Thickness (viscosity) | Bentonite, cellulose derivates, polyacrylates | Increase paint viscosity by creating networks between hydrophobic and hydrophilic parts of the paint |
| Dispersion additives | Tensides (surfactants) | Increase the wettability of pigments with the binder phase by formation of micelles |
| Siccatives (drying agents) | Metal salts of organic acids | Used in paints with oxygen-reactive binders. Improves the curing process |
| Cold stabilizers | Ethylene glycol, propylene glycol | Used in water-based paint to improve the stability to freezing |

## 8.3 Material-wise Classification of Surface Coatings

Surface coatings can be classified into two general types, convertible (thermosetting) and non-convertible (thermoplastic). Further classification is also done, particularly for convertible coatings based on (i) their curing mechanism (air-dry, stoving etc.), (ii) their performance characteristics (heat, corrosion resistance), and (iii) the order of application, i.e. based on the methods familiar to professional decorators, industrial finishers and the general public.

### *8.3.1 High-Solids Coatings*

High-solid coatings are environmentally friendly due to minimum emission of organic solvents compared to conventional (alkyd) paints. However, high-solid coatings have low binder contents, and therefore the chances of their auto-oxidation due to cross-linking are high. Also, the high solid contents result in a thicker applied coating and make thorough drying more difficult (Bieleman 2000).

### *8.3.2 Waterborne Coatings*

Different categories of waterborne coatings are water-reducing coatings, latex coatings, emulsion coatings and aquatic dispersions of resins in polyurethane latex. These coatings have no emission or disposal problems, and normally the water surface tension is more than that of an organic solvent.

### 8.3.3 High Build Coatings

High build coatings are intended to enable high dry thickness with a minimum number of coats. This is achieved by incorporation of a gelling agent in the coating composition, thereby allowing the application of thick layers without the risk of vertical sagging.

### 8.3.4 Solvent-Free Coatings

Solvent-free coatings are based on using low viscosity liquid resins when the presence of solvents is undesirable, e.g. coatings applied to the interiors of tanks. These coatings are also suitable for the submerged parts of steel and concrete structures. They have a high dry thickness and formulated to cure either by (i) chemical reaction (two-pack types such as epoxy and polyurethane) or (ii) drying slowly by solvent evaporation (non-convertible types such as chlorinated rubber) (Zovi et al. 2011).

The reasons for developing solvent-free paints are (a) reduction of atmospheric pollution, (b) savings in material and transportation costs, (c) savings in energy involved in paint production, (d) time saving owing to higher layer thicknesses per application cycle, and (e) improved safety.

## 8.4 Functional Classification of Coatings

Coatings are usually applied first as a primer followed by a topcoat. In some cases, e.g. motor vehicles, the entire coating can vary from four to six layers. Each layer has specific functions, although the overall performance of a multi-coat system also depends on the properties of the interfaces between the layers (Hegedus 2004; Verkholantsev 2003a).

### 8.4.1 Functional Coatings

Besides the normal paint properties, functional coatings have diverse special properties, depending on the intended applications and substrates (Wulf et al. 2002). There are various types of functional coating, e.g. based on inorganic, organic or ceramic materials, but the scope of the present chapter is limited to organic coatings. Typical examples are self-cleaning (Parkin and Palgrave 2005a; Nun et al. 2002), easy-to clean (anti-graffiti) (Kuhr et al. 2003), antifouling (Perez et al. 2003), soft feel, and antibacterial coatings (Tiller et al. 2001).

The coating functionalities depend on their physical, chemical, mechanical and thermal properties. Examples are coatings resistant to scratches, friction (wear) and thermal effects. Some applications and types of functional coatings are discussed in subsections 8.4.1.1–8.4.1.9.

#### 8.4.1.1 Camouflage Coatings

Use of camouflage and stealth technologies is especially to protect military manpower and strategic equipment from detection and identification by reconnaissance and surveillance, as well as prevent destruction from programmable weapon systems with advanced sensors. Stealth technology aims to manipulate the signature of the targets by the following methods:

- Divert emissions and reflections from target surfaces.
- Reduce the energy reflected or emitted from the target by surface treatment.
- Alter the frequency response using carefully selected materials.
- Absorb or deflect away incident radiations coming from active sensors by using surface coatings.
- Attenuate electromagnetic radiation within the atmosphere using obscurants.
- Use of target covers that blend with the surroundings and background.
- Alter the apparent shape of a target or its key revealing features with add-on devices, and blind sensors, e.g. by dazzling.

#### 8.4.1.2 Visual–Laser Camouflage Coatings

The visible electromagnetic spectrum (wavelengths between 0.4–0.7 μm) interacts with the surface of an object and is partially absorbed. The rest is reflected, transmitted or scattered. The scattered energy incident on a sensor is used for sensing and identifying the object. The human eye is the best-known imaging sensor in the visible spectrum. However, other imaging sensors are frequently used (i) in passive mode, relying on sunlight or diffuse illumination, or (ii) with active mechanisms, e.g. laser-designated weapon systems. The spectral response of an object owing to its surface condition, shape, size, and shadow, is the key factor determining the overall surface characteristics. Figure 8.8 is an example of modifying the spectral response using camouflage.

***Visual and near-infrared stealth efficiency***: The stealth efficiency of any coating is highly dependent on the scattering power. For a fixed value of the scattering power m, the wavelength ($\chi$) most efficiently scattered by a pigment particle of diameter (d) is given by Eq. (8.1):

(a) (b)

**Fig. 8.8** **a** Defence strategic objects show contrast with the surroundings, and **b** objects visually merged with the surroundings

$$\lambda = \frac{d}{\kappa} \tag{8.1}$$

$$where\ \kappa = \frac{0.90(m^2 + 2)}{n\pi(m^2 - 1)} \tag{8.2}$$

and m is the scattering power, d is the particle diameter, $K$ is the coefficient of absorption.

Optimal HID capacity (ability to mask the substrate) is achieved when the relative refractive index is as high as possible and the pigment's particle size is approximately 1/3 to 1/2 the wavelength of light. In the case of sunlight, it is important to optimize the particle size distribution to increase the scattering related to the long-wavelength component of the incident light, i.e. up to 1.2 μm for covering the near-infrared area of the electromagnetic spectrum.

***Laser threat avoidance***: Modern military scenarios envisage extensive use of lasers to detect targets. It is therefore necessary to develop laser-absorbing materials and coating systems to counter laser threats. The relation between the range and reflectivity ability of lasers (Pr) under clear air the conditions is given by Eq. (8.3):

$$\mathbf{Pr} = \mathbf{P_t e} - ^{\mathbf{2\alpha R}} . \mathbf{A_T} . \rho . \mathbf{A_R} . \mathbf{T_R} / \mathbf{A_B R^2} \tag{8.3}$$

where $P_t$ = transmitted power (W), R = detection range in metres (m), α = atmospheric attenuation per kilometre ($km^{-1}$), ρ = target reflectivity, $T_R$ = receiver's transmission, $A_T$ = target area ($m^2$), $A_R$ = receiver area ($m^2$), and $A_B$ = beam area ($m^2$)

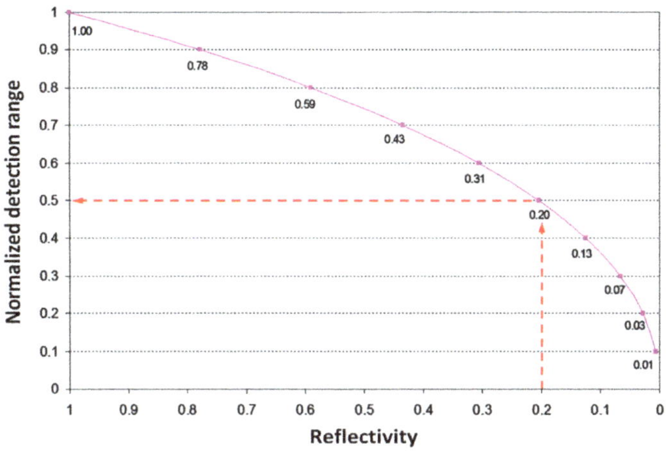


**Fig. 8.9** Normalized range versus laser reflectivity under clear weather conditions. [The arrows show 80% reduced reflectivity is needed to reduce the detection range by 50%]

$$\text{Then} \quad R^2 \propto \rho e^{-2\alpha R} \quad or \quad \rho \propto R^2 e^{2\alpha R}$$
$$\text{and} \quad \frac{\rho_2}{\rho_1} = \frac{R_2^2 e^{2\alpha R2}}{R_1^2 e^{2\alpha R1}} \tag{8.4}$$

$$\text{For } \rho_1 = \mathbf{1}, \ \text{ and } \mathrm{R1} = \mathbf{1}, \ \rho_2 = \left(\mathbf{R2^2 e^{2\alpha R2}}\right) / \mathbf{e^{2\alpha}} \tag{8.5}$$

Figure 8.9 shows normalized detection range values as functions of laser reflectivity under clear weather conditions (α = 0.2). The curve shows that a reflectivity reduction of about. 80% is required for a set target of 50% reduction in the detection range.

#### 8.4.1.3 Corrosion Protection Coatings

The term 'corrosion' usually refers to metals, although non-metallic substrates, such as plastics, concrete and wood also degrade by prolonged exposure to the environment. Metallic corrosion is an electrochemical process involving an anode (corrosion site), an electrolyte (corrosive medium), and a cathode (Verkholantsev 2003b). In general, organic coatings are applied to metal substrates to avoid the harmful effects of corrosion. The performance of corrosion protection coatings depends on many parameters, e.g. adhesion between coating and metal, and the coating's thickness and permeability to oxygen. In most cases a primer is used before applying the topcoat; and before that the metal surface is often pre-treated by physical etching and chemical treatments, both for corrosion protection and good adhesion with the paint system. A summary of the mechanisms by which organic coatings offer corrosion protection to metals and alloys is given below:

(1) *Sacrificial*: The electrochemical sacrificial behaviour of zinc in contact with steels is well-known. Various zinc-rich biological resin based coatings have been reported to protect metal substrates (Kouloumbi et al. 2005, 2003).
(2) *Barrier effect*: Polymeric coatings are impenetrable and act as barriers to corrosive species. Even so, additional protection can be obtained by adding certain pigment and clay particles to the paint formulation.
(3) *Inhibition*: Chromate-and lead-based pigments have been-and still are-the most common corrosion inhibitors for metal substrates. However, these substances are toxic and new anti-corrosive compounds are being developed. These pigments form a protective oxide layer on metallic substrates e.g., aluminum zinc phosphate, calcium zinc moderate; zinc molidut phosphate, calcium borozelate and strontium phosphor silicate.

Organo-functional silanes have emerged recently as an alternative to chromate treatment, owing to their environmental friendliness and good anti-corrosion properties (Palanivel and Ooij 2004). However, use of a silane requires that the substrate must have surface hydroxyl groups, either intrinsic or by surface modification.

Other recent developments are the use of (i) intrinsically conductive polymers (ICPs) for metal protection (Kendig et al. 2003; Kraljic et al. 2003); (ii) core–shell types of materials, e.g. a ferric oxide core with a shell of zinc phosphate or anticorrosive titanium dioxide coated with an organic polymer primer (Kumar et al. 2004); (iii) smart features, such as colour-change pH indicators, in the paint, and (iv) nanoclay, which can exchange anticorrosive agents with the corrosive species, thereby preventing corrosion.

#### 8.4.1.4 Thermal Resistant and Fire Resistant Coatings

***Thermal resistance***: Commonly used thermal resistant coatings for metal substrates are fluorine or silicon-based compounds. Their applications include non-stick cookery ware, barbecues and boilers. Phenolic-or epoxy-resin-based thermal-resistant coatings are also popular. However, fluorinated coatings are toxic above 300°C, and for high temperature applications (up to 1000°C) silicon resins or inorganic silicates dominate the market. More recent reports on thermal-resistant coatings, cheaper than silicones, include the use of titanium esters and titanium–aluminum combinations to increase thermal resistance up to 800°C.

***Fire resistance***: Indefinite protection against fire is impossible, but fire-resistant coatings can delay the spread of a fire. There are two main types:

- Intumescent coatings. These rely on expansion of carbon-coated layers which act as protective barriers against heat transfer, and hinder the dissemination of combustible gases and melted polymers. These coatings are composed of three components: (i) an inorganic acid (dehydration agent); (ii) the material that creates a carbonaceous char; and (iii) a blowing agent (Camino and Delobel 2000; Labuschagne 2003). Expandable graphite is also available: this contains chemical compounds, including acids, trapped between organic layers. In contact with high temperatures, the graphite is partitioned and provides an insulating layer for the substrate.

- Fire retardants. There are many types of phosphorus-and halogen-based fire-retardant systems. The combination of polyurethane and phosphate is a common one. However, the solubility of phosphate in water leads to a leaching-out problem. This difficulty can be avoided by chemically anchoring phosphate (D-ammonium hydrogen phosphate) into the polyurethane chains/shell (Giraud et al. 2002). Silicon or inorganic hydroxide-based fire-retardant coatings as well as polymer clay (layered silicates) nanocomposites have also been reported to have superior fire retardation (Kashiwagi and Gilman 2000; Horn 2000; Zanetti et al. 2000).

#### 8.4.1.5 Scratch and Abrasion Resistant Coatings

Coatings are susceptible to damage by scratches and abrasion. Paints and primers used on automobiles must have good scratch and abrasion resistances to prevent damage to the metal substrate. Scratch resistance can be improved by higher cross-linking in the binder resin. However, a highly cross-linked hard film has poor impact resistance, hence a compromise is required between hardness and flexibility. In this context, organic–inorganic hybrid coatings are paving the way for improved scratch-resistance. Recent advances in nanotechnology also play an important role in the development of scratch-resistant coatings (Chantarachindawong et al. 2012; Baer et al. 2003): Gläsel et al. have used siloxane encapsulated $SiO_2$ nanoparticles to develop scratch and abrasion-resistant coatings (Gläsel et al. 2000; Mehnert et al. 2001). Coatings with good abrasion and scratch resistant properties have also been reported by others (Thomas et al. 2001).

#### 8.4.1.6 Self-Cleaning Coatings

Self-cleaning functional coatings do not need manual cleaning. A shower of rain is sufficient to carry out the cleaning process. The self-cleaning action of such surface coatings is shown schematically in Fig. 8.10.

Discovery of the “Lotus Effect”, i.e. the self-cleaning property of lotus leaves owing to their specialized surface morphology and hydrophilicity, has inspired the development of self-cleaning coatings (Nakajima et al. 2001; Blossey 2003; Parkin and Palgrave 2005b). During the past few years, self-cleaning coatings using photo-catalytic titanium dioxide ($TiO_2$) have attracted considerable attention. When $TiO_2$ particles are illuminated with ultraviolet light, electrons are promoted from the valence band (VB) to the conduction band (CB), which creates positive charge ($h+$) holes in the VB and a free electron in the CB. These charge carriers can either recombine or migrate to the surface, while the holes can react with hydroxyl or adsorbed water molecules on the surface and produce different radicals such as

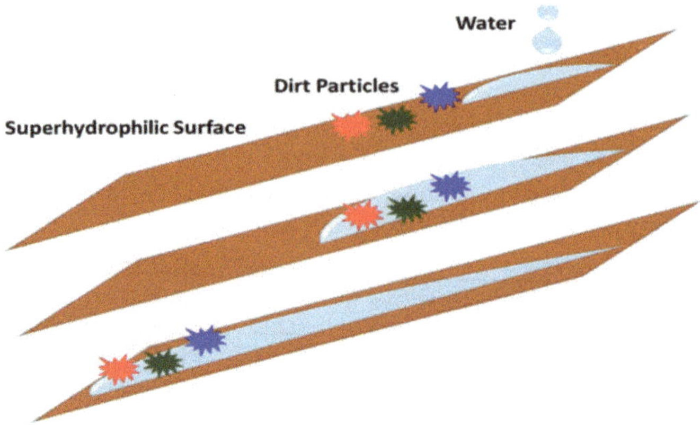


**Fig. 8.10** Self-cleaning action of coatings

hydroxyl radicals (OH·) and hydroperoxy radicals ($HO_2$) (Cheng 2005). These radicals are powerful oxidizing species and cause deterioration of organic contaminants or microbial species on the coating surface. The other beneficial effect of $TiO_2$ is its superhydrophilic behaviour, commonly known as the "water sheathing effect". This allows contaminants to be easily washed away with water or rainfall, as already shown in Fig. 6.10. The addition of silicon oxide together with $TiO_2$ has shown synergistic self-cleaning properties (Guan 2005).

#### 8.4.1.7 Antibacterial Coatings

Organic coatings are susceptible to microbial attack, and microbial growths on them have adverse consequences, including aesthetics (discoloration) and risks to health and hygiene. A wide variety of organic and inorganic biocides are available to prevent this type of attack. For example, biocides containing heavy metal ions function by penetrating cell walls and inhibiting the bacteria's metabolic enzymes; whereas antimicrobial agents with cationic surfaces cause rupture of the bacteria's cytoplasmic membranes. Examples of organic biocides include polymers, tertiary alkyl amines and organic acids (Tholmann et al. 2003; Sauvet et al. 2000). Inorganic biocides include silver, zinc oxide (ZnO), copper oxide (CuO), titanium oxide ($TiO_2$), and selenium (Trogolo et al. 2003; Mathiazhagan and Joseph 2011). Microcapsules containing biocides have also been developed in order to increase the longevity and efficiency of antimicrobial coatings (Xu et al. 2003; Shan et al. 2011).

#### 8.4.1.8 Antifouling Coatings

Marine organisms represent a major threat to ships and boats, owing to their attachment and growth to the hulls, which is known as "fouling". This is more prominent in coastal waters, where vessels are either docked or travel slowly. Depending upon the organisms involved, the fouling can be of two types, namely micro-fouling and macro-fouling. Both biocidal and non-biocidal coatings are used to prevent fouling.

Biocide antifouling coatings function by slow leaching of the biocides from the coatings. Owing to their toxicity the use of biocides is restricted: for example, tributyl tin (TBT) is a highly efficient marine biocide but it is no longer used. For long-term antifouling effects, either controlled-release or contact-active biocides are required (Wolfrum et al. 2002). Recently, a number of antifouling products have been developed using micro-encapsulation technology (Kamtsikakis et al. 2017; Kim et al. 2001).

The non-biocidal approach uses polymers with low surface energy to avoid the adhesion of marine organisms. Silicone elastomers are widely used for this purpose. However, this approach is effective only when the vessels move at relatively high speeds.

The applications for antifouling coatings are wide-ranging, and developments of special effect pigments offer new possibilities for their design (Edge et al. 2001; Thouvenin et al. 2002).

#### 8.4.1.9 Thermal Imaging Paints and Coatings

Thermal imaging technology has become one of the most valuable diagnostic tools for industrial applications. A thermal imager measures very small relative temperature differences in the infrared spectrum and converts otherwise invisible heat patterns into clear, visible images to be seen via a viewfinder or monitor. Currently these imaging systems are used in military as well as in industrial sectors. According to the type of radiation used, electro-optical imaging systems can be divided into two distinct groups: (i) imaging using the radiation emitted in wavelength between 3 and 15 μm by the targets, and (ii) imaging using the radiation reflected by the targets in the visible or NIR spectral range up to 1 μm.

To overcome the problems of confusion in detection, recognition and identification of strategic objects belonging to foes and friends, all strategic objects of defence are painted with general service camouflage paints that merge them with the immediate surroundings. Hence it is difficult to detect them. A suitable thermal imaging paint has been developed by using special grades of pigment and resin to provide a paint coating with different reflectivity and emissivity from general service camouflage coatings. The resultant differences in reflectivity and emissivity between the special and general service camouflage coatings enable discrimination between them.

## 8.5 Test Methods

Paints and coatings have to meet established quality standards with respect to viscosity, colour, gloss, and other required physical and functional properties. Finished coating tests are used to evaluate the physical properties of the cured coatings, and these tests are based on the American Society for Testing and Materials (ASTM) International Standards. Some performance tests are also based on other sources such as the American Architectural Manufacturers Association (AAMA) or European Union (EU), Bureau of Indian Standard (BIS) or as per International Standard Organization (ISO). Summaries of many of these tests are given in this Section.

### *8.5.1 Paint Viscosity*

Viscosity is determined by measuring the time required for a given quantity of paint to flow through a hole in the bottom of a metal cup. A number of cups, e.g. the Ford and Zahn types, with varying sizes and drain hole diameters are available for different viscometers. The recommended temperature for spray viscosity is 70ºF (21ºC). Figure 8.11 shows a schematic diagram and photograph of a Ford cup viscometer.

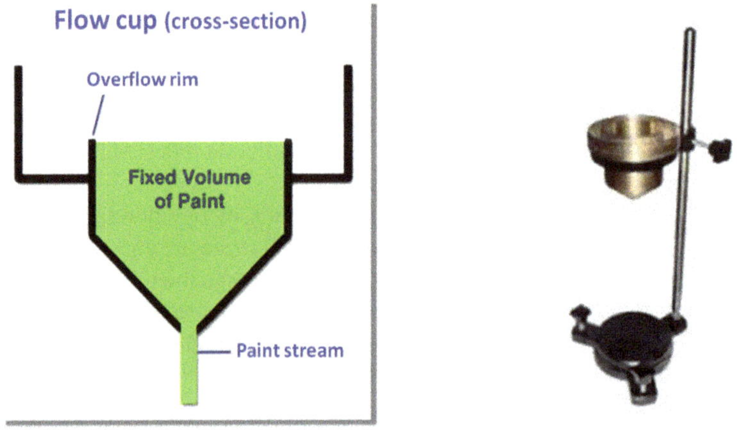


**Fig. 8.11** Schematic diagram and photograph of a ford cup. *Source* Sheen Instruments Co., Surrey, UK

### 8.5.2 Colour Matching

Colours must be consistent so that the same products are identical in appearance. In addition, many products have multiple components, sometimes with different substrates, and each component must be the same colour. Colour matching can be difficult because of varying and different lighting (e.g. fluorescent lamps versus outdoor light). Two light sources are often used to examine test panels, one predominantly blue and the other yellow. (It is important that the buyer and vendor agree upon the colour standards and comparison methods.) If standardized colour chips are used, they should be stored in darkness the dark to minimize fading. Elaborate colour measuring equipment is commercially available for determining the colour composition of a painted surface.

### 8.5.3 Gloss

The amount of light reflected from a paint or coating determines the gloss: more reflected light appears as higher gloss. Also, a smoother the surface reflects more incident light. The gloss is often measured with a photoelectric device. The reflected light is converted to an electric signal by a photovoltaic tube, and the strength of the signal is proportional to the amount of reflected light. This can be compared with a signal from the incident light. The angle of reflection may be varied to 20°, 45°, 60°, 90° or some other values, depending on the requirements.

The other important, but commonly measured properties for both specification as well as application, as also for selection of the paints are: (i) Coating Thickness, (ii) Hardness and (iii) Impact resistance. On the other hand, adhesion plays an important role in determining the which paint is suitable for some of the advanced applications:

### 8.5.4 Tape Adhesion Testing

Adhesion of a paint film to its substrate is often measured by abruptly pulling on an adhesive tape stuck over a scribed "X" or grid on the paint surface:

- Scribed "X" test. A numerical rating system from 0 (complete paint detachment, 0% adhesion) to 5 (no paint removal, 100% adhesion) is determined from this test.
- Scribed grid test (ASTM D 3359–97). The adhesion value is expressed by the percentage of squares showing loss of paint from aluminium substrates. The crosshatch pattern is subjected to boiling demineralized water for 20 min before being tested with the tape pull. This test method (AAMA 2604, 7.4.1.2) is used for outdoor environment evaluation, where the paint film would be subjected to water permeability and experience possible oxidation effects.

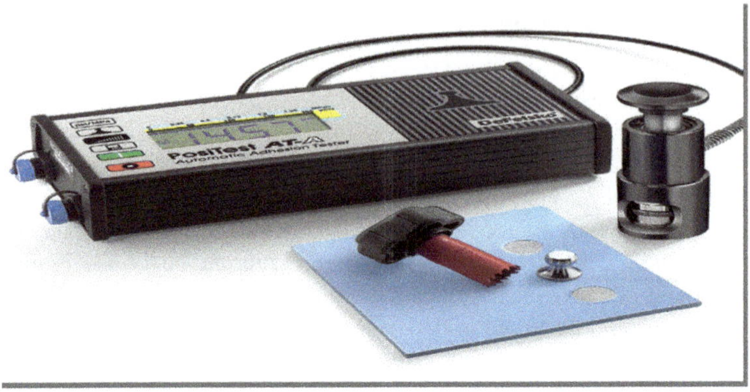


**Fig. 8.12** Automatic pull-off adhesion tester

Figure 8.12 illustrates an automatic pull-off adhesion tester that provides a consistent pulling-off of the adhesive tape.

## *8.5.5 Environmental Testing*

There are a number of environmental test methods, including humidity tests, salt spray tests, cyclic salt spray tests, UV + moisture tests, and outdoor exposure. These are discussed in this subsection.

### 8.5.5.1 Humidity Testing

Moisture can penetrate especially easily through pigmented primer coatings. When warmed, the moisture tends to vaporize and exert a pressure that causes the coating to swell. If the coating is flexible and the swelling is severe, there is a possibility of adhesion failure. Two examples of humidity tests are given here:

- Condensing humidity test. Water vapour is allowed to condense on sample panels. The condensed water drips off the panel and is revaporized by an evaporative heater at the bottom of the test chamber. Typical test conditions arewater temperature of 60ºC and a test duration of 24 h. After testing, the panels are checked for blistering, colour change, and loss of gloss.
- Constant humidity test. This involves exposing sample panels to 100% relative humidity at 38ºC for 24 h.

#### 8.5.5.2 Salt Spray Testing

The use of salt solution spray testing is intended to accelerate the corrosion process and cause early failure of the paint. Painted steel panels with "X" scribe lines are typically exposed for up to 14 days to a mist of 5% (w/v) aqueous sodium chloride solution at 33–36°C. The mist is produced by blowing hot saturated air through the salt solution. The panels are evaluated for two types of corrosion:

(1) Rust-through: The percentage of the surface that has rust visible through the paint.
(2) Creep: The distance in 1/32″ (0.8 mm) from the centres of the scribe lines to which the paint film breaks down and separates from the substrate. The results are measured on a scale with a predetermined distance adjudged to represent failure after the chosen exposure time.

Acetic acid is sometimes added to the salt spray solution to further accelerate the corrosion. Both neutral and acidified salt spray testing are used as a standard for paint and coating performance in service. However, one potential problem with salt spray testing is the steel substrate. Steel qualities vary substantially, and these variations may affect the paint and coating performances. Thus it is very important to always test standard steel panels as controls. Another important factor is the correct pretreatment of the paint film, which if not fully cured can cause premature failure.

#### 8.5.5.3 Cyclic Corrosion Testing

A cyclic corrosion test measures the behaviour of a paint or coating film under a combination of test conditions, e.g. salt, humidity, and changing temperature. The painted steel test panels are then rated against control tests on the basis of the percentages of rusted areas. Cycle corrosion testing is considered to be a more accurate way of predicting the comparative qualities of paints and coatings than straightforward salt spray testing. A typical cycle sequence (to be repeated as required) is 4 h in 5% neutral salt spray; 18 h at 38°C with 100% relative humidity; 2 h at–23°C.

#### 8.5.5.4 Quality UV (QUV) Testing

A quality ultraviolet (QUV) test chamber reproduces the damage caused by sunlight, rain, and dew on coated parts or test panels placed inside the chamber: the parts or test panels are subjected to alternating cycles of light and moisture at an elevated temperature. QUV testing measures the colour fastness and resistance to chalking, fading, cracking, blistering, embrittlement, strength loss, and oxidation. This test is more helpful than salt spray testing in evaluating the effects of outdoor exposure on different coating materials.

#### 8.5.5.5 Outdoor Exposure Testing

In outdoor exposure tests the test panels and samples are exposed to an outdoor environment for three months. The tests are slow but the best way to predict weatherability, *provided that the outdoor conditions represent the actual service environment*. The most common indications of paint and coating degradation caused by outdoor exposure are fading, cracking, chalking, blistering and peeling. Portions of panels are often buffed-up after testing to see how much of the original appearance can be restored.

## 8.6 Recent Developments in Paint and Coating Technologies

### *8.6.1 Volatile Organic Compound (VOC) Emissions*

Some of the major pollutants emitted by surface coatings are the organic solvents used during their processing and application. These are called volatile organic compounds (VOCs). It has been stated that the paint and coating industry contributes about 47% of total VOC emissions (Rengasamy and Mannar 2012). The detrimental effects of VOCs on the environment via chemical/photo-chemical reactions are well described in the literature (Annual Book of ASTM Methods 1993).

With the advent of sophisticated technology and functional and engineering materials, there is a trend to use more environmentally friendly alternatives to VOCs. New paint systems such as water-based coatings, UV and electron beam (EB) radiation curable coatings, high solids coatings, and powder coatings have come to the fore. UV curable coatings contain a photo-initiator (or photo-sensitizer), whereas EB curable coatings employ high energy electrons for ionization and excitation of resins.

The development of cost-effective technologies while staying abreast of environmental regulations is the main challenge for resin and paint manufacturers. Many possibilities to develop eco-friendly organic coating and paint systems are being investigated in the efforts to lower VOC emissions and reduce the risk to human health. Table 8.5 gives a summary of the basic technology concerned with eco-friendly coating systems, and also their relative merits.

**Table 8.5** Control of VOC emissions by eco-friendly coatings (Mathiazhagan and Joseph 2011)

| Coating types | % solvent reduction | Remarks |
|---|---|---|
| Water-borne coatings | 60–90 | Adapted to most equipment |
| High-solid coatings | 60–90 | Viscosity problem, limited systems |
| Powder coatings | 100 | High cost, need new equipment |
| Radiation curable coatings | Up to 100 | High cost, limited to flat surfaces |

#### 8.6.1.1 Camouflage and Stealth Coatings

Camouflage and stealth coatings are suitable for various regions of the electromagnetic spectrum (multi-spectral camouflage), from ultraviolet (UV) to near infrared (NIR), laser, thermal infrared, microwave and the radar region. Stealth and camouflage technology, also termed as LO (low observable) technology, is a subdiscipline of military tactics and passive electronic countermeasures for personnel, depots, tanks, aircrafts, ships, submarines, missiles and satellites to make them less detectable.

Development of camouflage/stealth technology aims to protect strategic military manpower and equipment from growing threats of detection, identification from reconnaissance and surveillance sensors, and destruction from programmable weapon systems using advanced sensor technologies. One of the greatest threats is posed by laser detection. Hence a priority is the development of laser absorbing paint and coating systems to counter the threats. The required visual camouflage shades giving 82–98% absorption of laser light have been successfully developed, and are suitable for a wide variety of terrains, including desert, semi-desert, green belts, snowbound areas and coastal regions. Some detailed information about visual and laser detectability is given in subsection 8.4.1.2.

## 8.7 Concluding Remarks

Development of functional paints and coatings is an important technological area of both commercial and strategic interest. This chapter surveys the science and technology of functional paints and coatings as well as some of their applications. The chapter also points out areas where more research is required to overcome new challenges, especially the development of eco-friendly paints and coatings. Military applications—for example, vehicles, artilleries and invisible radars—and aerospace products such as aircraft, satellites and solar panels, all involve the widespread use of coated materials. As worldwide demands for functional coatings continue to increase, new and cost-effective microencapsulation techniques will be developed, thus ensuring that functional paint and coating technology will remain at the forefront of research. At present, the major problem is to provide functional coatings which are easy to apply and have long-term stability. Consequently, it is expected that attention will be focussed on this topic.

## References

Annual Book of ASTM Methods. ASTM International (100 Barr Harbor Dr, Conshohocken, PA 19428, USA) Sec. 6, vol. 06.01, pp. 517–521 (1993)

Awasthi, S., Agarwal, D.: Preparation and characterisation of polyurethane coatings based on polyester polyol. Pigm. Resin Technol. **39**(4), 208–213 (2010)

Azcan, N., Demirel, E., Yılmaz, O., Erciyes, A.T.: Microwave heating application to produce dehydrated castor oil. Ind. Eng. Chem. Res. **50**, 398–403 (2011)

Baer, D.R., Burrows, P.E., El-Azab, A.A.: Enhancing coating functionality using nano science and nanotechnology. Prog. Org. Coat. **47**(3), 342–356 (2003)

Bajpai, P., Bajpai, M.: Development of a high performance hybrid epoxy silicone resin for coatings. Pigment. & Resin Technol. **39**(2), 96–100 (2010)

Bieleman, J.H.: Additives for Coatings, WILEY-VCH Verlag GmbH, D-69469 Weinheim (Federal Republic of Germany) (2000)

Blossey, R.: Self cleaning surfaces virtual realities. Nat. Mater. **2**, 301–306 (2003)

Camino, G., Delobel, R.: Intumescence. In: Grand, A.F., Wilkie, C.A. (eds.) Fire Retardance of Polymeric Materials, pp. 217–243. Marcel Dekker, New York NY 10016 (2000)

Chantarachindawong, R., Luangtip, W., Chindaudom, P., Osotchan, T., Srikhirin, T.: Development of the scratch resistance on acrylic sheet with basic colloidal silica ($SiO_2$)—methyltrimethoxysilane (MTMS) nanocomposites films by sol–gel technique. Can. J. Chem. Eng. **90**(4), 888–896 (2012)

Chen, L., Duan, Y., Liu, L., Guo, J., Liu, S.: Influence of $SiO_2$ fillers on microwave absorption properties of carbonyl iron/carbon black double-layer coatings. Mater. Des. **32**, 570–574 (2011)

Cheng, Y.T.: Microscopic observations of condensation of water on lotus leaves. Appl. Phys. Lett. **87**(19), 194112 (2005)

Cherian, B., Thachil, E.T.: Synthesis of unsaturated polyester resin–Effect of choice of reactants and their relative proportions. Int. J. Polym. Mater. Polym. Biomater. **53**(10), 829–845 (2004)

Cherian, A.B., Abraham, B.T., Thachil, E.T.: Modification of unsaturated polyester resin by polyurethane prepolymers. J. Appl. Polym. Sci. **100**(1), 449–456 (2006)

Crawford, D.M., Escarsega, J.A.: Dynamic mechanical analysis of novel polyurethane coating for military applications. Thermochim. Acta **357–358**, 161–168 (2000)

Edge, M., Allen, N.S., Turner, D., Robinson, J., Seal, K.: The enhanced performance of biocidal additives in paints and coatings. Prog. Org. Coat. **43**(1–3), 10–17 (2001)

Giraud, S., Bourbigot, S., Rochery, M., Vroman, I., Tighzert, L., Delobel, R.: Microencapsulation of phosphate: applications to flame retarded coated cotton. Polym. Degrad. Stab. **77**(2), 285–297 (2002)

Gläsel, H.J., Bauer, F., Ernst, H., Findeisen, M., Hartmann, E., Langguth, H., Mehnert, R., Schubert, R.: Preparation of scratch and abrasion resistant polymeric nanocomposites by monomer grafting onto nanoparticles. Macromol. Chem. Phys. **201**(18), 2765–2770 (2000)

Guan, K.: Relationship between photocatalytic activity, hydrophilicity and self-cleaning effect of $TiO_2/SiO_2$ films. Surf. Coat. Technol. **191**, 155–160 (2005)

Hegedus, C.R.: A holistic perspective of coating technology. JCT Res. **1**(1), 5–19 (2004)

Jr Horn, W.E.: Inorganic hydroxides and hydrocarbonates: their function and use as flame retarded additives In: Grand, A.F., Wilkie, C.A. (eds.) Fire Retardance of Polymeric Materials, pp. 285–332. Marcel Dekker, New York NY 10016 (2000)

Kamtsikakis, A., Kavetsou, E., Chronaki, K., Kiosidou, E., Pavlatou, E., Karana, A., Papaspyrides, C., Detsi, A., Karantonis, A., Vouyiouka, S.: Encapsulation of antifouling organic biocides in poly (lactic acid) nanoparticles. Bioeng. **4**(81), 1–18 (2017)

Kashiwagi, T., Gilman, J.W.: Silicon-based flame retardants. In: Grand, A.F., Wilkie, C.A. (eds.) Fire Retardance of Polymeric Materials, pp. 353–389. Marcel Dekker, New York NY10016 (2000)

Kendig, M., Hon, M., Warren, L.: Smart corrosion inhibiting coating. Prog. Org. Coat. **47**, 183–189 (2003)

Kim, J.C., Song, M.E., Lee, E.J., Park, S.K., Rang, M.J., Ahn, H.J.: Preparation and characterization of triclosan-containing microcapsules by complex coacervation. J. Dispersion Sci. Technol. **22**(6), 591–596 (2001)

Kouloumbi, N., Pantazopoulou, P., Moundoulas, P.: Anticorrosion performance of epoxy coatings on steel surface exposed to deionized water. Pigment. & Resin Technol. **32**(2), 89–99 (2003)

Kouloumbi, N., Ghivalos, L.G., Pantazopoulou, P.: Determination of the performance of epoxy coatings containing feldspars filler. Pigment. & Resin Technol. **34**(3), 148–153 (2005)

Kraljic, M., Roković, K., Kvastek, V., Radosevic, H., Duic, L.J.: Poly (ortho-ethoxyaniline) in corrosion protection of stainless steel. Corros. Sci. **49**(6), 2567–2580 (2003)

Kuhr, M., Bauer, S., Rothhaar, U., Wolff, D.: Coatings on plastics with the PICVD technology. Thin Solid Films **442**(1–2), 107–116 (2003)

Kumar, A., Stephenson, L.D.: Accelerated testing of self healing coatings. In: Ghosh, S.K. (ed.) Functional Coatings: By Polymer Microencapsulation: A General Perspective (2006). Wiley-VCH Verlag Gmbh & Co. KGaA, Weinheim (2004)

Labuschagne, F.J.W.J.: Metal catalysed intumescence of polyhydroxyl compounds. PhD Thesis, University of Pretoria, South Africa (2003)

Mathiazhagan, A., Joseph, R.: Nanotechnology-a new prospective in organic coating–review. Int. J. Chem. Eng. Appl. **2**(4), 225–237 (2011)

Mehnert, R., Hartmann, E., Gläsel, H.J., Rummel, S., Bauer, F., Sobottka, Elsner, C.: UV (EB) curable functional coatings. Mater. Sci. & Eng. Technol. **32**(10), 774–780 (2001)

Nakajima, A., Hashimoto, K., Watanabe, T.: Recent studies on super-hydrophobic films. Chem. Mon. **132**(1), 31–41 (2001)

Niihama, F., Yoshioka, N.: (to S.K. Kaken) Co., Ltd., Japan). Jap. Pat. JP. **308**, 407 (2004)

Nun, E., Oles, M., Schleich, B.: Lotus effect surfaces. Macromol Symp. **187**, 677–682 (2002)

Oldring, P.K.T., Lam, P.K.H.: Waterborne and Solvent Based Acrylics and Their End User Applications. John Wiley & Sons /SITA Technology Limited. London, UK (1996)

Palanivel, V., Van Ooij, W.J.: Silanes and other coupling agents. In: Mittal, K.L. (ed.), vol. 3, pp. 135–159. VSP, Utrecht, Netherlands (2004)

Parkin, I.P., Palgrave, R.G., Self-cleaning coatings. J Mater Chem. **15**(17), 1689–1695 (2005a)

Parkin, I.P., Palgrave, R.G. Self cleaning coatings. J. Mater. Chem. **15**(17), 1689–1695 (2005b)

Perez, M., Garcia, M., Del, A.B., Blustein, G., Stupak, M.: Core-shell pigments in antifouling paints. Surf Coat Int Part B Coat Trans **86**(4), 259–262 (2003)

Popov, S.Y., Popova, T.T., Sovina, O.A.: Rus Pat RU2. **036**, 265 (2005)

Qian, M., Andrew, M.S., Tan, X.H., Zeng, X.T., Wijesinghe, S.L.: Two-part epoxy-siloxane hybrid corrosion protection coatings for carbon steel. Thin Solid Films **517**(17), 5237–5242 (2009)

Rengasamy, S., Mannar, V.: Development of soy-based UV-curable acrylate oligomers and study of their film properties. Prog. Org. Coat. **76**(1), 78–85 (2012)

Sauvet, G., Dupond, S., Kazmierski, K., Chojnowski, J.: Biocidal polymers active by contact. V. Synthesis of polysiloxanes with biocidal activity. J. Appl. Polym. Sci. **75**(8), 1005–1012 (2000)

Shan, C.A.O., JiaDao, W., HaoSheng, C., DaRong, C.: Progress of marine biofouling and antifouling technologies. Chin. Sci. Bull. **56**(7), 598–612 (2011)

Shi, L., Zhong, Q., Zhang, J., Fang, J., Zhang, D.: China Pat. CN 1, 569. 992.16 (2005)

Tholmann, D., Kossmann, B., Sosna, F.: Polymers with antimicrobial properties. Eur. Coat. J. **1**(2), 16–33 (2003)

Thomas, F.A.P., Stoks, J.A.W., Buegman, A.: US Patent 6291054 (2001)

Thouvenin, M., Peron, J.J., Charreteur, C., Guerin, P., Langlois, J.Y., Valle-Rehel, K.: A study of the biocide release from antifouling paints. Prog. Org. Coat. **44**(2), 75–83 (2002)

Tiller, J.C., Liao, C.J., Lewis, K., Klibanov, A.M.: Designing surfaces that kill bacteria on contact. Proc. Natl. Acad. Sci. **98**(11), 5981–5985 (2001)

Trogolo, A.J., Rossitto, C.F., Welch, K.E.: World Patent WO 03/055941 July 10 (2003)

Verkholantsev, V.V.: Functional variety effects and properties in surface-functional coating systems. Eur. Coat. J. **9**, 18–25 (2003a)

Verkholantsev, V.V.: Self-stratifying coatings for industrial applications. Pigment. & Resin Technol. **32**(5), 300–306 (2003b)

Wolfrum, J., Huang, J., Blake, D.M., Maness, P.C., Huang, Z., Fiest, J., Jacoby, W.A.: Photocatalytic oxidation of bacteria, bacterial and fungal spores, and model biofilm components to carbon dioxide on titanium dioxide-coated surfaces. Environ. Sci. Technol. **36**(15), 3412–3419 (2002)

Wulf, M., Wehling, A., Reis, O.: Coatings with self cleaning properties. Macromol. Symp. **187**, 459–467 (2002)

Xu, T., Xie, C.S., Prog.: Tetrapod-like nano-particle ZnO/Acrylic resin composite and its multi-function property. Prog. Org. Coat. **46**(4), 297–301 (2003)

Zanetti, M., Lomakin, S., Gamino, G.: Polymer layered silicate nanocomposites. Macromol. Mater. Eng. **279**(1), 1–9 (2000)

Zhan, C., Liu, S., Zhao, J.Q., Huang, J.Y.: Synthesis and properties of a modified unsaturated polyester resin with phosphorus-containing pendant groups. Polym. Bull. **70**(4), 1097–1111 (2013)

Zovi, O., Lecamp, L., Loutelier-Bourhis, C., Lange, C.M., Bunel, C.: A solvent less synthesis process of new UV-curable materials based on linseed oil. Green Chem. **13**, 1014–1022 (2011)

# Chapter 9
# Hydrogels: A Unique Class of Soft Materials

**Dibyendu S. Bag, Akansha Dixit, and Eswara Prasad Namburi**

**Abstract** Hydrogels are well-known soft materials that are used to develop soft and wet technology. A solid cross-linked polymer remains in a solvent-entrapped swollen state, and this swollen mass is called a hydrogel. Hydrogel materials having high water absorption and retention capacity are specifically called superabsorbent hydrogels. Such hydrogels are widely used in baby diapers, sanitary towels and agricultural applications. Many hydrogels exhibit the phenomenon of sudden and reversible phase transitions under the influence of external stimuli such as temperature, pressure, electric and magnetic field, light intensity, pH and ionic strength of the medium and chemical triggers: these are called smart hydrogels (also stimuli-responsive gels or intelligent hydrogels). Smart hydrogels have been 'tailored' to challenging technological applications such as artificial muscles and organs, drug delivery systems, smart sensors and actuators. The synthesis of *functional* smart hydrogels having extraordinary activity like sensing, healing, actuation, and other functions, to fulfil the present technological demand of functional soft materials is a challenging task. Syntheses of fast stimuli-responsive and also strong and stretchable hydrogels are two other aspects to consider in the development of smart hydrogels. The present chapter surveys all aspects of hydrogel materials, including their synthesis, characterization, property evaluations, and recent trends in their technological applications.

**Keywords** Hydrogels · Shape memory and self-healing hydrogels · Double network hydrogels · Nanocomposite hydrogels · Applications of hydrogels

---

D. S. Bag
Defence Materials and Stores Research and Development Establishment (DMSRDE), Defence Research and Development Organisation (DRDO), Kanpur, Uttar Pradesh, India
e-mail: ds_bag@rediffmail.com

A. Dixit
Indian Institute of Technology, Kanpur, Uttar Pradesh, India

E. P. Namburi (✉)
Former Outstanding Scientist and Ex-Director, Defence Materials and Stores Research and Development Establishment (DMSRDE), Defence Research and Development Organisation (DRDO), Kanpur, Uttar Pradesh, India
e-mail: neswarap@gmail.com

 
E. P. Namburi et al. (eds.), *Novel Defence Functional and Engineering Materials (NDFEM) Volume 1*, Indian Institute of Metals Series,
https://doi.org/10.1007/978-981-99-9791-6_9

## 9.1 Introduction

When a cross-linked polymer is immersed in an aqueous medium, solvent molecules slowly enter the polymer network. Since the polymeric chains are cross-linked to each other, the chain molecules are not separated and the material swells instead of dissolving. This solvent-entrapped swollen and soft mass is called a hydrogel (Bag and Saxena 2014; Bag and Rao 2006; Zohuriaan-Mehr and Kabiri 2008). It is neither solid, nor liquid or gas (three states of matter). A hydrogel is also neither a plasma (the 4th state of matter), nor the Bose–Einstein condensate (BEC) of matter (the 5th state of matter). This unique state of soft hydrogel material (swollen state), with properties not exhibited by the other well-known five states of materials, may be designated as the 6th state of matter! Hydrogels have unique properties and are used to develop so-called soft and wet technology.

Superabsorbent hydrogels (SAHs) (also known as superabsorbent polymers, SAPs) are a class of cross-linked polymer network hydrogels with very high water absorption and retention capacities (Ahmed 2015). There is an increasing demand for superabsorbent hydrogels for baby diapers and sanitary towels, and agricultural and horticultural use to facilitate and improve plant growth.

The phenomenon of sudden and reversible volume changes in hydrogel materials under the influence of external stimuli such as temperature, pressure, electric and magnetic field, light intensity, pH and ionic strength of the medium, and chemical triggers are well-known. Such materials are called smart hydrogels (also stimuli-responsive or intelligent hydrogels) (Bag and Saxena 2014). Smart hydrogels have been 'tailored' to challenging technological applications such as artificial muscles and organs, drug delivery systems, smart sensors and actuators (Sun et al. 2012a). The synthesis of *functional* smart hydrogels having extraordinary activity like sensing, healing, actuation, and other capabilities to fulfil the increasing technological demand of functional soft materials is a challenging task. Two additional aspects are very important for smart hydrogels: (i) fast stimuli-response and (ii) strong and stretchable properties, instead of the usually brittle behaviour in their swollen state.

This chapter discusses all aspects of hydrogel materials including their synthesis, characterization and properties. The chapter also includes the recent development trends for various technological applications.

## 9.2 Hydrogel Swelling Mechanism

By definition, hydrogels are swollen 3D networks of cross-linked polymers in aqueous media. The swelling mechanism is the penetration of water molecules into the cross-linked polymer network; see Fig. 9.1. By absorbing an aqueous solvent the cross-linked polymer network achieves a fully swollen equilibrium state beyond which no more solvent can be absorbed. The equilibrium amount of swelling depends on the nature of the monomers, the cross-link density, and polymer–solvent interaction parameters.

Hydrogels with covalently cross-linked polymer networks are called "chemical hydrogels". There are also "physical hydrogels" formed by physical cross-linking between the polymeric chains and also physical cross-linking with the water molecules. In these hydrogels the polymer networks are held together by secondary forces and molecular entanglements.

The fundamental forces responsible for the behaviour of hydrogels are (i) ionic interaction, (ii) hydrogen bonding, (iii) hydrophobic interaction and (iv) Van der Waals forces. Since all these forces (or interactions) are reversible and can be disrupted by changing the physical conditions, physical hydrogels are also called "reversible hydrogels". These forces govern the ability of hydrogels to interconvert between their swollen and squeeze states exhibiting smart behaviour. For example, if the environmental conditions cause repulsive forces to predominate within the polymeric networks, then the hydrogels expand by absorbing water (Bag and Rao 2006). On the other hand, if the environmental conditions cause attractive forces to be stronger between the polymeric networks, then the hydrogels collapse by expelling the water.

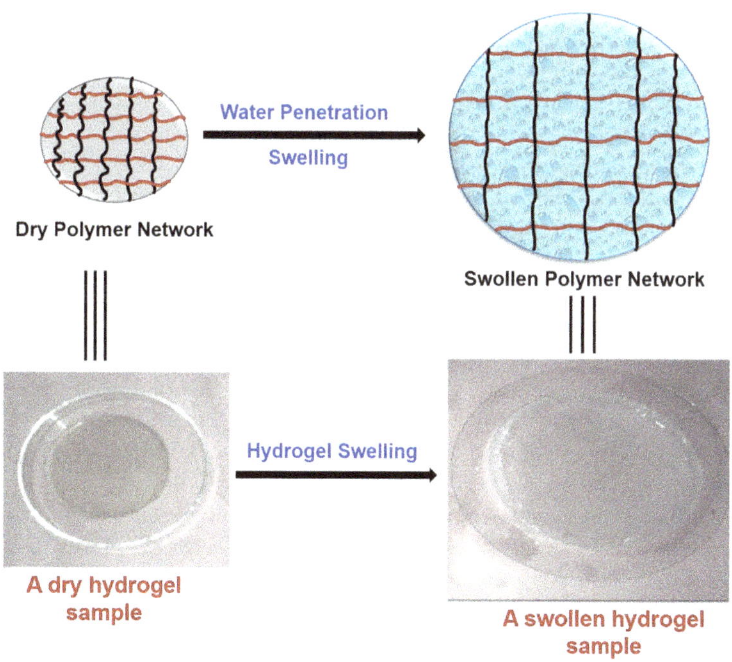


**Fig. 9.1** Mechanism of hydrogel swelling

## 9.3 Different Classes of Hydrogels

Hydrogels may be classified in different ways; see Table 9.1. Some additional information is given in the following text:

- According to their origin, hydrogels may be classified as natural, semi-synthetic and synthetic hydrogels. The vitreous humour in eyes is an example of a natural hydrogel. Polyacrylamide and modified chitosan based hydrogels are examples of synthetic and semi-synthetic hydrogels, respectively.
- Water is usually taken as solvent, and it forms hydrogels. However, gels can be prepared using organic solvents. These are called organic gels.
- Cross-linking (covalent or physical) determines the classification into chemical or physical hydrogels; see also Sect. 9.2 of this chapter.
- The nature of the polymers determines the classification into two groups: (i) uncharged polymers and (ii) charged polymers. Uncharged polymers capable of forming H-bonding with water molecules result in hydrogels. Charged polymers imbibing water molecules form swollen hydrogels owing to ionic repulsion.
- Uncharged polymeric hydrogels: any effect destroying the H-bonding (increasing temperature or changing the ionic strength of the solvent) induces hydrophobic–hydrophilic imbalance that causes hydrogel shrinking.
- Charged polymeric hydrogels: a reduction in net charge of the hydrogel, owing to changing the solvent, pH or by adding an efficient counter-ion (low molecular weight counter-ion or oppositely charged polyelectrolyte) decreases the repulsion between polymer segments and results in hydrogel shrinking.
- The physical and chemical characteristics of uncharged and charged polymeric hydrogels can be 'tailored' to provide smart hydrogels with various smart functions; see the list in Table 9.1. Examples of smart hydrogels are discussed in Sect. 9.6 of this chapter.
- Certain hydrogels have very high water absorption and retention capacities. As mentioned in Sect. 9.1, they are designated as superabsorbent hydrogels (SAHs) to distinguish them from ordinary hydrogels.
- Porous hydrogels and also hydrogel particles with high surface areas can be synthesized to obtain fast-swelling characteristics. Depending on the type of porosity, these hydrogels may be further classified as microporous or superporous. These types of hydrogels are discussed in Sect. 9.8.1 of this chapter.
- The sizes of synthesized hydrogel particles provide another classification, namely into microgels or nanogels. These are discussed in Sect. 9.8.2 of this chapter.
- Since hydrogels are normally brittle in their swollen states, different strategies have been adopted to improve the mechanical properties of hydrogels. Two important classes of such hydrogels are (i) double network hydrogels and (ii) nanocomposite hydrogels. These are discussed in Sect. 9.9 of this chapter.

**Table 9.1** Different classes of hydrogel materials

| Basis of classification | Different classes of hydrogels |
|---|---|
| Origin of materials | Natural, semi-synthetic and synthetic hydrogels |
| Nature of solvent | Hydrogels (water as a solvent) and organic gels (organic solvent, mineral or vegetable oil) |
| Nature of cross-linking | Chemical and physical hydrogels |
| Nature of polymers | Uncharged (capable of forming H-bonding) and charged polymers |
| Smart functions | Smart hydrogels<br>• Thermo-responsive hydrogels (stimulus: temperature)<br>• Electro-responsive hydrogels (stimulus: electric field)<br>• Magneto-responsive hydrogels (stimulus: magnetic field)<br>• Photo-responsive hydrogels (light or photon energy)<br>• pH-responsive hydrogels (change of pH)<br>• Chemically activated hydrogels (chemical trigger) etc. |
| Absorption capacity | Ordinary hydrogels and superabsorbent hydrogels (SAHs) |
| Porosity in hydrogels | Microporous and superporous hydrogels |
| Size of particles | Microgels and nanogels |
| Strategies of improving strength properties | Strong and stretchable hydrogels<br>• Double network hydrogels<br>• Nanocomposite hydrogels |

## 9.4 Synthesis and Characterization of Hydrogels

### *9.4.1 Synthesis of Hydrogels*

Several polymerization mechanisms and techniques are used to synthesize the cross-linked polymer networks for hydrogels. Some of the polymerization mechanisms which are usually followed for the synthesis of hydrogels are listed here:

- Radical polymerization of functional monomers with the aid of crosslinkers.
- Cross-linking of preformed polymers/natural polymers (chemical cross-linking and radiation cross-linking).
- Graft copolymerization of functional monomers onto preformed polymers/natural polymers.

Radical polymerization is the most widely used technique for the synthesis of cross-linked polymers forming hydrogels. In this technique, vinyl monomers are polymerized in the presence of crosslinkers to obtain polymer networks. Some of the common functional monomers and crosslinkers used for the synthesis of hydrogels are given in Table 9.2.

Polymer networks of synthetic and natural polymers are also obtained by chemical or radiation cross-linking reactions and other important synthesis procedures like graft polymerization of functional vinyl monomers onto preformed polymers (synthetic or natural polymers). Some of the natural polymers widely used for hydrogels include cellulose and its derivatives: starch, guar gum, chitosan, alginate, gelatine, etc.

The above-mentioned three commonly used polymerization techniques for synthesis of cross-linked polymeric networks are described in more detail in Sects. 9.4.1.1–9.4.1.3. Some special polymerization techniques are used to synthesize special categories of hydrogel materials, such as (i) microporous and superporous hydrogels and (ii) microgels and nanogels. These are discussed in Sects. 9.4.1.4 and 9.4.1.5.

**Table 9.2** Functional monomers and crosslinkers used for hydrogels

| Monomers/crosslinkers | Structures | Monomers/crosslinkers | Structures |
|---|---|---|---|
| N-isopropylacrylamide | 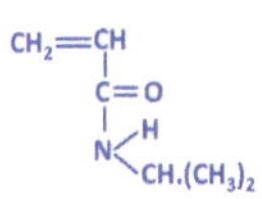 | Poly(N-ethylacrylamide) | 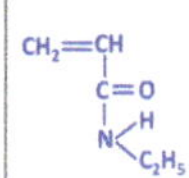 |
| Acrylamide | 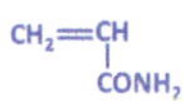 | Poly(N-methyl-N-ethyl acrylamide) | 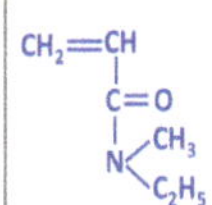 |
| Methacrylamide | 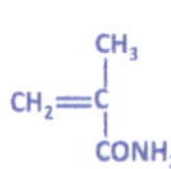 | Poly(N-vinyl caprolactam) | 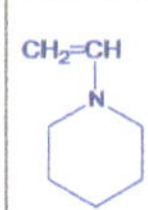 |
| Poly(N-methyl-N-ethyl acrylamide) | 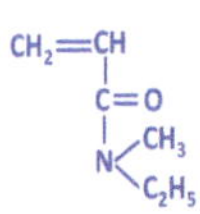 | Ethylene glycol dimethacrylate (EGDMA) | 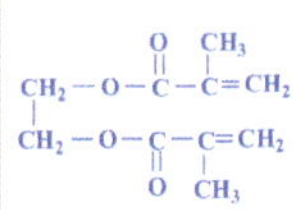 |
| Methacrylic acid | 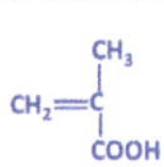 | Poly(vinyl methyl ether) | 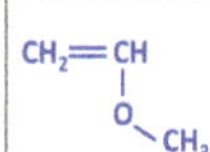 |
| Methyl acrylate | 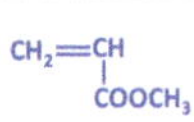 | N, N′-Methylene bis acrylamide (MBA) | 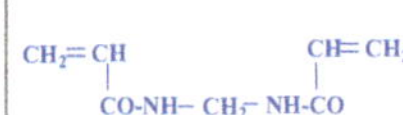 |
| Methyl methacrylate | 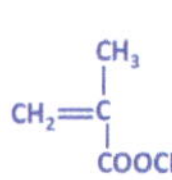 | 1,6-Hexane dioldiacrylate (HDDA) | 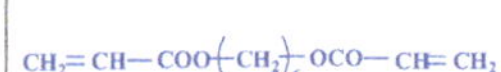 |
| Acrylic acid | 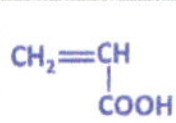 | 1,4-Butane dioldiacrylate | 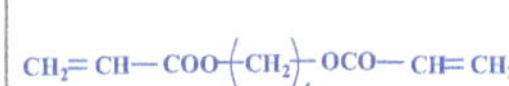 |

#### 9.4.1.1 Radical Polymerization of Functional Monomers with the Aid of Crosslinkers

Free radical polymerization is a very common polymerization mechanism used for polymerization of vinyl monomers (Bag 2013). Polymerization of functional hydrophilic vinyl monomers with the aid of crosslinkers produces cross-linked polymer networks that can form hydrogels, as is shown schematically in Fig. 9.2. Initiators like benzoyl peroxide (BPO), 2, 2′-azo-bis-isobutyronitrile (AIBN), ammonium persulfate (APS) and accelerators like tetramethylethylenediamine (TMEDA) are frequently used. Hydrophilic monomers like N-isopropylacrylamide (NIPAM), acrylamide, acrylic acid, methacrylic acid, 2-hydroxyethyl methacrylate, 4-vinyl pyridine, and/or their combinations, and crosslinkers such as N,N′-methylene-bis-acrylamide (MBA), ethylene glycol dimethacrylate (EGDMA) are widely used for the synthesis of hydrogels. It is important to remove the unreacted monomers and crosslinkers and other reagents from the network. This is usually carried out via the Soxhlet extraction technique.

#### 9.4.1.2 Cross-Linking of Preformed Polymers/Natural Polymers

Cross-linking is widely used to obtain hydrogel networks of various synthetic and natural polymers. In this method, chemical cross-linking of polymer chains (natural and synthetic polymers) is done via reaction of the functional groups (such as OH, COOH, and $NH_2$) present in polymers with crosslinker molecules like glutaraldehyde, adipic acid, dihydrazide, and epichlorohydrin (Chang et al. 2010). For example, carboxymethyl cellulose (CMC) chains are cross-linked using 1,3-diaminopropane to produce a CMC-hydrogel. Hydrogels can also be synthesized from cellulose in NaOH/urea aqueous solutions by using epichlorohydrin as a crosslinker. Similarly, cross-linking of corn starch or poly(vinyl alcohol) using glutaraldehyde as a crosslinker can also produce hydrogels; other combinations of polymers and crosslinkers are also possible.

Cross-linking of polymers could also be achieved using high energy radiation (such as UV-radiation, $\gamma$-rays, X-rays or electron beams). Several hydrogels have been synthesized by the radiation cross-linking of natural polymers such as methylcellulose, hydroxypropylcellulose, N-allylcarbamoylmethyl cellulose, carboxymethyl cellulose, carboxymethyl starch, gum arabic, carboxymethylated chitin and chitosan (Makuuchi 2010; Singh and Vashishth 2008).

The irradiation technique is especially used for the preparation of hydrogels and modification of biopolymers that have their end-use in biomedical applications, since this technique does not involve chemical additives and therefore retains the biocompatibility of the biopolymers.

Free radicals are generated in the polymers upon exposure to high energy radiation. Such generated radicals in close association with the original polymer and other secondarily formed polymer radicals form cross-links by way of radical–radical reactions and polymer–polymer radical reactions. The action of radiation depends

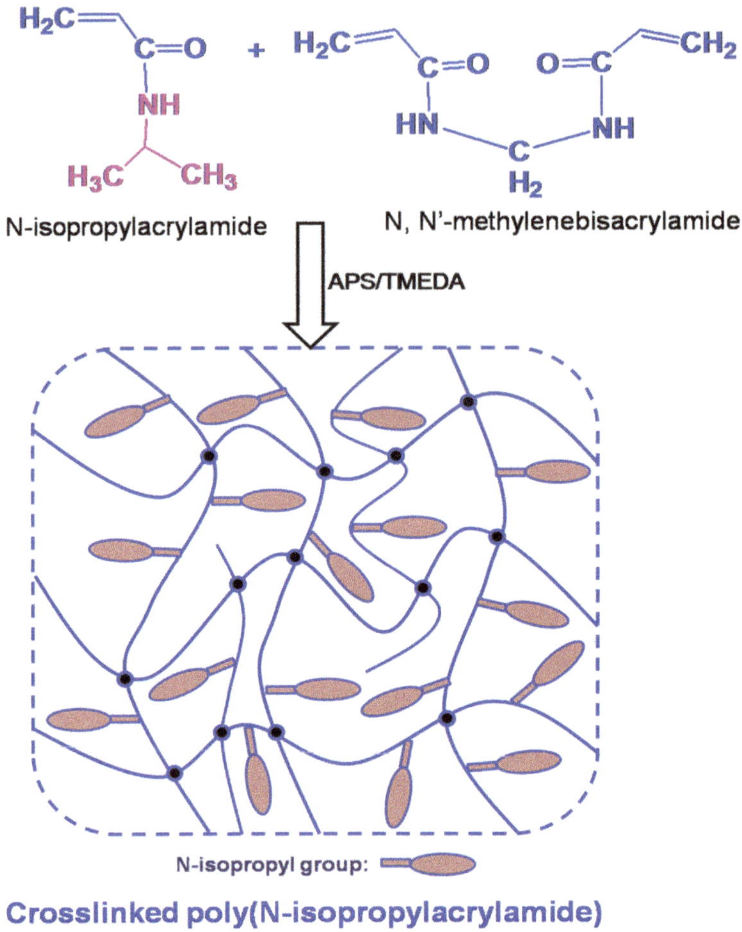


**Fig. 9.2** Schematic representation of synthesis of athermo-responsive cross-linked poly(N-isopropylacrylamide) hydrogel

on the polymer environment (i.e. dilute solutions, concentrated solutions, solid state). However, the presence of water promotes the diffusion of macro radicals to combine and form cross-linked hydrogel networks.

#### 9.4.1.3 Graft Copolymerization of Functional Monomers onto Polymers

Graft copolymerization refers to the polymerization of a functional monomer onto the backbone of a polymer in which active sites are created; see Fig. 9.3. In general, natural polymers like cellulose and its derivatives, chitosan, guar gum, etc. are used

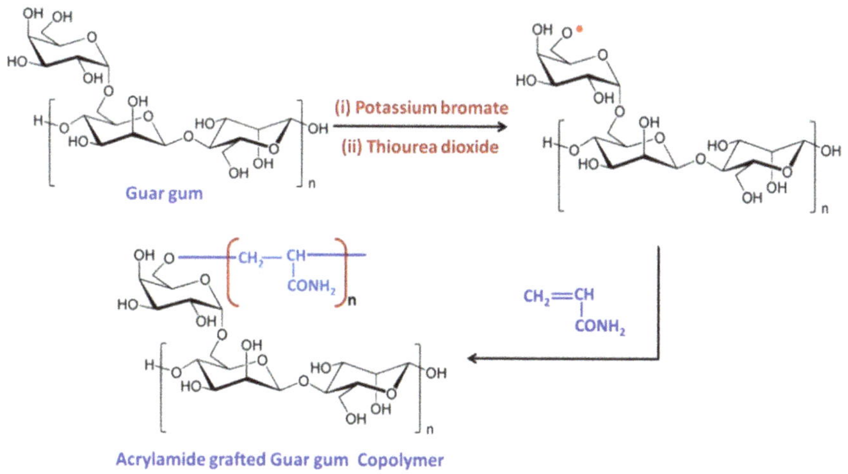


**Fig. 9.3** Schematic representation of graft copolymerization of acrylamide onto guar gum

for this type of synthesis. The active sites on the polymer chains are usually created by the action of either chemical reagents (e.g. redox initiator systems like $K_2S_2O_8$ and $Na_2SO_3$) or applying high energy radiation (e.g. γ-rays or X-rays) to the functional monomers. The polymerization of the functional monomers on the active chain radicals (macroradicals) leads to branching and finally cross-linking to produce the polymer hydrogels. Examples are as follows:

- Graft copolymerization of N-isopropylacrylamide (NIPAM) onto chitosan-obtained CS-g-NIPAM hydrogels, which enables thermo-responsive properties (Schuetz et al. 2008).
- Grafting carboxymethyl cellulose (CMC) with acrylic acid and using electron beam irradiation in aqueous solutions, resulting in pH-responsive hydrogels (Said et al. 2004).

#### 9.4.1.4 Synthesis of Microporous and Superporous Hydrogels

Porous structures are required to obtain fast responsive hydrogels. Such hydrogels are prepared with the usual synthesis techniques, but with the addition of a *foaming agent*. This enables in-situ generated gas to escape from the polymer matrix such that there is control over the homogeneous and interconnected porous structures in the hydrogels (Gemeinhart et al. 2000a; Omidian et al. 2005). Depending upon the foaming agent, foam stabilizer and polymerization kinetics, different types of porous structures with different water-absorbing networks are produced.

Microporous hydrogels can also be prepared using *pore-forming agents (porogens)* such as hydroxypropylcellulose and carboxy methyl cellulose. These are washed out during Soxhlet extraction, resulting in pores inside the hydrogels (Wu et al. 1992). Moreover, the unreacted monomers washed out during purification of the synthesized material by Soxhlet extraction can also be responsible for obtaining microporous/nanoporous structures in the hydrogels.

Porous hydrogels have also been synthesized without using any foaming agent or porogen. Microporous thermo-responsive hydrogels of poly(N-isopropylacrylamide) have been obtained when the hydrogels were synthesized above the lower critical solution temperature (LCST) and also by the γ-radiation technique (Kishi et al. 1997; Kabra and Gehrke 1991). In fact, using γ-radiation at a temperature above the LCST resulted in a sponge-like porous structure for poly(vinyl methylene), and this hydrogel exhibited larger thermal response than the hydrogel prepared below the LCST (Hirasa et al. 1991a).

#### 9.4.1.5 Synthesis of Microgels and Nanogels

Microgels and nanogels are usually synthesized using emulsion polymerization or the inverse micro-emulsion polymerization process, following the radical polymerization of vinyl monomers in the presence of crosslinkers. This technique yields narrow particle size distributions and enables preparation of very small microgel particles (i.e. particle diameters less than 150 nm) (Pelton et al. 2011). However, since *conventional emulsion polymerization* is carried out in the presence of an added surfactant, this technique suffers from the difficulty of completely removing residual surfactant from the synthesized microgels. Therefore *surfactant-free emulsion polymerization* is employed to obtain microgels (diameters 100–1000 nm) without residual surfactant contamination (Xiao et al. 2004). The surfactant-free emulsion polymerization technique is preferred for preparing microgels and nanogels used in biomedical applications.

In addition to emulsion polymerization, *free radical heterogeneous polymerizations* included in the dispersion, precipitation, inverse mini-emulsion, and inverse micro-emulsion preparation of hydrogels are used for microgel/nanogel synthesis (Cawse et al. 1997). Other strategies such as the photolithographic technique (Huang et al. 2007), micromoulding method (Tang et al. 2003) and microfluidics (Oh et al. 2009) approach have also been adapted for preparing well-defined microgels and nanogels.

### 9.4.2 *Characterization of Hydrogels*

#### 9.4.2.1 Structural Characterization

Spectroscopic techniques are employed for structural characterization of hydrogel materials. Thus, Fourier Transform Infrared (FTIR) spectroscopy and Nuclear Magnetic Resonance (NMR) spectroscopy are used for identification of chemical structures and functional groups present in the cross-linked polymeric networks of hydrogels. These techniques are also widely used to investigate the structural arrangement and hydrogen bonding in swollen hydrogels by comparison with the starting dry materials.

The solid-state cross-polarization magic angle spinning nuclear magnetic resonance ($^{13}$C-CP/MAS NMR) technique is used to characterize dry hydrogels and determine the compositions of monomers and crosslinker fractions in them. Since $^{1}$H-NMR can give information about the interchange of water molecules between the so-called free and bound states, this technique is also used to characterize and quantify the amount of free and bound water in hydrogels. Differential Scanning Calorimetry (DSC) analysis is another method used to characterize and quantify the amount of free and bound water in hydrogels.

Scanning Electron Microscopy (SEM) and Transmission Electron Microscopy (TEM) are powerful techniques for morphological characterization of materials. Both are extensively used to understand the characteristic 'network' structures, surface topographies and presence of porosity (microporous and superporous) in hydrogels. TEM is also used to examine the actual morphologies of self-assembled micellar aggregates of thermo-responsive amphiphilic copolymers, and especially to detect the presence of nanoparticles in nanocomposite hydrogels. X-ray diffraction (XRD) analysis and thermo-gravimetric analysis (TGA) are also used to confirm the formation of the cross-linked network structures of hydrogels.

#### 9.4.2.2 Properties

The main characteristic of hydrogels is their swelling behaviour. To study this behaviour, an accurately-weighed amount of a dry hydrogel sample is usually immersed in distilled water at room temperature. Water molecules start to penetrate the sample, and owing to the presence of a crosslinker and the occurrence of cross-linking the sample swells by absorbing water. The swollen hydrogel sample is taken out of the water after a fixed time period and allowed to drain on a sieve. Then any excess water on the surface of the sample is removed by blotting, and the sample is weighed. The degree of swelling ($S_t$) at different times ($t$) is then calculated from the two weights of the sample (dry weight and weight after swelling) using the following equation (Dixit et al. 2017):

$$\text{Degree of Swelling}\,(S_t)\,\% = \frac{W_t - W_0}{W_0} \times 100 \tag{9.1}$$

where $W_t$ = weight of the swollen hydrogel at time $t$; and $W_0$ = weight of the dry sample.

The swelling of a sample increases with time and finally attains equilibrium with a maximum swelling. The minimum time to attain maximum swelling is considered the time taken to reach equilibrium swelling. The time to equilibrium swelling and the equilibrium degree of swelling are usually estimated from the plot of degree of swelling ($S_e$) versus time ($t$). The equilibrium swelling can also be determined from the initial ($W_0$) and equilibrium ($W_e$) weights of the sample using the following equation:

$$\text{Equilibrium Degree of Swelling } (S_e)\,\% = \frac{W_e - W_0}{W_0} \times 100 \tag{9.2}$$

It is also worth mentioning, although fairly obvious, that the water uptake by the hydrogel at any time $t$ can be obtained from the above equations.

## 9.5 Superabsorbent Hydrogels

As stated earlier, superabsorbent hydrogels (SAHs) have high water absorption and retention capacity (usually 100–500 times or more of their dry weight) (Ahmed 2015; Sun et al. 2012a). The cross-linked polymer networks of superabsorbent hydrogels have excellent properties like hydrophilicity, high swelling capacity and biocompatibility. Natural polymers like carbohydrates, starch and cellulose are primarily used in synthesizing superabsorbent hydrogels (Yan et al. 2011). Agricultural waste products like orange and jackfruit peels, corn cobs, hazelnut and peanut shells, and soybean hulls are also being investigated, using acrylate modifiers, for making hydrogels.

Recently, sodium humate (SH) and sodium alginate (NaAlg), and their derivatives, have also been investigated for superabsorbent hydrogels. Functional monomers like acrylic acid (AA), sodium acrylate (Na-AA) and acrylamide (AM) are generally used to synthesize superabsorbent hydrogels. Introducing sodium humate and sodium alginate into hydrogels increases the water absorption and water retention capacity in sandy soils (Zhang et al. 2006a). For example, a superabsorbent hydrogel based on poly(acrylic acid /acrylamide/sodium humate) results in water absorbency as high as 724 $gg^{-1}$ (Singh and Singhal 2012). Another example is the very high amount of water absorption (906 $gg^{-1}$) in AA/NaAlg/SH superabsorbent hydrogels (Agnihotri and Singhal 2017).

The water absorption capacity increases due to a large number of hydrophilic functional groups such as hydroxyl (–OH) and amino (–$NH_2$) groups in the hydrogels. For example, during polymerization SH reacts with AA/NaAlg and forms a polymer web which enhances the water absorbency. The porosity, increased surface area and roughness in hydrogels also play active roles, since pores are regions of solvent penetration in which the interaction of hydrophilic groups with solvent molecules occurs, resulting in high water absorbency.

The water absorbency of SAHs depends on the nature of the functional monomers and crosslinkers and their concentration. The performance of common superabsorbent hydrogels based on acrylic acid and acrylamide and their copolymer poly(acrylic acid-co-acrylamide) was tremendously improved by incorporating sodium humate (SH) and attapulgite (APT) functional components into them (Zhang et al. 2006a, 2006b; Maya and Havazelet 2010; Zheng and Wang 2007; Hua and

**Table 9.3** Superabsorbent hydrogels and their water absorption capacity

| Superabsorbent hydrogels (composition) | Polymerization techniques | Water absorbency (g/g) | References |
|---|---|---|---|
| Butyl methyl cellulose and acrylic acid | Graft copolymerization | 51 | Sadeghi et al. (2011) |
| Carboxymethyl cellulose and acrylamide | Radiation induced graft copolymerization | 190 | Ki-Bum and Duk-Geun (2014) |
| Cellulose, acrylic acid and acrylamide | Graft copolymerization | 490 | Wu et al. (2012) |
| Cellulose nanofibrils, chitosan and acrylic acid | Graft copolymerization | 381–486 | Spagnol et al. (2012) |
| Chitosan, poly(acrylic acid) and attapulgite | Graft copolymerization | 159 | Zhang et al. (2007) |
| Chitosan, poly(acrylic acid) and sodium humate | Graft copolymerization | 183 | Liu et al. (2007) |
| Wheat straw and attapulgite | Graft copolymerization | 186 | Xie et al. (2011) |
| Wheat straw and acrylic acid | Chemical method | 417 | Liu et al. (2009) |
| Acrylic acid, acrylamide and sodium humate | Graft copolymerization | 591 | Zheng and Wang (2007) |
| Poly(acrylic acid), acrylamide and sodium humate | Solution polymerization | 1100 | Zhang et al. (2006b) |
| Poly(acrylic acid-co-acrylamide) and sodium humate | Solution polymerization | 1184 | Zhang et al. (2006a) |
| Sodium alginate, sodium humate and acrylic acid | Graft copolymerization | 1380 | Hua and Wang (2009) |

Wang 2009). Sodium humate graft in an AA/NaAlg/SH copolymer resulted in water absorbency as high as 1380 g/g; see Table 9.3.

### *9.5.1 Application of SAHs*

Superabsorbent hydrogels have found applications in disposable diapers, sanitary towels, cosmetic and hygiene products, paper towels, surgical sponges, thickening agents, sealing of underground cables, artificial snow, bandages and wound dressing, and drug delivery systems (Meena et al. 2008). This class of high-water-containing hydrogels is also being used successfully as soil conditioners in the agricultural industry, for holding water and enhancing nutrition in sandy soils. They help improve the permeability, structure, texture and density of soils, and control evaporation from the soils and infiltration rates of water through them. Thus, these hydrogels reduce irrigation frequency and stop erosion and water runoff, as well as increasing the soil aeration and the supply of nutrition to the plants. They are also used in the controlled release of fertilizers.

## 9.6 Smart Hydrogels

Smart hydrogels respond adaptively to a variety of external stimuli such as pH and ionic strength of the medium, temperature, pressure, electric and magnetic fields. The details can be obtained from Bag and Saxena 2014; Bag and Rao 2006. The phenomenon of rapid hydrogel swelling and shrinking, with large volume changes, may be brought about by small changes in external conditions. The drastic volume transition is reversible. The size of the hydrogel is changed but the shape of the gel usually does not alter due to this transition. This unique property of smart gels is utilized for various technological applications such as drug delivery systems, thermo-responsive cell culture dishes, microfluidic gel photovoltaics, sensors and actuators.

Smart hydrogels are classified in Table 9.1, depending on the behaviour response towards external stimuli (smart functions). The response of smart hydrogels owing to stimuli may be (i) volume change or strain development or (ii) reversible changes in any of the properties like viscosity, optical properties, etc. Table 9.4 gives examples of the chemical structures of polymers which exhibit two kinds of smart function responses: thermo-responsive or electro-responsive. A thermo-responsive hydrogel undergoes a sudden volume transition owing to the action of thermal energy or changing the temperature. Analogously, an electro-responsive hydrogel exhibits reversible deformation in size and shape by the application of an electric field. The main classes of smart hydrogels are discussed in Sects. 9.6.1–9.6.6.

The term "multi-responsive hydrogels" is also found in the literature (Bag and Rao 2006). These are hydrogels that can respond to more than one external stimulus. Copolymers/terpolymers synthesized by copolymerizing various types of monomers (having hydrophilic, hydrophobic, ionic moieties) enable the formation of multi-responsive smart gels (Bag et al. 2004). Poly(dimethylaminoethylmethacrylate) (PDMAEMA) exhibits both pH-responsive and thermo-responsive behaviour; and polyelectrolyte gels are both pH-responsive and electro-responsive.

Table 9.4 Smart hydrogels and their structures

| Polymers for hydrogels | Chemical structures |
|---|---|
| *Thermo-responsive hydrogels* | |
| Poly(N-isopropylacrylamide) (LCST = 32 °C) | 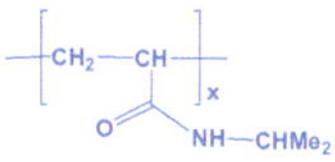 |
| Poly(N-methyl-N-ethyl acrylamide) (LCST = 56 °C) | 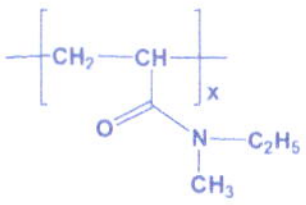 |
| Poly(N-ethylacrylamide) (LCST = 72 °C) | 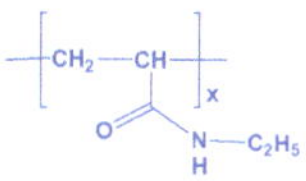 |
| Poly(vinyl methyl ether) (LCST = 34 °C) | 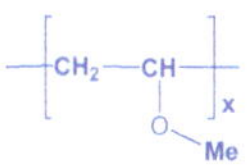 |
| Poly(N-vinyl caprolactam) (LCST = 40 °C) | 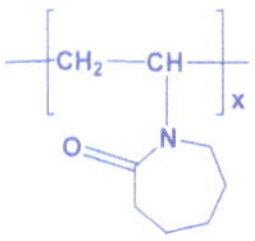 |
| Poly(N-vinyl piperidine) (LCST = 4–5 °C) | 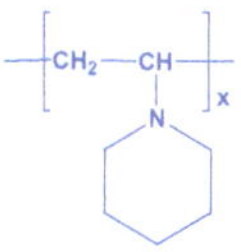 |
| *Electro-responsive hydrogels* | |
| Sodium acrylate-acrylamide copolymer | 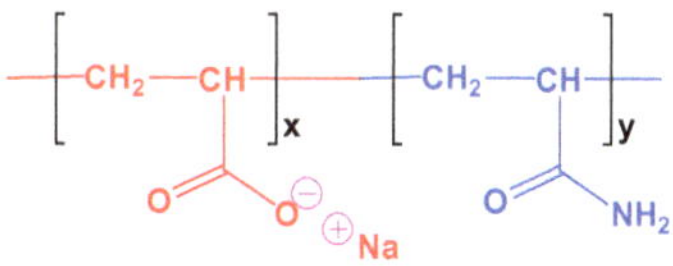 |
| Ionized poly (N-isopropyl acrylamide) | 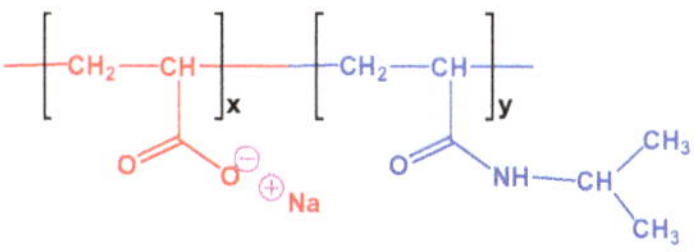 |

(continued)

**Table 9.4** (continued)

| Polymers for hydrogels | Chemical structures |
|---|---|
| Poly (2-acrylamido-2-methylpropane 1-sulphonic acid) (PAMPS) | 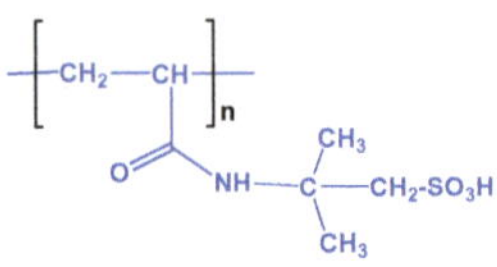 |
| Perfluorosulphonateionomer (Nafion) | 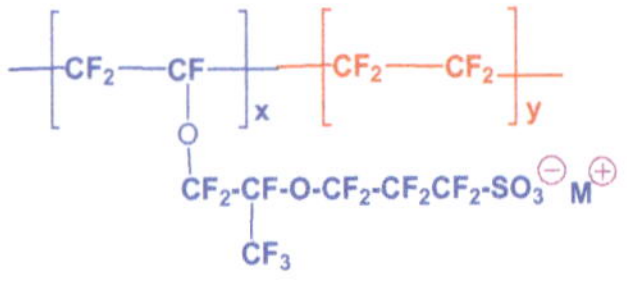 |
| Perfluorocarboxylateionomer (Flemion) | 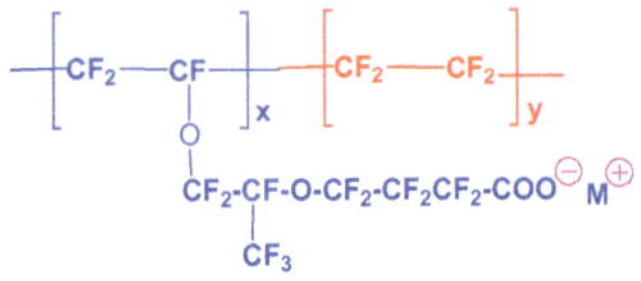 |

## 9.6.1 *Thermo-Responsive Hydrogels*

Thermo-responsive hydrogels have a minimum temperature at which their swelling and de-swelling responses are observed, and this is called the lower critical solution temperature (LCST) or cloud point (Taylor et al. 1975). These hydrogels are swollen in aqueous solution up to the LCST and are squeezed when the temperature is raised above the LCST. This thermal response is due to the sudden expulsion of water from the hydrogels owing to disruption of the H-bonding between the polymer networks and water molecules.

Poly(N-isopropylacrylamide)—or polyNIPAM for short—and its copolymers are the most widely investigated thermo-responsive hydrogels (Schild 1992). The LCST of polyNIPAM is 32 °C (at pH = 7.2), which is just below human body temperature (37 °C). However, the LCST varies according to the nature of the co-monomers forming the copolymer networks in the hydrogels:

- In general, hydrophilic co-monomers lead to LCSTs above 32 °C, while hydrophobic co-monomers lower the LCST.
- Introduction of ionic co-monomers into polyNIPAM not only influences the LCST but also makes the LCST pH-dependent (Bag et al. 2004; Schild 1992). For example, the LCST of poly(NIPAM-co-DMAEMA) is 40 °C at pH = 7.2 while it is 30 °C at pH = 8.0.
- Incorporation of nanofillers like fullerenes (which are hydrophobic) into the polyNIPAM network also lowers the LCST, but only marginally.

Other well-known polymers of thermo-responsive hydrogels are poly(*N*-alkyl acrylamides); poly(*N*-acryloylpiperidine) (Gan et al. 2000); poly(*N*-acryloylpyrrolidine); poly(dimethylaminoethylmethacrylate) (Plamper et al. 2007); poly(oligo(ethyleneglycol) methyl ether methacrylate) (POEGMA); and their copolymers (Lutz et al. 2007). POEGMA analogues have attracted special attention due to their biocompatibility and 'tuneable' LCSTs, ranging from 26 to 90 °C.

Besides conventional cross-linked hydrogel structures, comb-type grafted thermo-responsive hydrogels of polyNIPAM have been synthesized in special strategies, whereby the comb structure is achieved by tailoring the hydrogel architecture at the molecular level (Corcione et al. 2008; Yoshida et al. 1995). The grafted polyNIPAM chains are freely mobile ends that are not found in conventional hydrogel structures. Increasing temperature causes the grafted polyNIPAM chains to collapse from their expanded (hydrated) form to the compact (unhydrated) form before the polyNIPAM network begins to shrink. This behaviour is responsible for temperature-induced rapid swelling and de-swelling response of the comb-type hydrogels and is schematically illustrated in Fig. 9.4. Although this temperature-induce response is much faster than that of conventional polyNIPAM hydrogels, the LCST is the same (32 °C).

Comb-type architectures have also been reported for the hydrogels (i) poly(N-isopropylacrylamide-co-acrylic acid), P(NIPAM-co-AA) with grafted polyNIPAM chains: and (ii) comb-type grafted poly(N,N-diethylacrylamide-co-acrylic acid), P(DEAM-co-AA), in which polyDEAM acts as the grafted freely mobile chains (Corcione et al. 2008). Such hydrogels also exhibit rapid thermo-responsive behaviour compared with their conventional analogues. [**N.B**: Comb-type hydrogels having an acrylic acid component also exhibit rapid pH-responsive behaviour].

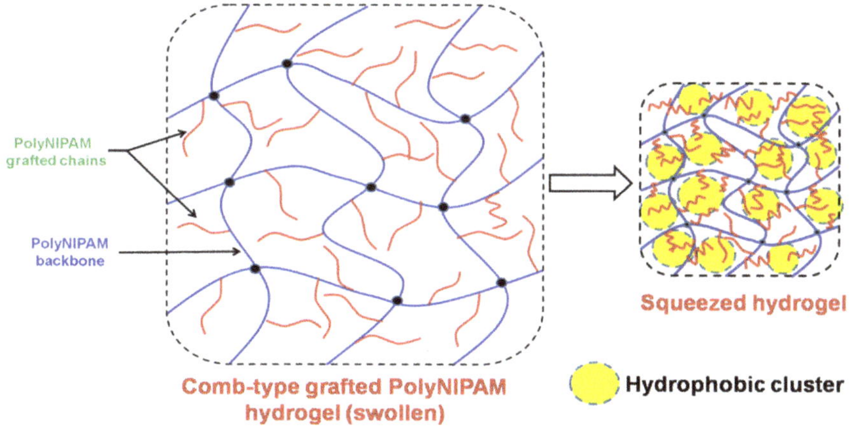


**Fig. 9.4** Schematic representation of a comb-type architecture of grafted polyNIPAM, in which grafted polyNIPAM mobile chain ends are responsible for fast swelling and de-swelling behaviour owing to forming hydrophobic zones (Yoshida et al. 1995) (*Adopted/modified with the permission of Nature*)

#### 9.6.1.1 Applications of Thermo-Responsive Hydrogels

Thermo-responsive smart hydrogels have been investigated for various applications such as controlled drug release, enzyme immobilization, gene carriers, and solute separation processes (Yoshida et al. 1995). In particular, polyNIPAM hydrogels are being widely investigated for controlled drug release (Li et al. 2013). This is because the polymer solution has a low viscosity at room temperature, enabling it to be injected into the body as a liquid: it then forms a gel in situ when the solution temperature rises above the LCST (32 °C). Hence liquid polyNIPAMs can serve as carrier matrices for a wide range of biomedical and pharmaceutical applications. Another clinical application of injectable hydrogels is the treatment of irregularly shaped tissue locations.

### *9.6.2 Electro-Responsive Hydrogels*

In general, "electro-responsive hydrogels" are so-called if any of their properties are reversible by an electric field. However, if hydrogels possess electro-mechanical behaviour they are specifically called electro-active hydrogels (Bar-Cohen 2001). The response to an electric field stimulus in such hydrogels is strain development. Because the electric field stimulus is easily controlled, this class of smart hydrogels has attracted special attention for developing hydrogel actuators. Polyelectrolyte gels (ionic polymeric gels) are examples of this class of smart hydrogels; see the right-hand columns in Table 9.4, in particular poly(sodium acrylate-acrylamide) copolymers and partially ionized polyNIPAM. Also, poly(2-acrylamido-2-methyl propane sulphonic acid) (PAMPS) hydrogels show worm-like mobility under the influence of an electric field stimulus (Osada et al. 1992).

#### 9.6.2.1 Mechanism of Electro-Active Behaviour and Applications

The ionic electro-active (polyelectrolyte) hydrogels are active only in their charged state. The electro-mechanical response may result in hydrogel deformation (gel swelling, or shrinking, or bending), and the extent of deformation depends on the electric field intensity and the concentration of the mobile ions inside the hydrogels. The deformed electro-active gels may be returned to their original shapes by removal of the applied electric field or by changing the polarity of the field.

When a DC electric field is applied across an electro-active hydrogel, mobile ions move towards their counter-electrodes. This ion drift causes two substantial changes: (i) the osmotic pressure changes and (ii) there is a conformational change in the hydrogel network. Thus, a hydrogel swells when the osmotic pressure increases and shrinks when it decreases. Figure 9.5 illustrates measurement (a)–(c) of the electro-responsive behaviour of an ionic hydrogel (based on acrylamide, N-isopropyl

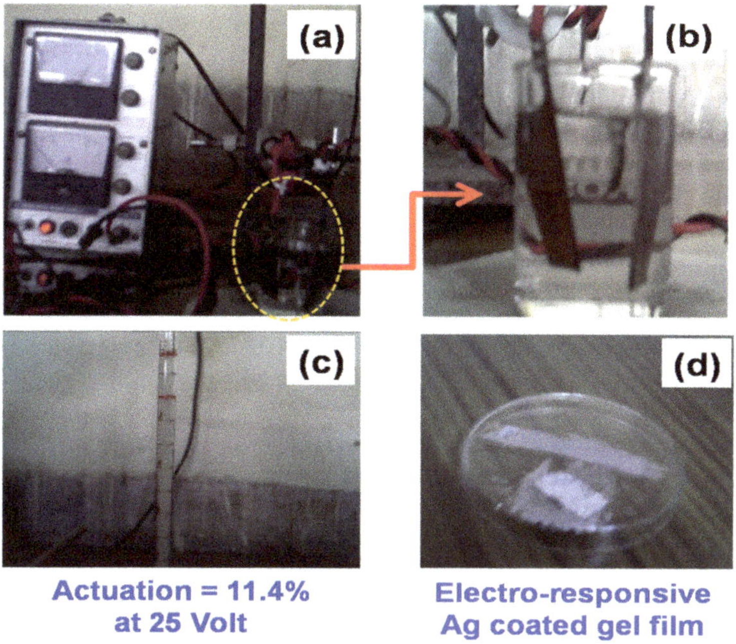


**Fig. 9.5** Measurement of electro-responsive behaviour of an ionic hydrogel: **a** DC voltmeter, **b** platinum electrodes, **c** swollen hydrogel in capillary tube (measured actuation volume 11.4% at 25 V) and **d** electro-responsive silver-coated hydrogel film (Bag et al., unpublished work)

acrylamide and sodium acrylate) under a DC electric field; and also shows an electro-responsive silver-coated gel film.

#### 9.6.2.2 Applications of Electro-Responsive Hydrogels

The electric field induced deformation of electro-active hydrogels has been used to develop actuators. Such soft gel actuators are used for artificial muscles and in micro- or nano-machines (Fang et al. 2010). For example, a "hydrogel hand" with gel fingers controlled by electric field stimuli can grasp an egg without breaking it. However, since hydrogels are usually brittle, ionomers (or polyeletrolytes) such as perfluorosulphonate ("Nafion", DuPont, USA) and perfluorocarboxylate ("Flemion" Asahi Glass, Japan) have been used to make electro-active hydrogel/metal composites. The seionomeric polymer–metal composites (IPMCs) are now used to develop soft sensors and actuators (Shahinpoor 2003).

### 9.6.3 Electro-Conductive Hydrogels

Electro-conductive hydrogels are also simply called conducting hydrogels and are defined as hydrogels with electrically-conducting components (Lu et al. 2005). These hydrogels are generally prepared by incorporating conducting fillers or conducting polymers into hydrogels. They can also be prepared by in situ polymerization of conducting monomers (such as aniline, pyrrole, thiophene and modified thiophenes) within the hydrogel matrices or by the usual hydrogel synthesis procedures in the presence of conducting fillers like graphite, conducting carbon, carbon nanotubes, graphene oxide and graphene.

Electro-conductive composite hydrogels based on polyaniline–poly(2-acrylamido-2-methyl propane sulphonic acid) (PAMPS), polyaniline–polyacrylamide, and polyaniline–poly(2-hydroxyethylmethacrylate) have been reported (Lira and Cordoba 2005). Such composite hydrogels are electro-conductive owing to conducting polyaniline entrapped in the hydrogel pores. Because these hydrogels respond to an applied electric field, they can transform chemical free energy directly into mechanical work.

Polyaniline–polyacrylamide conductive hydrogels have been used in electrochemically controlled drug delivery devices (Lira and Cordoba 2005). Another possibility is the use of polyaniline–poly(2-hydroxyethylmethacrylate) as bio-recognition membranes for immobilization of antibodies in mycotoxins. A composite hydrogel composed of acrylamide and acrylic acid copolymer doped with polypyrrole/carbon black has been used in developing an artificial muscle (Moschou et al. 2006).

### 9.6.4 Magneto-Responsive Hydrogels

Hydrogels that respond reversibly under a magnetic field are called magneto-responsive hydrogels. The responsive behaviour is provided by magnetic nanoparticles present inside the hydrogels. The degree of swelling of magneto-responsive hydrogels containing magnetic particles is usually lower than that of the corresponding particle-free hydrogels, owing to additional cross-links formed by nanoparticles interacting with the polymer chains (Starodubtsev et al. 2005). However, incorporation of magnetic particles strengthens the hydrogels.

This class of hydrogels may be prepared by three general strategies: (i) synthesizing magnetic particles inside hydrogel matrices, (ii) synthesizing cross-linked hydrogel networks in the presence of magnetic particles, or (iii) adding (by mixing or dispersing) magnetic particles to pre-formed polymers or hydrogels.

A magneto-responsive hydrogel obtained by the synthesis of magnetic particles in the presence of the hydrogel and its behaviour under a magnetic field are schematically represented in Fig. 9.6. A magnetic field stimulus aligns the magnetic particles, leading to a change in shape of the hydrogel, such that it can exert a strain (force). A

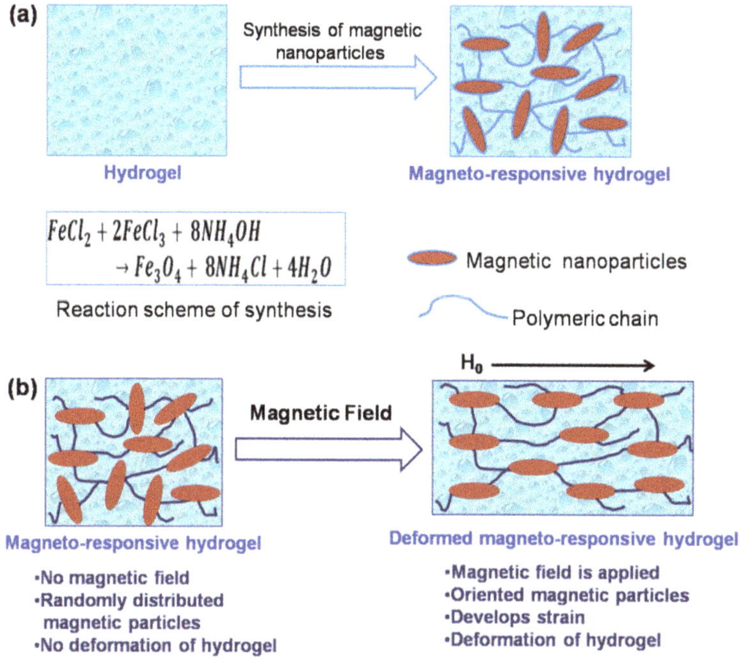


**Fig. 9.6** Schematic representation of **a** synthesis of a magneto-responsive hydrogel (magnetic nanoparticles embedded in the hydrogel) and **b** its deformation under a magnetic field (*Adapted by the permission of John Wiley and Sons*)

balance between the magnetic and elastic interactions decides the final shape of the hydrogel. The mechanical shape change is, of course, reversible.

Examples of magneto-responsive hydrogels are as follows:

- Ferrogels which are colloidal dispersions of ferrite nanoparticles (~10 nm) in chemically cross-linked polymeric networks. Ferrogels have very interesting properties due to both their magnetic response and their fluidity (Liu et al. 2006).
- Magnetic-nanoparticle-containing hydrogels based on acrylamide, alginate (i.e. a polysaccharide bearing one carboxyl group per monomer), polyurethanes and silicones (Zhao et al. 2011).
- Magnetic-nanoparticle-containing hydrogels of polyNIPAM, poly(N-vinylcaprolactam-co-acetoacetoxyethylmethacrylate), copolymers of maleilatedcarboxymethyl chitosan with NIPAM (Pich et al. 2004). These have also been investigated for their magnetic field sensitivities with respect to temperature and/or pH. Interestingly, the nanoparticles in such hydrogels do not alter these sensitivities or the lower critical solution temperature (LCST). These magneto- and thermo-responsive hydrogels could be used in novel remote-controlled drug release systems, whereby the release of an encapsulated drug results from collapse of thermo-responsive gels owing to the magnetic particles being heated by a magnetic field (Gelbrich et al. 2010).

#### 9.6.4.1 Applications of Magneto-Responsive Hydrogels

In general, magneto-responsive hydrogels may find applications as smart dampers for structural vibration control, actuators, sensors, and in biomedical use for targeted drug release upon the application of an alternating magnetic field. An additional advantage of some magnetic hydrogels is their biodegradability under physiological temperatures and pH. Furthermore, the release of drugs can be achieved by adjusting the amounts of embedded magnetic nanoparticles and the magnetic field intensity.

### *9.6.5 Photo-Responsive Hydrogels*

Photo-responsive hydrogels exhibit reversible property changes in response to light (Roy et al. 2010). The mechanism of photo-responsiveness is associated with the shape changes of certain chromophores like azobenzene, stilbene, and spirobenzopyran present in the hydrogels either in the polymer backbones or as pendant groups. These chromophores can transfer light energy into changes in polymer chain shapes, resulting in property changes, e.g. in viscosity, solubility, pH, conductivity, and spectral response; and also resulting in mechanical effects such as reversible expansion and contraction of polymer films (Fig. 9.7). Examples of photo-responsive hydrogels are as follows:

- Poly(*p*-N, N-dimethylamino)-N-$\gamma$-D-glutamanilide) film deformation by ultraviolet (UV) illumination up to 35% (Aviram 1978).
- PolyNIPAM changes its dimensions when stimulated by a laser beam. When the beam is switched off the gel returns to its original size (Fig. 9.7) (Juodkazis et al. 2000).
- A copolymer gel of N-isopropyl acrylamide and a photosensitive molecule, bis(4-(dimethylamino)phenyl)((4-vinylphenyl)methyl leucocyanide sensitive to ultraviolet light (Mamada et al. 1990).
- A soft, optically transparent polymer gel based on *N*-isopropylacrylamide, sodium acrylate, and a photoactive chromophore, which was activated by visible light (switched on) but also deactivated (switched off) by altering the local environment using three different means: *p*H, temperature, and light (Suzuki et al. 1996).

Photo-responsive smart hydrogels are used in several applications such as in the construction of photoactive devices and transducers, actuators, photosensors, and information optical-storage devices, photo-responsive artificial muscles, switches and memory devices and biomedical applications (Itsuro et al. 2011).

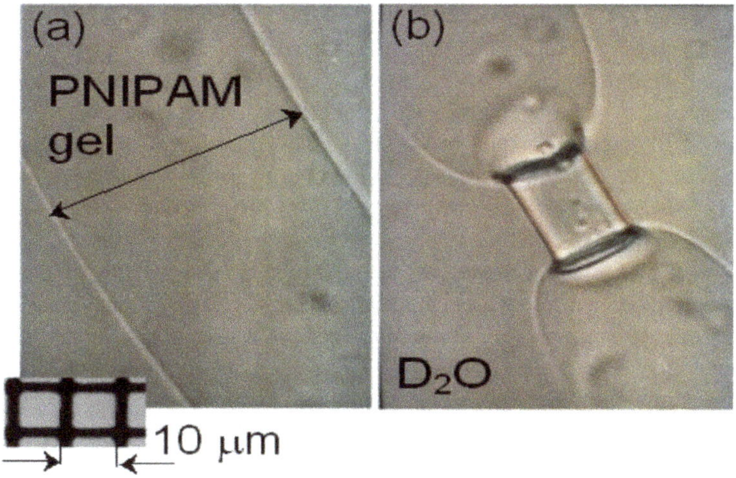


**Fig. 9.7** Poly(*N*-isopropylacrylamide) gel rod in $D_2O$ before (**a**) and after (**b**) illumination by 0.75 W power laser illumination at l = 1064 nm wavelength (Juodkazis et al. 2000) (*Photographs by permission of Nature*)

### *9.6.6 pH-Responsive Hydrogels*

pH-responsive hydrogels generally contain acidic and alkaline functional groups, and they respond to a change in pH of a medium (Sethuraman et al. 2006). The response is reversible and may be a change in volume and shape. Examples of such hydrogels are poly(acrylic acid) (PAA), poly(2-ethylacrylic acid), poly(methacrylic acid) (PMAA), poly(ethylene imine) (PEI), poly(propylene imine), poly(*N*, *N*-dimethylaminoethylmethacrylate)(PDMAEMA), poly(L-lysine) and poly(L-histidine), chitosan and other modified natural polymers.

The change of pH leads to the ionization (protonation or deprotonation) of the functional groups present in this class of hydrogels. The generated ionized groups create electrostatic repulsive force which leads to hydrogel swelling. Therefore, pH-responsiveness depends on the degree of ionization of the functional groups in hydrogels. Uses of these pH-responsive hydrogels are in controlled drug delivery systems and also as pH sensors (Gerlach et al. 2005).

## 9.7 Self-healing Hydrogels

As their name implies, self-healing hydrogels can autonomically repair any damage to them (Jones et al. 2010). A variety of hydrogels can do this, and others need an external stimulus such as temperature, light and pH (Sun et al. 2012b; Liu et al. 2012). These hydrogels have the ability to form new bonds within the gel materials while old bonds are broken. The reversible bond breaking and bond making phenomenon prevents fracture of the molecular backbones in such materials.

The self-healing phenomena are also associated with the reversible molecular interactions such as H-bonding, electrostatic interactions, π–π stacking, molecular recognition, dynamic chemical bonds, and metal-coordinating bonds. Secondary interactions and hydrophobic association are employed for healing in supramolecular networks and non-covalent hydrogels. Self-healing phenomenon has also been reported as a result of strong hydrophobic interactions in polymeric hydrogels (Akay et al. 2013). These hydrogels are hydrophobically modified and prepared by copolymerization of hydrophilic monomers (e.g. acrylamide) and a small amount of a hydrophobic co-monomer via the micellar polymerization without a chemical crosslinker.

Self-healing ability has also been explored for chemically cross-linked hydrogels having an optimal balance of hydrophilic and hydrophobic forces in their pendant side chains (Phadke et al. 2012). The dangling flexible pendant side chains, with minimal steric hindrance and hydrophobic collapse, mediate between H-bonds across the hydrogel interfaces, thereby exhibiting a self-healing capacity.

For example, a polymer hydrogel, obtained from the polymerization of N-acryloyl-6-aminocaproic acid (A6ACA) using N, N′-methylenebisacrylamide as a crosslinker, exhibited rapid healing (Phadke et al. 2012). In this case, the hydrogel was able to mediate hydrogen bonding across two hydrogel interfaces by way of the amide and carboxylic functional groups. The self-healing is observed in an aqueous environment at low pH ($pH \leq 3$), but at high pH ($pH \geq 9$) the healing is prevented because carboxyl groups become deprotonated, leading to electrostatic repulsion between the side chains. Recently, self-healing hydrogels have also been prepared by mixing ionically cross-linked alginate and covalently cross-linked polyacrylamide (Sun et al. 2012b).

Self-healing hydrogels have potential applications as scaffolding in reconstructive tissue engineering. The high porosity of hydrogels controls the diffusion of cells during migration and also the transfer of nutrients and waste products away from cellular membranes. A very different application is their use as sealants for vessels containing corrosive acids. In such cases, hydrogels have the ability to selectively cross-link under acidic conditions. However, it is very challenging to impart self-healing ability in permanently cross-linked hydrogels, owing to then irreversibility of chemical cross-links (Taylor and Panhuis 2016).

## 9.8 Fast-Response Hydrogels

In order to achieve fast swelling and fast response behaviour of smart hydrogels, the following strategies are generally adopted: (i) formation of porous structures, (ii) obtaining cross-linked polymeric materials with high surface area in the form of microgels and nanogels, and (iii) making hydrogels with smaller gel particle sizes as well as porous structures.

### 9.8.1 *Microporous and Superporous Hydrogels*

Porous (microporous and superporous) hydrogels are specially designed hydrogels with pore sizes much larger than the typical mesh size of conventional hydrogels. Synthesis of porous hydrogels has been described in Sect. 9.4.1.4. Since porous structures have high surface area, such hydrogels are fast-responsive, and their swelling and de-swelling take place much faster than for conventional hydrogels. Porous hydrogels also exhibit superabsorbent properties. The pore size of this class of hydrogels ranges from less than 1 μm to more than 1000 μm, whereas the typical mesh size of a conventional hydrogel is below 100 nm (Chen and Park 1999). Some examples of microporous and superporous hydrogels are as follows:

- Fast-responsive microporous thermo-responsive polyNIPAM hydrogels (Omidian et al. 2005; Wu et al. 1992; Kishi et al. 1997). Thermo-responsive behavior was also studied for the porous hydrogels of poly(N-isopropylacrylamide-co-acrylamide) (Gemeinhart et al. 2000b).
- Collagen grafted poly(acrylamide-co-acrylic acid)microporoushydrogels with very good pH-responsive behaviour (Pourjavadi and Kurdtabar 2007)
- Microporous cross-linked copolymer, P(NIPAM-co-Na-AA) has been reported to have a very high surface area (304.6 $m^2/g$)(Bag and Alam 2012).
- Porous smart hydrogels of poly(acrylamide-co-acrylic acid), P(AM-co-AA) exhibited pH-sensitivity in alkaline solutions and fast-swelling properties in acidic solutions, both effects being caused by ionization–deionization of the carboxyl functionalities.
- Sponge-like porous poly(vinylmethyl ether) (PVME) gel fibres have been reported to have higher thermal responsiveness than conventional hydrogels (Hirasa et al. 1991b). The fibre diameters changed from 400 mm at 20 °C to 200 mm at 40 °C.

Porous hydrogels achieve their fast response at the expense of strength. Hence they are modified into composites and/or hybrid structures to obtain mechanically strengthened superporous hydrogels. Sodium alginate, chitosan, gelatine, poly(vinylalcohol) etc. are used as reinforcing fillers or hybridizing agents that can be cross-linked in situ during the synthesis of superporous hydrogels (Gemeinhart et al. 2000a). Along with their fast swelling response, such superporous hydrogel hybrids/composites exhibit reasonably good elastic and rubbery properties in their fully swollen state. For example, a hydrogel hybrid of alginate and acrylamide was stretchable up to 2–3 times its original length and it was able to resist a static mechanical pressure of about 10 N (Omidian et al. 2006). For more information see Sect. 9.9 on strong and stretchable hydrogels.

### 9.8.2 *Microgels and Nanogels*

Microgels and nanogels are swollen polymeric micron-sized and submicron-sized hydrogel particles. Hydrogels with particles in the size range 100 nm to 1 μm are called microgels, and those with particle sizes in the range 1–100 nm are called nanogels. The hydrogel particles are uniformly dispersed in a continuous aqueous medium with characteristics that depend on the synthesis method, the monomer(s), the cross-linked density and the solution condition (Seiffert 2013). Microgels and nanogels have similar properties to water-swollen conventional hydrogels, but they respond faster owing to their very small sizes. Special properties may also be inducted into such smart microgels/nanogels via incorporation of active functional groups in them (Smith and Lyon 2012). Some examples of these types of hydrogels are given here:

- Thermo-responsive microgels and nanogels of polyNIPAM, poly(*N*-isopropylmethacrylamide, poly(*N*-ethylacrylamide), poly(*N*-ethylmethacrylamide) poly(*N*-acryloylpyrrolidine), poly(2-(dimethylamino)ethyl methacrylate) and others have been studied widely. The thermo-responsive behaviours of microgels and nanogels are reported to be similar to those of conventional (macroscopic) gels (Oh et al. 2008; Wei et al. 2014).
- PolyNIPAM microgels are swollen at room temperature, but collapse above 32 °C (the LCST). The swelling capacity and volume phase transition temperature of polyNIPAM nanogels (diameter about 100 nm) were also reported to have similar characteristics to those of their macroscopic counterparts (Kratz et al. 2001).

Since smart microgels and nanogels can swell and shrink reversibly with changes in external environment, this means that the gel particle sizes are altered by the external stimuli. Because of their large surface areas and fast response, and also because the particle sizes can be 'adjusted' over the nanometre to micrometre range, microgels and nanogels have attracted special attention, especially for drug delivery systems (DDS). Smart microgels and nanogels of naturally occurring carbohydrate-based biopolymers such as chitosan (CS), hyaluronam (HA) and dextran are being investigated extensively for various biomedical applications.

## 9.9 Strong and Stretchable Hydrogels

As stated previously, hydrogels are mechanically very weak (usually brittle), particularly in their swollen state. The poor mechanical properties, especially lack of toughness and flexibility, are severe limitations on their applicability and use. Many applications require hydrogels to endure significant mechanical loads in aggressive environments, including artificial cartilage in tissue engineering, artificial nerves and muscles and swellable packets in the oil industries and other engineering applications

(Martinez et al. 2014; Hu and Chen 2014; Du et al. 2016). Two important approaches have been adopted in developing mechanically strong and highly stretchable hydrogels. These are (i) Double Network (DN) hydrogels and (ii) Nanocomposite (NC) hydrogels.

### 9.9.1 Double Network (DN) Hydrogels

Double network (DN) hydrogels have attracted much recent interest in improving the strength properties of hydrogels. These hydrogels have high molecular mass and double polymer networks, in which one type of monomer is polymerized and cross-linked in the presence of another swollen network (Zhang et al. 2014). Figure 9.8 is a schematic of the synthesis of a DN hydrogel. The mechanical properties of DN hydrogels are always superior to those of single network hydrogels owing to mutual support by each network. Figure 9.9 compares the tensile fracture strengths and strains of DN hydrogels with the equivalent properties of single network (SN) hydrogels and other soft materials (plastics and rubbers). The DN hydrogels are significant improvements on SN hydrogels and can reach fracture *strains* exceeding those of rubbers and plastics. However, DN hydrogels cannot match rubbers and plastics in terms of fracture *stress*. This is hardly surprising, given that the hydrogels are 90% water.

More specific examples of DN hydrogel properties are listed here:

- A tough (DN) hydrogel of (i) poly(2-acrylamid-2-methyl-propane sulphonic acid) (PAMPS) as the tightly cross-linked first network, and (ii) polyacrylamide (PAM) as the loosely cross-linked (or even uncross-linked) second network has been reported (Omidian et al. 2006). This PAMPS/PAM hydrogel consisted of 90 wt% water and yet exhibited relatively high compressive strength (more than 20 MPa), strain, toughness and Young's modulus (0.1 MPa) (Gong and Katsuyama 2003).
- An extremely stretchable and tough DN hydrogel has been obtained, consisting of (i) an ionically cross-linked alginate (through divalent and multivalent cations) as the first network, and (ii) a covalently cross-linked polyacrylamide (Sun et al. 2012b; Yang et al. 2013). The hydrogels had much better mechanical properties than those of the parent alginate and polyacrylamide gels. The fracture stress and strain of the DN hydrogel with divalent cation were 156.0 kPa and 23%; whereas the individual parent fracture stresses and strains were (i) 3.7 kPa and 1.2% for the alginate gel and (ii) 11.0 kPa and 6.6% for the polyacrylamide gel (Sun et al. 2012b).
- The authors have recently reported DN hydrogels consisting of a PVA-borax double strand network and another network of a cross-linked copolymer of acrylamide (AM) and 2-hydroxyethylmethacrylate (HEMA) (Dixit and Bag 2016). Such a DN hydrogel having 80% water content exhibited a tensile strength of 32.2 kPa and a truly remarkable elongation of 206%.

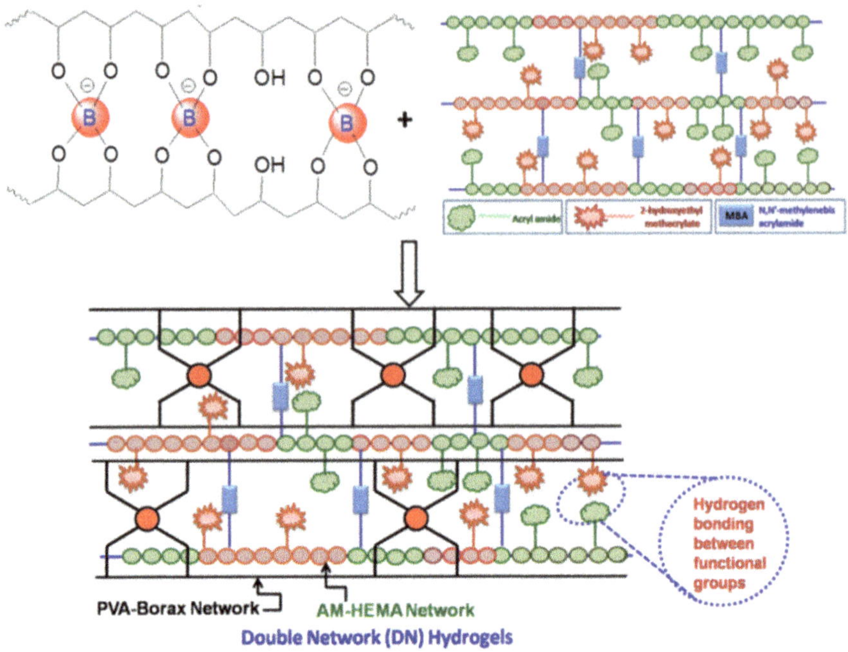


**Fig. 9.8** Schematic representation of DN hydrogel synthesis: **a** formation of the first network of poly(vinyl alcohol) and borax, **b** formation of the second network of acrylamide and 2-(hydroxyl ethylmethylacrlate) in the presence of the first network, leading to **c** formation of the DN network (Dixit et al. 2017) (*Adopted with the permission of Elsevier*)

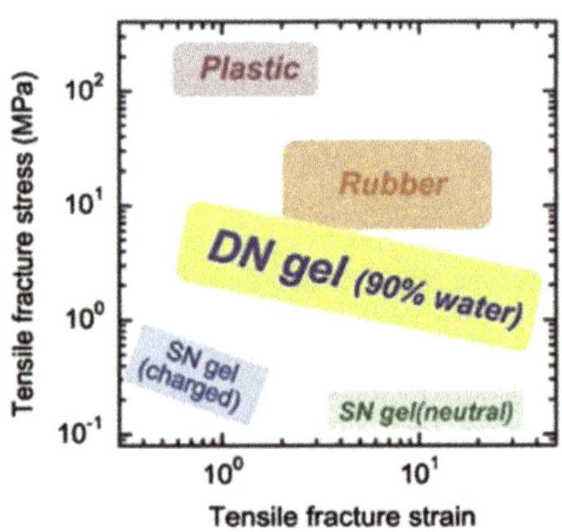


**Fig. 9.9** Comparison of the toughnesses of single (SN) and double (DN) network hydrogels with other polymer materials (Haque et al. 2012) (*Adopted by the permission of Elsevier*)

**Table 9.5** Compressive strength of some DN hydrogels (Gong and Katsuyama 2003)

| Sl. no | First network* | Second network* | Water content (%) | Fracture stress (MPa) | Fracture strain (%) |
|---|---|---|---|---|---|
| 1 | PAMPS | PAA | 92 | 2.3 | 75 |
| 2 | PAMPS | PAM | 90 | 17.2 | 92 |
| 3 | PAA | PAM | 89 | 2.1 | 95 |
| 4 | Agarose | PHEMA | 66 | 2.4 | 87 |
| 5 | Cellulose | Gelatin | 78 | 3.7 | 37 |

* PAMPS: poly(2-acrylamido-2-methyl propane sulphonic acid); PAA: poly(acrylic acid); PAM: poly(acrylamide); PHEMA: poly(2-hydroxy ethylmethacrylate); TFEA-2,2,2-triflouroethyl acrylate

The mechanical strengths of DN hydrogels depend not only on composition but also significantly on the presence of H-bonding. Water content also has an influence. Table 9.5 is included here to illustrate the mechanical property results for several types of DN hydrogels.

### 9.9.2 Nanocomposite (NC) Hydrogels

Nanocomposite hydrogels (NCs) are usually nanofiller/nanoparticle-reinforced polymer networks swollen in an aqueous medium. They are usually prepared from polar monomers or polymers plus nanofillers that have at least one dimension less than 100 nm. Synthesis is by chemical and physical cross-linking of functional monomers in the presence of nanofillers. The nanofillers/nanoparticles may be carbon-based nanomaterials like fullerenes, carbon nanotubes, graphene or other inorganic and metal nanoparticles. They have large surface area to mass ratios owing to their small size, leading to high filler connectivity and small inter-particle separation when dispersed in a matrix.

Nanoscale dispersion of fillers in the composite hydrogels can introduce new physical properties and novel behaviour absent in unfilled hydrogels. The fillers can also improve some of the existing properties. Improvements on properties may depend on the nature of the nanoparticles and their sizes, shapes, aspect ratios, moduli, dispersion, and alignments and also interaction of the nanoparticles with the matrix polymers/hydrogels.

These composite hydrogels have various advantages such as reinforced strength and toughness, modified hardness and plasticity. Nanocomposite hydrogels are also reported to have other functional properties like catalytic activity, super paramagnetic and conducting properties. For example, fullerene-containing nanocomposite hydrogels are used for drug delivery systems (DDS) and also in solar cell applications. Other important applications of nanocomposite hydrogels are in sensors, smart actuators for chemical valves, artificial ‘muscles’, optical systems, soft biomimetic machines, 'on/off' switches for chemical reaction, scaffolds for tissue engineering, and matrices for bio separation (Nath et al. 2014; Zinchenko et al. 2015; Arslantunali et al. 2014; Davaran et al. 2014; Curcio et al. 2015).

#### 9.9.2.1 Fullerene-Containing NC Hydrogels

Fullerenes are less used in smart hydrogels than carbon nanotubes (CNTs) and graphene. Fullerene ($C_{60}$)-containing cross-linked poly(2-hydroxyethylmethacrylate) hydrogels have been reported to exhibit a lower equilibrium degree of swelling than that of the fullerene-free hydrogels, owing to the presence of hydrophobic fullerene nanomaterials inside the hydrogels (Katiyar et al. 2013). These nanomaterials act as inherent hydrophobic nanofillers during water permeation inside the polymers and thereby slow the swelling process. Other effects induced by the fullerene nanofillers are as follows:

- Fullerene-containing thermo-responsive poly(N-isopropylacrylamide) hydrogels also show a lesser degree of swelling (Katiyar et al. 2014). However, the LCSTs of these hydrogels were hardly altered. On the other hand, the glass transition temperature ($T_g$) and thermo-stability increased with increasing fullerene content in the cross-linked polymers.
- Fullerene-containing hydrogels have also been studied for biomedical applications. Glycol-chitosan nanogels containing fullerene were investigated for efficient tumour therapy (Kim et al. 2014).

#### 9.9.2.2 CNT-Containing NC Hydrogels

Carbon nanotubes (CNTs) are huge cylindrical molecules built up from a hexagonal arrangement of carbon atoms ($sp^2$ hybridized). The walls of CNTs are composed of one or more rolled-up layers of graphene sheets. Thus, one may distinguish between single-walled nanotubes (SWCNTs) and multi-walled nanotubes (MWCNTs). Both types are being investigated for biomedical, functional and mechanical applications.

CNTs are hydrophobic, and there is limited interaction with hydrophilic groups because of $\pi$–$\pi$ interactions between the CNTs (Li et al. 2014; Gaharwar et al. 2014). This hydrophobic nature requires pre-treatments (such as functionalization with a strong acid or surface modification with surfactants, polymers or proteins) for improving the dispersion of CNTs during hydrogel synthesis. CNT-containing NC hydrogels combine properties of hydrogels and CNTs, resulting in improved functional (Ye et al. 2012; Zhang et al. 2011), biological (Manek et al. 2016; Liu et al. 2014) and mechanical properties (Shin et al. 2013). Some information on their mechanical properties is given in the following paragraphs:

(1) The mechanical properties of CNT-containing NC hydrogels are reported to be highly dependent on the nanotube dispersion, which directly influences the molecular tube–tube and tube–polymer interfacial bonding in the composites,

and hence the NC strength and other properties. Variations in nanotube dispersion could be the major cause of changes in swelling properties, compressive strength and elastic modulus (Shin et al. 2012; Dong et al. 2013).

(2) Polyacrylamide hydrogels can be significantly improved by incorporating the hydrophilic MWCNTs with poly(acrylic acid) brushes. The tensile strength of the composite hydrogel PAM/ MWCNT-PAA390 with 1.4 wt% of MWCNTs is 3.9 times that of the original hydrogel, reaching 82.7 kPa (Shin et al. , 2013).

(3) CNT nanocomposite DN hydrogels with very good mechanical strength have been reported (Dong et al. 2013). An individual poly(2-acrylamido-2-methylpropanesulfonic acid) (PAMPS) hydrogel containing CNTs was fabricated as the first nanocomposite network, and allowed to swell to equilibrium in an acrylamide solution, which was then polymerized to form the second network. The CNT additions to the PAMPS/AM nanocomposite DN hydrogel resulted in improved mechanical properties. For example, the compressive stress of PAMPS-PAAM DN gels was 19 MPa at a fracture strain of 0.94, whereas the nanocomposite DN hydrogel with 4 wt% CNTs reached 78 MPa at a strain of 0.98 and without breaking at this strain level (Dong et al. 2013).

#### 9.9.2.3 Graphene-Based NC Hydrogels

Graphene is a truly two-dimensional material consisting of a single atomic layer of graphite and has drawn immense attention in science and technology due to its excellent mechanical properties and superior thermal and electrical conductivities (Rao et al. 2009). Among various carbon-based nanoparticles (NPs), graphene-based materials including graphene, graphene oxide (GO) and reduced graphene oxide (rGO), have been extensively explored as novel reinforcing fillers for nanocomposite hydrogels. This is because of their distinctive properties, such as large surface area (2620 $m^2/g$), high aspect ratio, outstanding mechanical properties (Young's modulus of 1 TPa and intrinsic strength of 130 GPa), excellent thermal and electrical conductivity, and also good optical transparency (Chung et al. 2013; Limbu et al. 2017).

The incorporation of graphene-based materials into hydrogels has proved extremely valuable for improvement of the mechanical, thermal and conductive properties of hydrogels (Zhoo et al. 2017; Sayyar et al. 2015; Chen et al. 2013). Some results are listed here:

- NC hydrogels with graphene oxide were found to be more ductile and capable of sustaining large deformation and complex shear force fields. The simultaneous increase in strength and ductility was attributed to the strength and flexibility of the graphene oxide components (Zhoo et al. 2017).
- Swelling of a graphene/chitosan composite hydrogel decreases with increasing graphene content, presumably due to the interaction between the polymer matrix and the hydrophobic graphene nanosheets (Ye et al. 2012). Also, incorporation of only 0.5 wt% graphene improved the tensile strength by 58%; whereas the addition of 3 wt% graphene improved the tensile strength by more than 223% and the Young's modulus by more than 135%. These improvements indicate

good dispersion of graphene sheets in the composite. Another improvement is that swollen samples show much better elongation at break compared to dried samples (Chen et al. 2013).
- Graphene oxide (GO) has been incorporated into DN (double network) hydrogels of sodium alginate/polyacrylamide. Details of this complex process are given in Zhu et al. (2014). The GO additions increased the tensile and compressive strengths, and doubled the elastic modulus from 127.5 to 202.3 kPa. Such GO-containing hydrogels also maintained good ductility. The improvement in properties is associated with the effective cross-link of GO nanosheets with the polymer chains of hydrogels through surface anchoring and hydrogen interaction.
- Graphene oxide (GO) incorporated nanocomposite cationic hydrogels of acrylamide (AM) and 2-(dimethylamino)ethylacrylatemethochloride (DAC) exhibited good mechanical properties without sacrificing their self-healing ability (Pan et al. 2017). Also, both properties were tunable by the GO content and the mass ratio of AM and DAC. Such GO-containing nanocomposite hydrogels possess high stiffness (Young's modulus: ~1.1 MPa) and toughness (~9.3 $MJm^{-3}$) as well as high self-healing efficiency (>92% of tensile strength).

## 9.10 Applications of Hydrogels

### *9.10.1 General Applications*

Soft hydrogel materials have been extensively used in a variety of applications. Some of the important applications are as follows:

- Super absorption, e.g. in diapers, sanitary towels, and as soil conditioners for plant growth in sandy and desert lands.
- Stimuli-responsive hydrogels for biomedical application like drug delivery systems (DDS), cell sheet engineering, and wound healing.
- Smart hydrogels for windows, sensors and actuators.
- Hydrogels in contact lenses and smart skin.
- Hydrogels for microfluidic and photovoltaic applications.
- Self-healing hydrogels for applications as scaffoldings in reconstructive tissue engineering.

As early as 1960, a hydrogel based on poly (2-hydroxyethylmethacrylate) (PHEMA) was described as a synthetic biocompatible material useful for contact lens applications. Later on, the discovery of the unique stimuli-responsive properties of smart hydrogels became extensively used to develop soft and wet smart technology. In 1996, Gel Sciences/Gel Med commercialized its first "smart gel" which was based on a thermo-responsive hydrogel and used as a shoe-insert to provide comfort. Since then, various applications of smart hydrogels have been explored which include "cloud gels" in smart windows and display devices, smart hydrogels in drug delivery systems, chemical memories, valves, sensors, hydrogel actuators for artificial muscles and organs, and hydrogels for reversible separation processes.

A more detailed summary of some of the more important hydrogel applications follows:

(1) Thermo-responsive and pH-responsive smart hydrogels, and hydrogels responding to specific biochemical triggers (particularly glucose-sensitive hydrogels and antigens- and enzyme-sensitive hydrogels), have been widely investigated for potential biomedical applications; see point (3) also.
(2) The chemical-, solvent- and pH-dependent swelling–deswellng abilities of hydrogels are used to develop appropriate chemical and pH sensors. Hydrogel-based chemical sensors are also being considered for 'artificial noses' and 'artificial tongues'.
(3) Electrically-stimulated and magneto-responsive hydrogel swelling–deswelling is favoured for drug delivery systems (DDS), since an external electric or magnetic field stimulus is easily controlled, and it is desirable to have precise control over drug release from devices in the body. In addition, thermo-responsive and pH-responsive smart hydrogels are also useful for drug deliveries. Another, more recent, development is the possibility of using smart microgels and nanogels for more precise drug deliveries, although there are health concerns with respect to using nanotechnology.
(4) Magneto-responsive hydrogels, especially magnetic particles embedded in thermo-responsive hydrogels, are widely studied for targeted drug delivery and cancer therapies, since the drugs can be magnetically guided to the appropriate location in the body.
(5) Electro-responsive and electro-conductive hydrogels are very effective soft materials for making sensors and actuators, since the electric field stimulus is easily controlled. Also, the sensing and actuation responses of hydrogels can be tailored by choosing an appropriate electrical input signal.
(6) Electro-conducting hydrogels are also used in fuel cells, dye sensitive solar cells, super capacitors, and rechargeable lithium batteries. They are also being investigated for high-speed valves and pumps in microfluidic applications in toys, valves and switches, and heat-shrinkable tubing; and they can also be used to build micro-electro-mechanical systems (MEMS) both in (or as) actuators and sensors.
(7) Hydrogels are now being actively researched for functional textiles and smart clothing. One of the best methods to produce textiles with more functionality is smart finishing of textiles by surface-modifying stimuli-responsive hydrogels. For example, a smart skin hydrogel, using a combination of hydrophilic/hydrophobic copolymers can be embedded in an open-cell foam layer bonded to the inside of a closed-cell neoprene layer; and this forms part of a composite wet suit with nylon outer and inner layers.

## 9.10.2 Defence Applications

### 9.10.2.1 Actuators

Modern technological actuator applications require improved performance, ease of fabrication, reduced size and mass, low power consumption, and low-cost actuator materials. These challenges are being met by polymeric hydrogel materials, which are superior to conventional electro-active (piezoelectric and electro strictive) ceramic materials. Also, electro-active ceramics are inherently (and unavoidably) brittle materials.

***Advantages of electro-active hydrogels***: These have the advantages of light weight and ease of fabrication in various shapes and are also more suitable for small-scale systems like micro- and nano-robots and devices. Thus it is possible to fabricate low-power-consuming and inexpensive miniature gel actuators. Moreover, the actuation capability of electro-active polymers may be more than two orders of magnitude higher than that of electro-active ceramics. This high actuator capability of hydrogel actuators enables them to be designed for the operation of biological muscles. Such actuators also have the unique characteristics of high toughness and inherent vibration damping.

***Disadvantages of electro-active hydrogels***: Polyelectrolyte hydrogel actuators have certain inherent disadvantages. They require the presence of water/solvent and salts for their actuation, and the actuating electrolysis is accompanied by generation of a large electric current and heat. Moreover, highly cross-linked hydrogels are brittle, especially in their swollen state, and hence unable to provide good strength. In order to improve the strength and overcome the shortcomings of ionic hydrogels, ionomeric polymer-metal composites (IPMCs) are now being studied for actuator materials. In IPMCs metal ions (platinum, gold, etc.) are dispersed throughout the hydrophilic film surface of the ionomer films. An example of an IPMC actuator is shown in Fig. 9.10.

### 9.10.2.2 Other Major Applications

Some of the applications of hydrogel materials discussed in in Sect. 9.6 and also in 9.10.1 and 9.10.2 may be useful for the defence sector:

(1) Hydrogel-based sensors and actuators may be useful for components in different instruments and kits required for defence applications, including aero structures.
(2) Biomedical application: hydrogels are essential for military personnel, including combat medical care and wound-healing applications. For example, haemorrhaging is a leading cause of death on the battle field: up to 90% of preventable combat deaths occur due to uncontrolled bleeding. Conventional techniques to stop arterial bleeding must pack gauze directly into the wound cavity, sometimes as deep as 5 inches into the body. However, this may not always work properly.

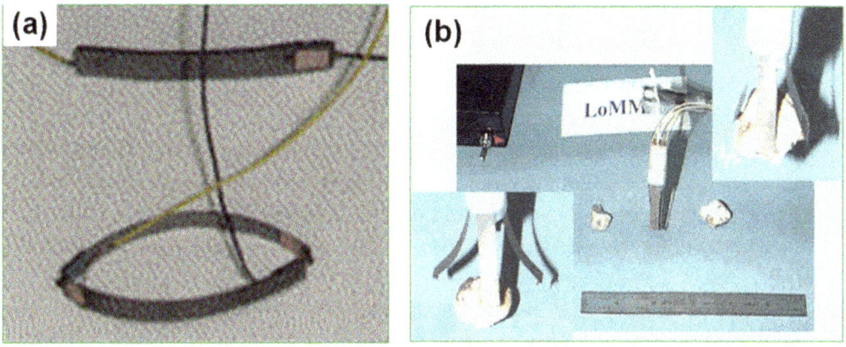


Film pair linear actuator: 5 Volt, 11% actuation

Gripper lifting 10.3 gm rock under 5 Volt using four 0.1gm fingers

**Fig. 9.10** IPMC actuator: **a** IPMC based on nafion and platinum composite (film (thickness: 0.18 mm, strips: 1 × 0.125 inch, weight: 0.1 gm), and **b** actuator lifting a rock (*Photographs by the permission of Proceedings of SPIE's 5th Annual International Symposium on Smart Structures and Materials, 1–5 March, 1998, San Diego, CA. Paper No. 3324-27 SPIE Copyright © 1998*)

If bleeding is not stopped after three minutes of applying direct pressure, all the gauze must be pulled out, a very painful process, and the procedure must start again.

(3) Currently, hydrogels can be used to stop severe bleeding almost instantly; see Fig. 9.11.This could save the lives of soldiers on the battlefield. The hydrogels may be used in different forms such as tablets, in packets or on gauze. Hydrogel therapy also provides physical protection while maintaining a moist environment to aid long-term healing and prevent infection. A healing gel implanted with stem cells will encourage migration of new cells into the wound site to promote healing.

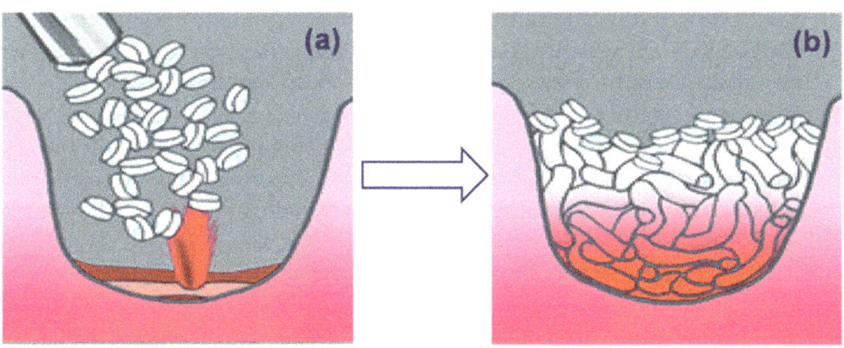


**Fig. 9.11** Combat medical care to stop bleeding using hydrogels **a** hydrogel tablets are deposited in the wound, and **b** they swell and prevent bleeding (*Photographs from* https://www.meddeviceonline.com/doc/fda-clears-xstat-hemostatic-dressing-for-civilian-use-0001)

## 9.11 Summary and Concluding Remarks

1. Hydrogel materials are already being used in commercial applications such as contact lenses, hygiene products and wound dressings. However, commercial hydrogel products for tissue engineering and drug delivery are still limited. Many hydrogel-based drug delivery devices and scaffolds, soft gel actuators, sensors and others have been designed, studied and demonstrated, but not many have reached the marketplace. This situation is due, to some extent, to high production costs.
2. There are thus many challenges in achieving commercial success in developing and commercialising hydrogel applications. Fast swelling hydrogels with high swelling capacity and fast controlled response to external stimuli are required for most successful applications of smart hydrogels. Moreover, smart hydrogel materials having macro- and nano-dimensions are necessary for certain areas, especially biomedical applications inside the human body. High strength and stretchable hydrogels are desirable for gel-based actuators, artificial cartilage in tissue engineering, artificial nerves and muscles.

Some of the strategies for the successful use of soft and novel hydrogel materials are as follows:

(1) To counter too-slow responses in conventional hydrogels, research is concentrating on size decreases to micro- or nano-dimensions (micro- and nanogels) and also the formation of porous structures (microporous and superporous hydrogels).
(2) Double network (DN) and also nanocomposite hydrogels have been focussed on as strategies to develop hydrogels with sufficient strength and stretch ability for many actuators and other applications.
(3) Development of self-healing hydrogels is one of the most interesting areas of current hydrogel research. Such hydrogels would be beneficial in all applications in general and especially as scaffolding in reconstructive tissue engineering.
(4) Electro-responsive and magneto-responsive hydrogels have been much studied because they can be easily controlled by electric or magnetic fields. Such smart hydrogels can be used for making gel actuators and artificial muscles.
(5) Magneto-responsive hydrogels and multifunctional nanogels for small interfering RNA (siRNA) delivery have shown potential promise in cancer therapy. Targeted drug delivery using magneto-responsive hydrogels and magnetic field stimulus is an innovative approach for controlled drug release in the treatment of cancers.
(6) Thermo-responsive hydrogel-based cell culture dishes are advantageous for cell sheet engineering.
(7) Hydrogels also have potential use in energy applications, e.g. regenerable microfluidic hydrogel photovoltaics mimic the functionality of a plant leaf to accumulate and store solar energy.

Some specific shortcomings (discussed earlier) of hydrogels have restricted many commercial applications. Much effort will be required to overcome these limitations, but the challenges are, and will be, very worthwhile. Perhaps the most outstanding challenge is to properly understand the true mechanisms of stimuli-responsive hydrogels, in order to improve their performance and fully utilize their potentials for smart devices and biological systems.

**Acknowledgements** The authors gratefully acknowledge the Defence Research and Development Organization (DRDO), Ministry of Defence, Government of India, for all the necessary support and funding to carry out this research. They would like to particularly thank Professor GN Mathur, Former Director, DMSRDE, Dr. G. Satheesh Reddy, Former Chairman, DRDO & Secretary, DD R&D and Dr. Samir V. Kamat, Chairman, DRDO and Secretary, DD R&D for their valuable association and constant encouragement.

## References

Agnihotri, S., Singhal, R.: Effect of sodium humate on the swelling characterizateristic and agricultural application of superabsorbrnt hydrogels of poly (acrylic acid/sodium humate). J. Polym. Mater. (in December, 2017) (2017)

Ahmed, E.M.: Hydrogel: preparation, characterization and applications: a review. J. Adv. Res. **2**(6), 105–121 (2015)

Akay, G., Hassan, R.A., Tuncaboylu, D.C., Orakdogen, N., Abdurrahmanoglu, S., Oppermann, W., Okay, O.: Self-healing hydrogels formed in catanionic surfactant solution. Soft Matter **9**(7), 2254–2261 (2013)

Arslantunali, D., Budak, G., Hasirci, V.: Multiwalled CNT-PHEMA composite conduit for peripheralnerve repair. J. Biomed. Mater. Res. Part A **102**(3), 828–841 (2014)

Aviram, A.: Mechanophotochemistry. Macromolecules **11**(6), 1275–1280 (1978)

Bag, D.S., Alam, S.: Chiral chemical absorption property of a cross-linked poly(*N*-isopropyl acrylamide-*co*-sodium acrylate) thermo-responsive smart gel. Chirality **24**, 506–511 (2012)

Bag, D.S., andRao, KUB,: Smart polymers and their applications. J. Polym. Mater. **23**, 225–228 (2006)

Bag, D.S., Alam, S., Mathur, G.N.: Terpolymer smart gels: synthesis and characterizations. Smart Mater. Struct. **5**(13), 1258–1262 (2004)

Bag, D.S.: Principles of Polymers-An Advanced Book, pp. 532–542. Nova Science Pub., New York (2013)

Bag, D.S., Saxena, A.K.: Smart gels. In: Kirk-Othmer (ed.) Encyclopedia of Chemical Technology, pp. 1–50. Wiley, New York, USA (2014)

Bar-Cohen, Y. (ed.).: Electroactive Polymer (EAP) Actuators as Artificial Muscles: Reality, Potential and Challenges. SPIE Press, Bellingham, USA (2001)

Cawse, J.L., Lovell, P., El-Aasser, M.S.: Emulsion Polymerization and Emulsion Polymers, pp. 743–761. Wiley, West Sussex, UK (1997)

Chang, C., Zhang, L., Zhou, J., Zhang, L., Kennedy, J.F.: Structure and properties of hydrogels prepared from cellulose in NaOH/urea aqueous solutions. Carbohyd. Polym. **1**(82), 122–127 (2010)

Chen, J., Park, K.: Superporous hydrogels: fast responsive hydrogel systems. J. Macromol. Scie. Part A: Pure Appl. Chem. **36**, 917–930 (1999)

Chen, Y., Chen, L., Bai, H., Li, L.: Graphene oxide–chitosan composite hydrogels as broad-spectrum adsorbents for water purification. J. Mater. Chem. A **1**(6), 1992–2001 (2013)

Chung, C., Kim, Y.K., Shin, D., Ryoo, S.R., Hong, B.H., Min, D.H.: Biomedical applications of graphene and graphene oxide. Acc. Chem. Res. **46**(10), 2211–2224 (2013)

Corcione, E.C., Frigione, M., Maffezzoli, A., Malucelli, G.J.: Photo – DSC and real time – FT-IR kinetic study of a UV curable epoxy resin containing. Eur. Polym. J. **45**, 2060–2067 (2008)

Curcio, M., Spizzirri, U.G., Cirillo, G., Vittorio, O., Picci, N., Nicoletta, F.P., Iemma, F., Hampel, S.: On demand delivery of ionic drugs from electro-responsive CNT hybrid films. RSC Adv. **56**(5), 44902–44911 (2015)

Davaran, S., Alimirzalu, S., Nejati-Koshki, K., Nasrabadi, H.T., Akbarzadeh, A., Khandaghi, A.A., Abbasian, Mand Alimohammadi, S.: Physicochemical characteristics of $Fe_3O_4$ magnetic nanocomposites based on poly(N-isopropylacrylamide) for anti-cancer drug delivery. Asian Pacific J. Cancer Prevent. **15**(1), 49–54 (2014)

Dixit, A., Bag, D.S.: A novel double network hydrogel with high swelling and mechanical properties. J. Polym. Mater. **33**(2), 393–404 (2016)

Dixit, A., Bag, D.S., Kalra, S.J.S.: Synthesis of strong and stretchable double network (DN) hydrogels of PVA-borax and P(AM-co-HEMA) and study of their swelling kinetics and mechanical properties. Polymer **119**(1), 263–273 (2017)

Dong, W., Huang, C., Wang, Y., Sun, Y., Ma, P., Chen, M.: Superior mechanical properties of double-network hydrogels reinforced by carbon nanotubes without organic modification. Int. J. Mol. Sci. **14**, 22380–22394 (2013)

Du, J., Xu, S., Feng, S., Yu, L.: Tough dual nanocomposite hydrogels with inorganic hybrid crosslinking. Soft Matter **12**, 1649–1652 (2016)

Fang, B.-K., Lin, C.-C.K., Ju, M.-S.: Development of sensing/actuating ionic polymer-metal composite (IPMC) for active guide-wire system. Sens. Actuators: Phys. **158**(1), 1–9 (2010)

Gaharwar, A.K., Peppas, N.A., Khademhosseini, A.: Nanocomposite hydrogels for biomedical applications. Biotechnol. Bioeng. **111**(3), 441–453 (2014)

Gan, L.H., Gan, Y.Y., Roshan, D.G.: Poly(N-acryloyl-N′-propylpiperazine): a new stimuli-responsive polymer. Macromolecules **21**(33), 7893–7897 (2000)

Gelbrich, T., Marten, G.U., Schmidt, A.M.: Reversible thermo-flocculation of magnetic core shell particles induced by remote magnetic heating. Polymer **51**, 2818–2824 (2010)

Gemeinhart, R.A., Park, H., Park, K.: Pore structure of superporoushydrogels. Polym. Adv. Technol. **11**, 617–625 (2000a)

Gemeinhart, R.A., Chen, J., Park, H., Park, K.: pH-sensitivity of fast responsive superporous hydrogels. J. Biomater. Sci. Polym. Edn. **12**(11), 1371–80 (2000)

Gerlach, G., Guenther, M., Sorber, J., Suchaneck, G., Arndtand, K.-F., Richter, A.: Chemical and pH sensors based on the swelling behavior of hydrogels. Sens. Actuators, B Chem. **112**(11), 555–561 (2005)

Gong, J.P., Katsuyama, Y.: Double network hydrogels with extremely high mechanical strength. Adv. Mater. **15**, 1155–1158 (2003)

Haque, M.A., Kurokawa, T., Gong, J.P.: Super tough double network hydrogels and their application as biomaterials. Polymer **53**, 1805–1822 (2012)

Hirasa, O., Ito, S., Yamauchi, A., Fujishige, S., Ichijo, H.: Thermoresponsive polymer hydrogels. In: De Rossi, D., Kajiwara, K., Osada, Y., Yamauchi, A. (eds.) Polymer Gels. Fundamentals and Biomedical Applications, pp. 247–256. Plenum Press, New York (1991)

Hirasa, O., Ito, S., Yamauchi, A., Fujishige, S., Ichiji, H.: Thermoresponsive polymer hydrogels. In: Polymer Gel: Fundamentals and Biomedical Applications, Eds, DeRossi, D, Kajiwara, K, Osada Y and Yamauchi, A (Plenum Press, New York, USA) Pp.247–256 (1991)

Hu, Z., Chen, G.: Novel nanocomposite hydrogels consisting of layered double hydroxide with ultrahigh sensibility and hierarchical porous structure at low inorganic content. Adv. Mater. **26**(34), 5950–5956 (2014)

Hua, S., Wang, A.: Synthesis, characterization and swelling behaviors of sodiumalginate-g-poly(acrylic acid)/sodium humate superabsorbent. Carbohyd. Polym. **75**, 79–84 (2009)

Huang, J., Cusick, B., Pietrasik, J., Kowaleski, W.T., Lin, Q., Matyjaswski, K.: Synthesis and in situ atomic force microscopy characterization of temperature-responsive hydrogels based on

poly(2-(dimethylamino)ethyl methacrylate) prepared by atom transfer radical polymerization. Langmuir **23**, 241–249 (2007)
Itsuro, T., Peng, K., Alexander, K.: Photoresponsive hydrogels for biomedical applications. Adv. Drug Deliv. Rev. **63**(11–12), 1257–1266 (2011)
Jones, A.R., Sottos, N.R., White, S.R.: Full recovery of fiber/matrix interfacial bond strength using a microencapsulated solvent-based healing system. Compos. Sci. Technol. **79**, 1–7 (2010)
Juodkazis, S., Mukai, N., Wakaki, R., Yamaguchi, A., Matsuo, S., Misawa, H.: Reversible phase transitions in polymer gels induced by radiation forces. Nature **408**, 178–181 (2000)
Kabra, B.G., Gehrke, S.H.: Effect of the initiator on thermosensitive rate of poly(*N*-isopropylacrylamide) hydrogels. Polym. Commun. **32**, 322–323 (1991)
Katiyar, R., Bag, D.S., Nigam, I.: Fullerene ($C_{60}$) containing poly (N-isopropylacrylamide) thermo-responsive smart hydrogels and their swelling behavior. J. Polym. Mater. **30**, 15–26 (2013)
Katiyar, R., Bag, D.S., Nigam, I.: Synthesis and evaluation of swelling characteristic of fullerene ($C_{60}$) containing cross-linked poly(2-hydroxyethyl methacrylate) hydrogels. Adv. Mater. Lett. **5**(4), 214–222 (2014)
Ki-Bum, K., Duk-Geun, H.: Kinetic parameters, bleaching and radiation response of thermoluminescence glow peaks separated by deconvolution on Korean calcite. Radiat. Phys. Chem. **103**, 16–21 (2014)
Kim, S., Lee, D.J., Kwag, D.S., Lee, U.Y., Youn, Y.S., Lee, E.S.: Acid pH-activated glycol chitosan/fullerene nanogels for efficient tumor therapy. Carbohyd. Polym. **101**, 692–698 (2014)
Kishi, R., Hirasa, O., Ichijo, H.: Fast responsive poly(N-isopropylacrylamide) hydrogels prepared by γ-ray irradiation. Polym. Gels Netw. **5**, 145–151 (1997)
Kratz, K., Hellweg, T., Eimer, W.: Structural changes in PNIPAM microgel particles as seen by SANS, DLS, and EM Techniques. Polymer **42**(15), 6631–6639 (2001)
Li, W., Liu, M., Chen, H., Xu, J., Gao, Y., Li, H.: Phenylboronate-diol crosslinked polymer/SWCNT hybrid gels with reversible sol–gel transition. Polym. Adv. Technol. **25**(2), 233–239 (2014)
Li, P., Xu, R., Wang, W., Li, X., Xu, Z., Kelvin, W.K., Paul, K.C.: Platelet interaction with surface-immobilized IgG induces the secretion of surface bound microparticles and fragmentation of the cells. Colloids Surfaces B: Biointerfaces **101**, 251–262 (2013)
Limbu, T.B., Hahn, K.R., Mendoza, F., Sahoo, S., Razink, J., Katiyar, R.S., Weiner, B.R., Morell, G.: Grain size-dependent thermal conductivity of polycrystalline twisted bilayer graphene. Carbon **117**, 367–375 (2017)
Lira, L.M., Cordoba, T.S.I.: Conducting polymer-hydrogel composites for electrochemical release devices: synthesis and characterization of semi-interpenetrating polyaniline-polyacrylamide networks. Electrochem. Commun. **7**, 717–723 (2005)
Liu, T.-Y., Hu, S.-H., Liu, T.-Y., Liu, D.-M., Chen, S.-Y.: Magnetic-sensitive behavior of intelligent ferrogels for controlled release of drug. Langmuir **22**(14), 5974–5978 (2006)
Liu, F., Li, F., Deng, G., Chen, Y., Zhang, B., Zhang, J., Liu, C.Y.: Rheological images of dynamic covalent polymer networks and mechanisms behind mechanical and self-healing properties. Macromolecules **45**(3), 1636–1645 (2012)
Liu, J., Wang, Q., Wang, A.: Synthesis and characterization of chitosan-g-poly(acrylic acid)/sodium humate superabsorbent. Carbohydr. Polym. **70**, 166–173 (2007)
Liu, Z., Miao, Y., Wang, Z., Yin, G.: Synthesis and characterization of a novel superabsorbent based on chemically modified pulverized wheat straw and acrylic acid. Carbohydr. Polym. **77**, 131–135 (2009)
Liu, X.-W., Huang, Y.-X., Sun, X.-F., Sheng, G.-P., Zhao, F., Wang, S.-G., Yu, H.-Q.: Conductive carbon nanotube hydrogel as a bioanode for enhanced microbial electrocatalysis. ACS Appl. Mater. Interfaces **6**(11), 8158–8164 (2014)
Lu, J., Wang, W., Wu, D., Chen, G.: Fabrication and electrical and mechanical properties of carbon nanotube interconnections. Adv. Func. Mater. **15**, 1358–1361 (2005)
Lutz, J.F., Weichenhan, K., Akdemir, O., Hoth, A.: The phase transitions in aqueous solutions of thermoresponsive copolymers and hydrogels based on 2-(2-methoxyethoxy)ethyl methacrylate and oligo(ethylene glycol) methacrylate. Macromolecules **40**(7), 2503–2508 (2007)

Makuuchi, K.: Critical review of radiation processing of hydrogel and polysaccharide. Radiat. Phys. Chem. **3**(79), 267–271 (2010)

Mamada, A., Tanaka, T., Kungwatchakum, D., Irie, M.: Photoinduced phase transition of gels. Macromolecules **23**(5), 1517–1519 (1990)

Manek, E., Berke, B., Miklosi, N., Sajban, M., Doman, A., Fukuda, T., Czakkel, O., Laszlo, K.: Thermal sensitivity of carbon nanotube and graphene oxide containing responsive hydrogels. EXPRESS Polym. Lett. **10**(8), 710–720 (2016)

Martinez, R.V., Gelavan, A.C., Keplinger, C., Oyetibo, A.J., Whitesides, G.H.: Soft actuators and robots that are resistant to mechanical damage. Adv. Func. Mater. **24**(20), 3003–3010 (2014)

Maya, D.-P., Havazelet, B.-P.: A quantitative analysis of alginate swelling. Carbohyd. Polym. **80**, 1020–1027 (2010)

Meena, R., Chhatbar, M., Prasad, K., Siddhanta, A.K.: Development of a robust-hydrogels system based on agar and sodium alginate blend. Polym. Int. **57**(2)**,** 329–336 (2008)

Moschou, E.A., Madou, M.J., Bachas, L.G., Daunert, S.: Voltage-switchable artificial muscles actuating at near neutral pH. Sens. Actuators, B Chem. **115**, 379–383 (2006)

Nath, B.C., Gogoia, B., Boruaha, M., Sharma, S., Khannama, M., Ahmed, G., Dolui, S.K.: High performance polyvinyl alcohol/multi walled carbon nanotube/polyaniline hydrogel (PVA/MWCNT/PAni) based dye sensitized solar cells. Electrochim. Acta **146**, 106–111 (2014)

Oh, J.K., Drumright, R., Siegwart, D.J., Matyjaszewski, K.: The development of microgels/nanogels for drug delivery applications. Prog. Polym. Sci. **4**(33), 448–477 (2008)

Oh, J.K., Lee, D.I., Park, J.M.: Biopolymer based microgels/nanogels for drug delivery applications. Prog. Polym. Sci. **34**, 1261–1282 (2009)

Omidian, H., Rocca, J.G., Park, K.: Advances in superporoushydrogels. J. Control. Release **102**, 3–12 (2005)

Omidian, H., Rocca, J.G., Park, K.: Superporous hydrogel hybrids of polyacylammide and sodium alginate. Macromol. Biosci. **6**(9), 703–710 (2006)

Osada, Y., Okuzaki, H., Hori, H.: Stimuli-responsive poly(N-isopropylacrylamide): photo- and chemical induced phase transitions. Nature **355**, 242–244 (1992)

Pan, C., Liu, L., Chen, Q., Zhang, Q., Guo, G.: Tough, stretchable, compressive novel polymer/graphene oxide nanocomposite hydrogels with excellent self-healing performance. ACS Appl. Mater. Interfaces **9**(43), 38052–38061 (2017)

Pelton, R., Hoare, T., Fernandez-Nieves, A., Wyss, H.M., Mattsson, J., Weitz, D.A.: Microgel Suspensions: Fundamentals and Applications, Chap. 1. Wiley, New York (2011)

Phadke, A., Zhang, C., Arman, B., Hsu, C.-C., Mashelkar, R.A., Lele, A.K., Tauber, M.J., Arya, G., Varghese, S.: Rapid self-healing hydrogel. Proc. Natl. Acad. Sci. U.S.A. **109**(12), 4383–4388 (2012)

Pich, A., Bhattacharya, S., Lu, Y., Boyko, V., Adler, H.J.P.: Temperature-sensitive hybrid microgels with magnetic properties. Langmuir **20**(24), 10706–10711 (2004)

Plamper, F.A., Ruppel, M., Schmalz, A., Borisov, O., Ballauff, M., Muller, H.E.: Tuning the thermoresponsive properties of weak polyelectrolytes: aqueous solutions of star-shaped and linear poly(*N,N*-dimethylaminoethyl methacrylate). Macromolecules **40**(23), 8361–8366 (2007)

Pourjavadi, A., Kurdtabar, M.: Collagen-based highly porous hydrogel without any porogen: synthesis and characteristics. Eur. Polym. J. **43**, 877–889 (2007)

Rao, C.N.R., Sood, A.K., Subrahmanyam, K.S., Govindaraj, A.: Graphene: the new two-dimensional nanomaterial. Angew. Chem. Int. Edn. **48**(42), 7752−7777 (2009)

Roy, D., Camber, J.N., Sumerlin, B.S.: Future perspectives and recent advances in stimuli-responsive materials. Prog. Polym. Sci. **35**(1–2), 278–301 (2010)

Sadeghi, M., Heidari, B., Montazeri, K.: Investigation and evalution of swelling kinetics related to biocopolymers based on CMC-poly(AA-co BuMC). World Acad. Sci. Eng. Technol. **5**, 217–219 (2011)

Said, H.M., Alla, S.G.A., El-Naggar, A.W.M.: Synthesis and characterization of novel gels based oncarboxymethyl cellulose/acrylic acid prepared by electron beam irradiation. React. Funct. Polym. **61**, 397–404 (2004)

Sayyar, S., Murray, E., Thompson, B.C., Chung, J., Officer, D.L., Gambhir, S., Spinks, G.M., Wallace, G.G.: Processable conducting graphene/chitosan hydrogels for tissue engineering. J. Mater. Chem. B **3**, 481–490 (2015)

Schild, H.G.: Poly(*N*-isopropylacrylamide): experiment, theory and application. Prog. Polym. Sci. **2**(17), 163–249 (1992)

Schuetz, Y.B., Gurny, R., Jordan, O.: A novel thermoresponsive hydrogel based on chitosan. Eur. J. Pharm. Biopharm. **68**, 19–25 (2008)

Seiffert, S.: Small but smart: sensitive microgel capsule. Angew. Chem. Int. Ed. **52**, 11462–11468 (2013)

Sethuraman, V.A., Na, K., Bae, Y.H.: PH-responsive sulfonamide/PEI system for tumor specific gene delivery: an in vitro study. Biomacromol **7**(1), 64–70 (2006)

Shahinpoor, M.: Ionic polymer–conductor composites as biomimetic sensors, robotic actuators and artificial muscles—a review. Electrochem. Acta **48**(14–16), 2343–2353 (2003)

Shin, S.R., Jung, S.M., Zalabany, M., Kim, K., Zorlutuna, P., Kim, S., Nikkhah, M., Khabiry, M., Azize, M., Kong, J., Wan, K., Palacios, T., Dokmeci, M.R., Bae, H., Tang, X., Khademhosseini, A.: Carbon-nanotube-embedded hydrogel sheets for engineering cardiac constructs and bioactuators. ACS Nano **7**(3), 2369–2380 (2013)

Shin, S.R., Bae, H., Cha, J.M., Mun, J.Y., Chen, Y.C., Tekin, H., Shin, H., Farshchi, S., Dokmeci, M.R., Tang, S., Khademhosseini, A.: Carbon nanotube reinforced hybrid microgels as scaffold materials for cell encapsulation. ACS Nano, **6**(1), 362–372 (2012)

Singh, T., Singhal, R.: Poly (acrylic acid/acryl amide/ sodium humate) superabsorbrnt hydrogels for metal ion/ dye adsorption: effect of sodium humate concentration. J. Appl. Polym. Sci. **4**(125), 1267–1283 (2012)

Singh, B., Vashishth, M.: Development of novel hydrogels by modification of sterculia gum through radiation cross-linking polymerization for use in drug delivery. Nucl. Inst. Methods Phys. Res. B **266**, 2009–2020 (2008)

Smith, M.H., Lyon, L.A.: Multifunctional nanogels for siRNA delivery. Acc. Chem. Res. **45**(7), 985–993 (2012)

Spagnol, C., Rodrigues, F.H.A., Fajardo, R.A., Rubira, F.A., Muniz, F.E., Pereira, A.G.B.: Superabsorbent hydrogel composite made of cellulose nanofibrils and chitosan-graft-poly(acrylic acid). Carbohydr. Polym. **87**, 2038–2045 (2012)

Starodubtsev, S., Saenko, E., Dokukin, M.E., Aksenov, V.L., Klechkovskaya, V., Zanaveskina, I., Khokhlru, A.: Magnetic-sensitive behavior of intelligent ferrogels for controlled release of drug. J. Phys. Condens. Matter **22**(17), 1471–1480 (2005)

Sun, J.Y., Zhao, X., Illeperuma, W.R.K., Chaudhuri, O., Oh, K.H., Mooney, D.J., Vlassak, J.J., Suo, Z.: Highly stretchable and tough hydrogels. Nature **6**(489), 133–136 (2012a)

Sun, J.-Y., Zhao, X., Illeperuma, W.R.K., Chaudhuri, O., Oh, K.H., Money, D.J., Vlassak, J.J., Suo, Z.: Highly stretchable and tough hydrogels. Nature **489**, 133–136 (2012b)

Suzuki, A., Ishii, T., Maruyama, Y.: Optical switching in polymer gels. J. Appl. Phys. **80**(1), 131–136 (1996)

Tang, M.D., Golden, A.P., Tien, J.: Molding of three-dimensional microstructures of gels. J. Am. Chem. Soc. **125**, 12988–12989 (2003)

Taylor, L., Cerankowski, L.D.: Preparation of films exhibiting a balanced temperature dependence to permeation by aqueous solutions—a study of lower consolute behavior. J. Polym. Sci. Part A: Polym. Chem. **13**(11), 2551–2570 (1975)

Taylor, D.L., Panhuis, M.: Self-healing hydrogels. Adv. Mater. **28**(41), 9060–9093 (2016)

Wei, J., Chen, Q., Chu, L.Y.: Novel levan and pNIPA temperature sensitive hydrogels for **5**-ASA controlled release. Langmuir **30**(5), 1455–1464 (2014)

Wu, X.S., Hoffman, A.S., Yager, P.: Synthesis and characterization of thermally reversible macroporous poly (N-isopropylacrylamide) hydrogels. J. Polym. Sci. Part a: Polym. Chem. **30**, 2121–2129 (1992)

Wu, F., Zhang, Y., Liu, L., Yao, J.: Synthesis and characterization of a novel cellulose-g-poly(acrylic acid-co-acrylamide) superabsorbent composite based on flax yarn waste. Carbohydr. Polym. **87**, 2519–2525 (2012)

Xiao, X.-C., Chu, L.-Y., Chen, W.-M., Wang, S., Xie, R.: Preparation of submicrometer-sized monodispersed thermoresponsive core-shell hydrogel microspheres. Langmuir **20**(13), 5247–5353 (2004)

Xie, L., Liu, M., Zhang, M., Xu, N.B., Wang, Y.: Slow-release nitrogen and boron fertilizer from a functional superabsorbent formulation based on wheat straw and attapulgite. Chem. Eng. J. **167**, 342–348 (2011)

Yan, B., Ma, J., Li, N.: Synthesis and swelling behavior of sodium carboxymethyl cellulose-g-poly(AA-co-AM-co-AMPS)/MMT superabsorbent hydrogel. Carbohyd. Polym. **84**, 76–82 (2011)

Yang, C.H., Wang, M.X., Haider, H., Yang, J.H., Sun, J.-Y., Chen, Y.M., Zhao, J., Suo, Z.: Strengthening alginate/polyacrylamide hydrogels using various multivalent cations. ACS Appl. Mater. Interfaces **5**(21), 10418–10422 (2013)

Ye, Y., Mao, Y., Wang, H., Ren, Z.: Hybrid structure of pH-responsive hydrogel and carbon nanotube array with superwettability. J. Mater. Chem. **22**(6), 2449–2455 (2012)

Yoshida, R., Uchida, K., Kaneko, Y., Sakai, K., Kikuchi, A., Sakurai, Y., Okaro, T.: Temperature-sensitive drug delivery system composed of poly (N,N-dimethylaminoethyl methacrylate-co-acrylamide). Nature **374**, 240–242 (1995)

Zhang, J.L., Huang, W.M., Gao, G., Fu, J., Zhou, Y., Salvekar, A.V., Venkatraman, S.S., Wong, Y.S., Tay, K.H., Birch, W.R.: Shape memory/change effect in a double network nanocomposite tough hydrogel. Eur. Polymer J. **58**, 41–51 (2014)

Zhang, J., Ruifeng, L., Li, A., Wang, A.: Preparation, swelling behaviors, and slow-release properties of a poly(acrylic acid-co-acrylamide)/sodium humate superabsorbent composite. Ind. Eng. Chem. Res. **45**, 48–53 (2006)

Zhang, J., Li, A., Kang, A.: Synthesis and characterization of multifunctional poly(acrylic acid-co-acrylamide) sodium humate superabsorbent composite. React. Funct. Polym. **66**(7), 747–756 (2006)

Zhang, J., Wang, Q., Wang, A.: Synthesis and characterization of chitosan-g-poly(acrylic acid)/attempulgite superabsorbent composite. Carbohydr. Polym. **68**(2), 367–374 (2007)

Zhang, X., Pint, C.L., Lee, M.H., Schubert, B.E., Jamshidi, A., Takei, K., Ko, H., Gillies, A., Bardhan, R., Urban, J.J., Wu, M., Fearing, R., Javey, A.: Optically- and thermally-responsive programmable materials based on carbon nanotube-hydrogel polymer composite. Nano Lett. **11**(8), 3239–3244 (2011)

Zhao, X., Kim, J., Cezar, C.A., Huebsch, N., Lee, K., Bouhadir, K., Mooney, D.J.: Active scaffolds for on-demand drug and cell delivery. Proc. Natl. Acad. Sci. U.S.A. **108**(1), 67–72 (2011)

Zheng, Y., Wang, A.: Study on superabsorbent composites. XVIII. Preparation, characterization, and property evaluation of poly(acrylic acid-*co*-acrylamide)/ organomontmorillonite/sodium humate superabsorbent composites. J. Appl. Polym. Sci. **108**(1), 211–219 (2007)

Zhoo, L., Huang, J., Wang, T., Sun, W., Tong, Z.: Multiple shape memory, self-healable, and super tough PAA-Go-$Fe^{3+}$ hydrogels. Macromol. Mater. Eng. **302**(2), 1600359 (9 pages) (2017)

Zhu, Z., Song, G., Liu, J., Whitten, P.G., Liu, L., Wang, L.: Liquid crystalline behavior of graphene oxide in the formation and deformation of tough Nanocomposite hydrogels. Langmuir **30**(48), 14648–14657 (2014)

Zinchenko, A., Taki, Y., Sergeyev, V.G., Murata, S.: DNA-assisted solubilization of carbon nanotubes and construction of DNA-MWCNT cross-linked hybrid hydrogels. Nanomaterials **5**, 270–283 (2015)

Zohuriaan-Mehr, M.J., Kabiri, K.: Superabsorbent polymer materials: a review. Iranian Polym. J. **17**(6), 451–477 (2008)

# Chapter 10
# Ionic Liquids: New Functional Fluids as Lubricants

**Jyoti Srivastava, Tandra Nandi, Satyen Saha, and Eswara Prasad Namburi**

**Abstract** Ionic liquids (ILs) have emerged as new functional fluids that find a variety of applications as solvents, catalysts, electrolytes, solar thermal energy agents and lubricants. Their attributes of very low vapour pressure and high thermal stability fulfil the basic requirements of ionic liquids as lubricants. The selection of cations and anions in an ionic liquid and also the arrangements of different ionic side chains form the basis of development of different specific lubricants and lubricant additives. There are high expectations for the development of new ionic liquid applications as lubricants for extreme environments, such as high temperatures and high vacuum, where the use of conventional lubricants is limited. This chapter discusses the above mentioned aspects with emphasis on ionic liquids as lubricants.

**Keywords** Ionic liquids · Imidazolium · Lubricants · Synthesis and properties

J. Srivastava
Defence Materials and Stores Research and Development Establishment (DMSRDE), Defence Research and Development Organisation (DRDO), Kanpur, Uttar Pradesh, India

T. Nandi
Formerly With Defence Materials and Stores Research and Development Establishment (DMSRDE), Defence Research and Development Organisation (DRDO), Kanpur, Uttar Pradesh, India

S. Saha
Department of Chemistry, Institute of Science, Banaras Hindu University, Varanasi, India

E. P. Namburi (✉)
Former Outstanding Scientist and Ex-Director, Defence Materials and Stores Research and Development Establishment (DMSRDE), Defence Research and Development Organisation (DRDO), Kanpur, Uttar Pradesh, India
e-mail: neswarap@gmail.com



E. P. Namburi et al. (eds.), *Novel Defence Functional and Engineering Materials (NDFEM) Volume 1*, Indian Institute of Metals Series,
https://doi.org/10.1007/978-981-99-9791-6_10

## 10.1 Introduction

The science and technology of lubrication engineering has to meet new challenges and issues, including the reduction of wear, friction, emissions and energy losses and to increase the service life and durability of tribological systems (tribosystems). Tribology involves every industry directly or indirectly, and in a tribosystem one material slides/rubs against another or against itself. The size and shape of contact surface changes when one or both materials are worn during the sliding and rubbing processes.

A lubricant is a substance introduced essentially to reduce friction between surfaces in mutual contact and also reduces the heat generated when the surfaces move rapidly against each other. The outcome of machine parts is directly related to efficiency of lubricant. Normally, a lubricant formulation is based on a mineral or synthetic oil or grease with a mixture of different additives. These additives not only reduce friction and wear but also increase viscosity, thus improving the resistance to corrosion and oxidation, ageing or contamination of the surfaces that are lubricated (Bermudez et al. 2009a; Mang et al. 2007) Lubricants also have to function effectively in some specific conditions such as working under vacuum or in a micro-electro-mechanical-systems (MEMS) environment, for which the use of conventional lubricants does not suffice. Besides commercial lubricants mainly derived from petroleum, there is an urgent requirement for new, effective, and environmentally friendly lubricants.

Ionic liquids (ILs) are promising solutions to these challenges (Street et al. 2011; Palacio and Bhushan 2010; Doerr et al. 2010; Zhou et al. 2009). ILs are salts with melting points below 100 °C. They consist of organic cations in combination with a variety of organic or inorganic anions; see Fig. 10.1. Some of the most common cations are imidazolium, phosphonium, pyridinium and ammonium, while some common anions are $BF_4^-$, $PF_6^-$, $CF_3SO_3^-$, $NTf_2^-$ and $N(CF_3SO_2)^-$; see Tables 10.1 and 10.2.

ILs are unique solvents that consist entirely of ions. They usually contain large asymmetric organic cations and weakly coordinating anions. Owing to the large molecular sizes and the nature of the functional groups of anions, the charges on the ions are diffuse, resulting in reductions in electrostatic forces and the symmetry between the anions and cations. Also, these salts lose crystalline structure and attain a liquid state at room temperature; e.g. see Fig. 10.2.

A series of anions and cations can be used to obtain thousands of combinations of ILs, each with its own unique properties. ILs also have a large number of useful and unique properties including higher thermal stability, lower melting point, a wide liquid us range, and minimum vapour pressure. Therefore, ILs are highly 'tuneable' for particular applications (Ye et al. 2001a; Qu et al. 2006; Canter 2005), and the reduced solvent losses from volatilization have led to the development of 'green' solvents.

Ionic liquids (ILs) consist of large, asymmetric organic cations and usually an inorganic anion. Due to the large size of their molecules and the nature of the chemical

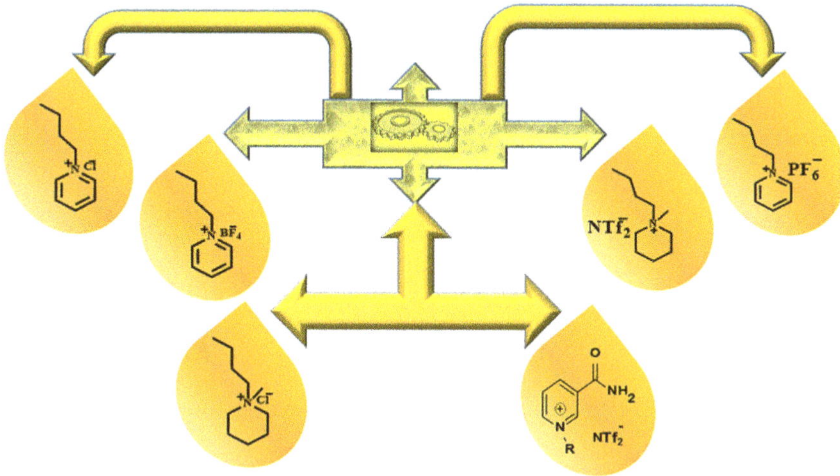


**Fig. 10.1** Ionic liquid as lubricant

groups of the anions, the charges on the ions of these salts are usually diffuse. As a result of reduced electrostatic forces between the anion and cation in these salts and their asymmetry, it is difficult to form a regular crystalline structure, and therefore, they can be liquid at room temperature.

## 10.2 Structures of Ionic Liquids: Common Cations and Anions

The structure of an IL consists of an organic cation with a weakly coordinating organic or inorganic anion. As mentioned in the Introduction, ILs are also featured as designer solvents because several combinations of anions and cations can result in ionic liquids for specific desired applications (Silvester and Compton 2006).

### *10.2.1 Cations*

The most frequently used cations are ammonium and imidazoliums. However, pyridinium and pyrrolidinium are also used. Table 10.1 shows some imidazolium-, pyridinium-, pyrrolidinium-, ammonium- and guanidinium-based cations with their abbreviations (the lengthy names make it convenient to use abbreviations). Imidazoliums are the most widely used cations and a variety of them are commercially available. The lubricating properties of imidazoliums-based ILs can be easily tailored by attaching different types of anions. For example, [BMIM][Cl] decomposes at 250

**Table 10.1** Full names, abbreviations and formulas of some cations used for ILs

| Abbreviations/name | Structure | Abbreviations/name | Structure |
|---|---|---|---|
| **[EMIM]**/1-ethyl-3-methylimidazolium | 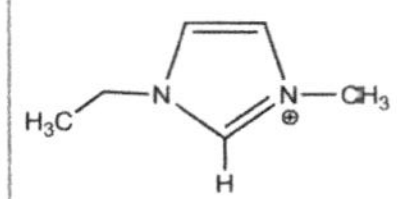 | **[Pip]**/Piperidinium | 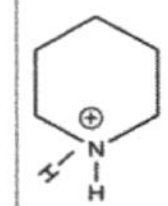 |
| **[EMMIM]**/1-ethyl-2,3-dimethylimidazolium | 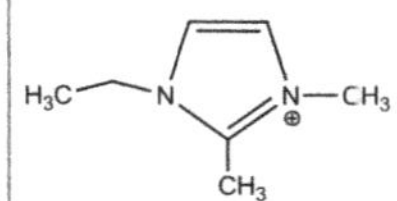 | **[Py]**/Pyridinium | 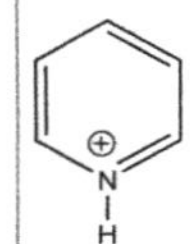 |
| **[BMIM]**/1-butyl-3-methylimidazolium | 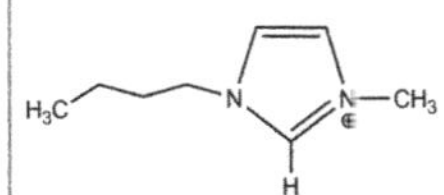 | **[nmp]**/Pyrrolidinium | 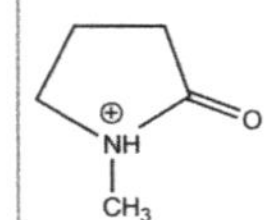 |
| **[HMIM]**/1-hexyl-3-methylimidazolium | 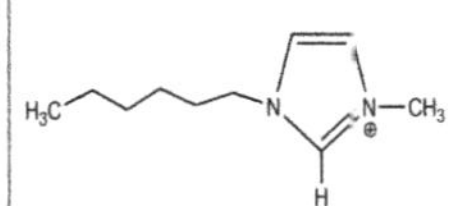 | **[EtNH$_3$]**/Ammonium | 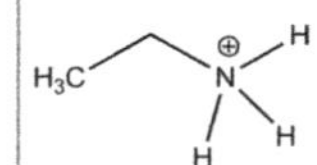 |
| **[OMIM]**/1-octyl-3-methylimidazolium | 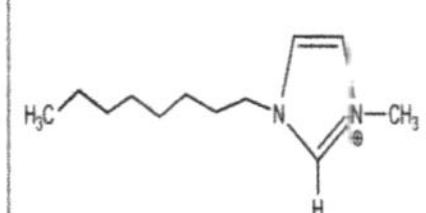 | **[tmg]**/Guanidinium | 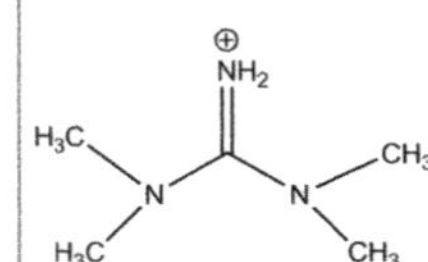 |

**Table 10.2** A few anions used for ILs

| Anions | Formula |
|---|---|
| Hexafluorophosphate | $PF_6^-$ |
| Tetrafluoroborate | $BF_4^-$ |
| Tetrachloroaluminate | $AlCl_4^-$ |
| Nitrate | $NO_3^-$ |
| Alkyl sulphate | $RSO_4^-$ |
| Acetate | $CH_3COO^-$ |
| Methane sulphonate | $CH_3SO_3^-$ |
| Trifluoromethanesulphonylamide | $N(CF_3SO_2)_2^-$ |
| Sachharine | $C_7H_5NO_3S^-$ |
| Lactate | $C_3H_5O_3^-$ |
| Trifluoromethylsulphonyl)imide | $NTf_2^-$ |
| (Perfluoroethyl)trifluorophosphate | $FAP^-$ |
| p-toluenesulphate | $SO_3C_6H_5CH_3^-$ |

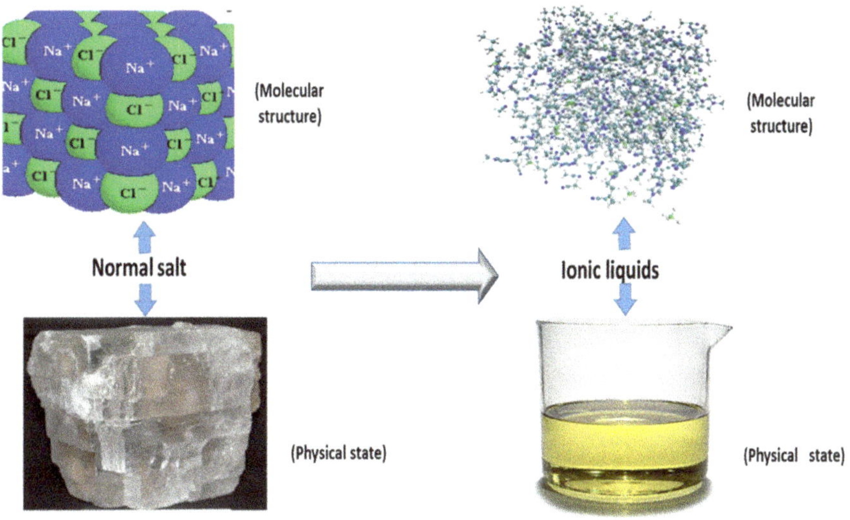


**Fig. 10.2** Schematic of molecular structure and physical state of common salt and an Ionic Liquid at room temperature

°C, while [BMIM][$PF_6$] and [BMIM][$NTf_2$] are stable up to 280 °C and 410 °C, respectively. Also, ILs containing $BF_4^-$, $PF_6^-$, and $NTf_2^-$ anions are more thermally stable than the corresponding halides (Sowmiah et al. 2009).

### *10.2.2 Anions*

The anions widely used in ionic fluids are listed in Table 10.2. The most commonly used anions are polyatomic inorganic moieties. Anion selection has a large influence on the properties of ILs. There are some issues with toxicity: ILs using $PF_6^-$ and $BF_4^-$ sometimes decompose to give off HF. To overcome this disadvantage, anions containing fluorine are bonded to carbon because the C-F bond is inert to hydrolysis.

## 10.3 Common Methods for IL Synthesis

There are two primary methods for preparing ILs: a metathesis reaction and acid–base neutralization; see (A) and (B) in Fig. 10.3. A metathesis reaction involves double displacement and acid–base neutralization (A), while acid–base neutralization is more direct (B). Two different methods are direct alkylation of imidazole (C) and the carbonate method (D).

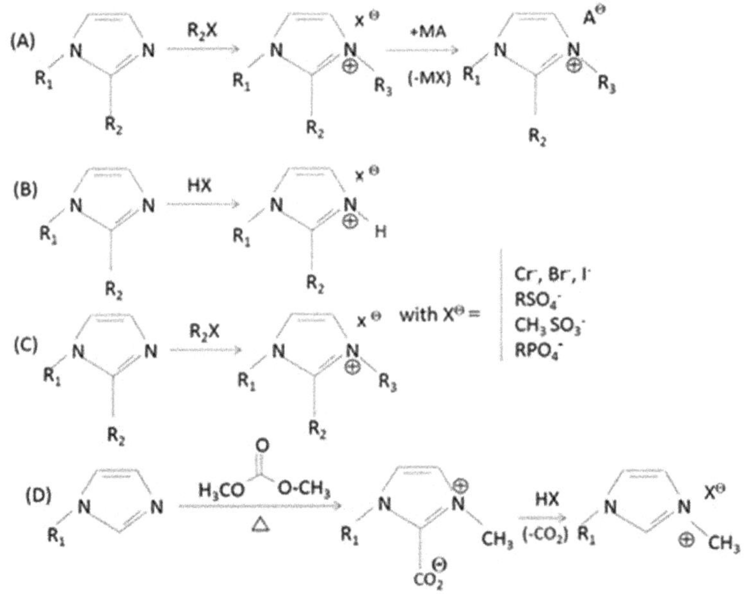


**Fig. 10.3** **a** Metathesis reaction, **b** acid–base neutralization, **c** direct alkylation of imidazole, and **d** the carbonate method: diagrams reproduced from Vekariya (2017)

## 10.4 Properties of ILs

The structures of ILs result in properties equivalent to those of conventional high performance lubricants. ILs have high thermal stability and negligible vapour pressures and are therefore suitable for harsh conditions. Their polarities are high compared to those of other lubricants: hence they form effective absorption films. Table 10.3 lists the IL properties advantageous for lubricants.

ILs contain tribologically interesting elements such as N, P, B and F for antiwear and extreme pressure use. The viscosity increases with cation type in the sequence imidazoliums < Pyridinium < Pyrolidinium; and for anions in the sequence$NTf_2^-$ < FAP < $BF_4^-$ < $PF_6^-$ < $Cl^-$ < $Br^-$. The anions change viscosity as well as influencing the formation of a boundary layer: thus with the same cation For same $FAP^-$ and $NTf_2^-$ give better antiwear properties than $BF_4^-$ and $PF_6^-$. It is also noteworthy that phosphonium ILs give better antifriction and antiwear properties owing to the active element phosphorus (Cabeza et al. 2014).

### *10.4.1 Disadvantages of ILs as Lubricants*

Although ionic liquids have unique properties like negligible volatility, non-flammability, high thermal stability, a broad liquid us range and polarity, as listed in Table 10.3, they do have some disadvantages. Fluorine-containing anions are very sensitive to the environment and can form HF, which attacks materials and pollutes the surroundings. Thermal oxidation can also occur: protons and 1-alkene are among the thermal oxidation products. However, there are solutions to these problems. The fluorine anions FAP and $Tf_2N$ can be replaced by other anions; and the use of anti-corrosion additives like Benzotriazole to the lubricants is always advisable (Bermudez et al. 2009a).

## 10.5 Tribological Behaviour and Lubrication Phenomena

Lubricants can be categorized into three classes: (i) liquid lubricants such as water or mineral oils; (ii) semi-solid lubricants like greases; and (iii) solid lubricants e.g. $MoS_2$ and graphite. Liquid lubricants have more advantages compared to greases and solid lubricants, including long lasting nature, low mechanical loudness, and very low friction in the hydrodynamic regime.

**Table 10.3** IL property advantages for Lubricants (Bermudez et al. 2009b)

| Property/characteristic | Ionic liquid advantages as lubricants |
|---|---|
| Viscosity and viscosity index ranges | Wide variations in viscosity and viscosity index are obtained by simply changing the anions in ionic liquids [see Table 10.4] |
| Lamellar-like liquid crystal structure | Structure comparable to most natural oils and conventional Lubricants, i.e. helpful in enhancing lubricity and wear resistance 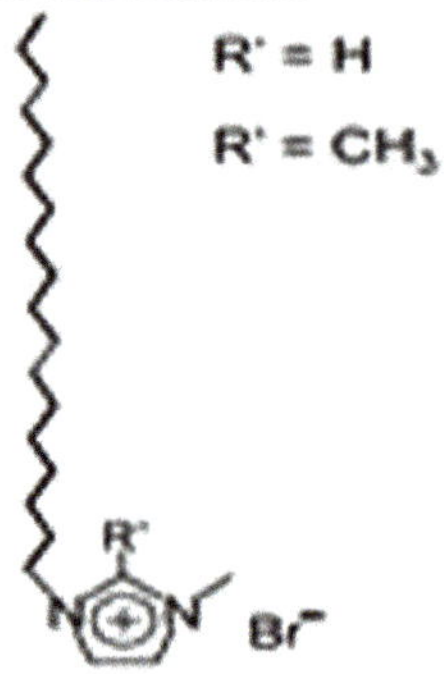  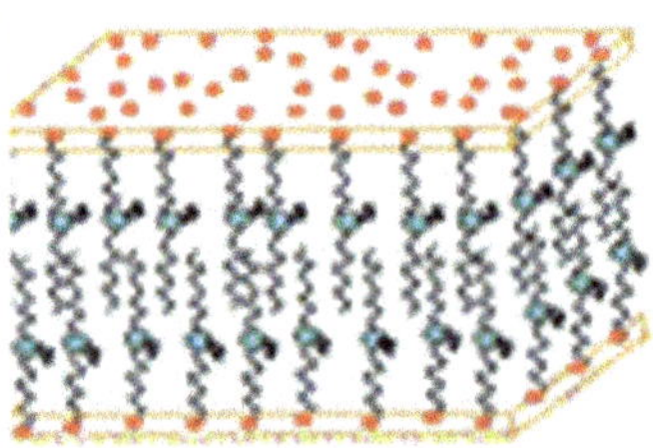 Zhou et al. (2012) |
| Good thermal stability | Can be used up to high temperatures: e.g. ≈ 410 °C for guanidinium based ILs (Huang et al. 2017) |
| Negligible vapour pressure | Evaporation rate much less than for most other lubricants |
| Environmentally friendly (non-toxic) | Promote energy conservation by decreasing friction; eco-friendly (non-toxic) ILs can be easily made and most can be obtained from non- petroleum products |
| Non-flammability | Safe transport and usage |
| Polarity | Easily adsorbed on positively charged worn metal surfaces because of bipolar molecular nature: an active boundary layer surface film forms to reduce friction |

**Table 10.4** Physical properties of ionic liquids and mineral and synthetic oil

| S.No. | Ionic liquids and base oils | | Kinematic Viscosity (Cst) | | Viscosity Index(VI) | Flash Point (°C) | Pour Point (°C) |
|---|---|---|---|---|---|---|---|
| | | | 40 °C | 100°C | | | |
| (A) Ionic Liquids | {BMIM}[B F$_4$]$^{+3}$ | 1-Butyl-3methylimidazolium Tetrafluoroborate | 43.4 | 9.03 | 187 | 288 | <-80 |
| | {BMIM}[P F$_6$]$^{+4}$ | 1-Butyl-3methylimidazolium hexafluorophosphate | 81.0 | 13.6 | 172 | >350 | <-50 |
| | {BMIM}[T FSI]$^{+6}$ | 1-Butyl-3methylimidazolium bis(trifluoromethylsul phonyl)imide | 29.0 | 6.88 | 211 | >200 | <-50 |
| | {EMIM}[D CN]$^{+7}$ | 1-Ethyl-3methylimidazolium dicyanamide | 8.6 | 3.01 | 252 | 221 | <-50 |
| (B) Base oils | Synthetic (PAO 6) | Poly alpha olefin | 30.5 | 5.92 | 136 | 245 | -68 |
| | Synthetic (PAO 4) | Poly alpha olefin | 16.8 | 3.94 | 128 | 226 | -73 |
| | Mineral oil | 100 SUS | 20 | 4.00 | 97 | 200 | -15 |
| | Synthetic (engine oil) | Shell Advance Ultra, 10W-40) | 90.2 | 14.2 | 163 | 230 | -33 |
| | Semi-synthetic (engine oil) | (Shell Advance AX7, 10W-40) | 98.6 | 15.8 | 172 | 230 | -36 |
| | Silicone oil | Texol Silicone Oil ™ (E50) | 50 | 11.5 | 216 | 318 | -70 |
| | Synthetic esters | Diisononylsebacate | 19 | 4.6 | 176 | 232 | -51 |
| | Synthetic esters | Alkylated aromatic | 29 | 5.3 | 115 | 229 | -54 |
| | Vegetable oil | Rapeseed oil | 36.0 | 8.19 | 214 | 31 8 | -20 |
| | Paraffinic oil | - | 30.0 | 5.26 | 105 | 21 5 | -15 |

### *10.5.1 The Stribeck Curve*

Liquid lubrication is classified into three regimes according to the distance between the moving surfaces. This is illustrated by the Stribeck curve (Sotres and Arnebrant 2009) shown in Fig. 10.4. Fundamentally the curve gives the relation between friction coefficient and bearing number, $\eta V/P$, where $\eta$ is the dynamic viscosity of the fluid, V is the sliding velocity and P is the applied load. The friction coefficient is plotted on the y-axis and the bearing number on the x-axis. There are three regimes shown in Fig. 10.4, corresponding to the sliding condition changing from boundary lubrication to mixed lubrication and then to hydrodynamic lubrication:

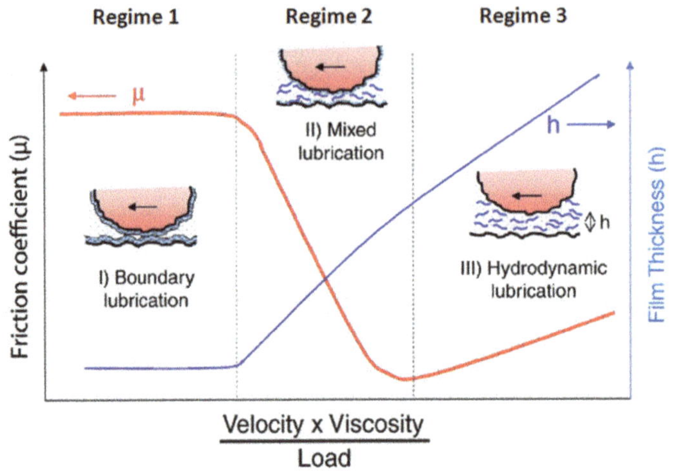


**Fig. 10.4** Stribeck curve: friction coefficient versus lubrication parameter ηV/P (Sotres and Arnebrant 2009)

**Regime 1**: In this regime the lubrication liquid film is beneficial, but may not prevent asperities on the sliding surfaces coming into contact. This causes 'stick–slip' and the breaking-off of some asperities. Obviously, friction and wear are more severe in this regime.

**Regime 2**: This is the mixed lubrication regime, where the lubrication film is thin such that there is still chance of contact between asperities.

**Regime 3**: This is the regime for hydrodynamic lubrication. In this regime a viscous liquid is introduced between the sliding surfaces and a stable liquid film is formed. The liquid film fully supports the applied load and the two surfaces do not make contact. Viscosity is the deciding factor for lubricant performance in this regime, since a thick liquid film can support and transfer some load under conditions of high pressure and low speed.

## *10.5.2 Physical Properties of ILs*

Most ionic liquids are liquid at room temperature, enabling them to act both as base oils and additives. Also they enable a variety of lubricants with a wide range of viscosities and other properties to be made. Table 10.4 compares some of these properties for ionic liquids (blue-shaded boxes) with those of some mineral and synthetic oils and lubricants.

## 10.6 Applications of ILs

In this section we shall discuss (i) applications of 'neat' IL lubricants, (ii) applications as additives and (iii) special considerations for using ILs as lubricants for micro electro-mechanical systems (MEMS). Two specific examples from Table 10.4 are worth noting here:

- 1-Butyl-3methylimidazolium tetrafluoroborate has a lower pour point (−80 °C) compared to silicone oil (−70 °C).
- 1-Butyl-3methylimidazolium hexafluorophosphate has the highest flash point (>350 °C).

These examples illustrate the flexibility of ILs as lubricants. As mentioned in the introduction to this chapter, many permutations and combinations are possible just by selecting desirable anions and cation groups. This leads to versatility of ILs for various applications, including both high temperature and cryogenic operations.

### *10.6.1 ILs as Neat Lubricants*

Wilkes and Zaworotko (1992) first reported that water-soluble ionic liquids could be used as lubricants. Liu et al. later studied alkylimidazoliumtetrafluoroborates with respect to steel contact surfaces, followed by investigation of combinations of steel surfaces with steel, aluminium, copper, $SiO_2$, Si(100), sialon and $Si_3N_4$ contact surfaces (Ye et al. 2001b). An efficient friction reduction was obtained for all these combinations. The application of ionic liquids in lubrication engineering are well reviewed by Minami (2009).

Hernandez et al. (2016) studied the tribological performance of phosphonium-containing ionic liquids like ([$P_{66614}$][BEHP]), i.e. Trihexyltetradecylphosphonium bis(2,4,4-trimethylpentyl) phosphonate; and [$P_{66614}$][$C_8C_8PO_2$], i.e. Trihexyltetradecylphosphonium bis(2,4,4-trimethylpentyl) phosphinate as neat lubricants for steel–steel contacts. The improvements in tribological performance were attributed to the formation of a phosphate tribofilm.

Totolin et al. (2013) studied halogen-free 1-Allylimidazoletributylborane-, 1-Ethylimidazole tributylborane-, and imidazoliumborane-containing ionic liquids also for steel–steel contacts. The tribological performance was improved in terms of friction and wear as compared to halogen-containing ionic liquids.

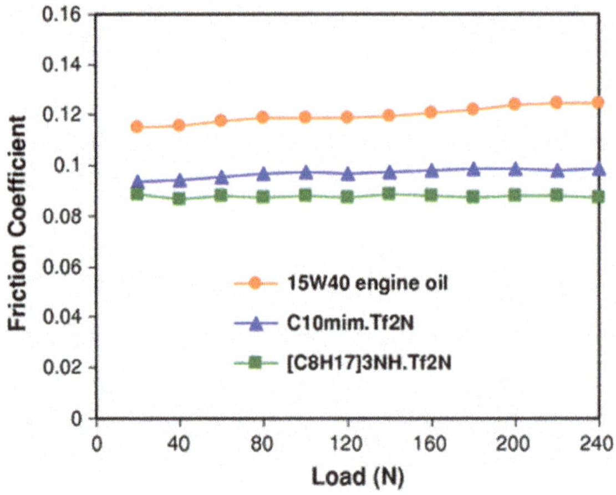


**Fig. 10.5** Comparisons of friction coefficients of IL and conventional engine oil 15W40: reproduced from Qu et al. (2009a)

Qu et al. (2009a) compared the friction coefficients of the ionic liquids 1-ethyl-3-decylimidazolium-bis(trifluoromethylsulfonyl) imide ($C_{10}mim{\cdot}Tf_2N$) and trioctylammoniumbis (trifluoromethylsulfonyl) imide ($[C_8H_{17}]_3NH{\cdot}Tf_2N$) with that of the conventional engine oil 15W40 in the range of 20–240 N load. They concluded that the IL friction coefficients were 20% and 30% lower than the 15W40 oil; see Fig. 10.5.

### *10.6.2 ILs as Neat Lubricants for Light Alloys (Mg, Al, Ti)*

The selection of lubricants for the light alloys magnesium (Forsyth et al. 2009), aluminium (Mu et al. 2005) and titanium (Jimenez and Bermudez 2009) seems to be difficult, but some ILs have been found suitable.

1. **Magnesium**: It has been reported that most lubricants are unsuccessful in preventing the wear of magnesium alloys. Forsyth et al. (2009) reported that (ILs) containing trihexyltetradecylphosphonium alongwith diphenylphosphate or bis(trifluoromethanesulfonyl) amide are used to form layers on magnesium alloy surfaces to enhance the corrosion resistance of the alloy. Espinosa et al. (2014) have studied phosphonateimidazolium-containing ILs for improving the abrasion resistance between magnesium alloy surfaces.
2. **Aluminium**: 1-(3′-O,Odiethylphosphonyl-*n*-propyl)-3-alkylimidazolium tetrafluoroborates were investigated as promising lubricants for aluminum–steel contacts (Mu et al. 2005). The tribological behaviour of these ILs was better than that of alkylimidazolium derivatives. Also for aluminum–steel contacts, ILs containing ammonium and phosphonium moieties gave better lubricating

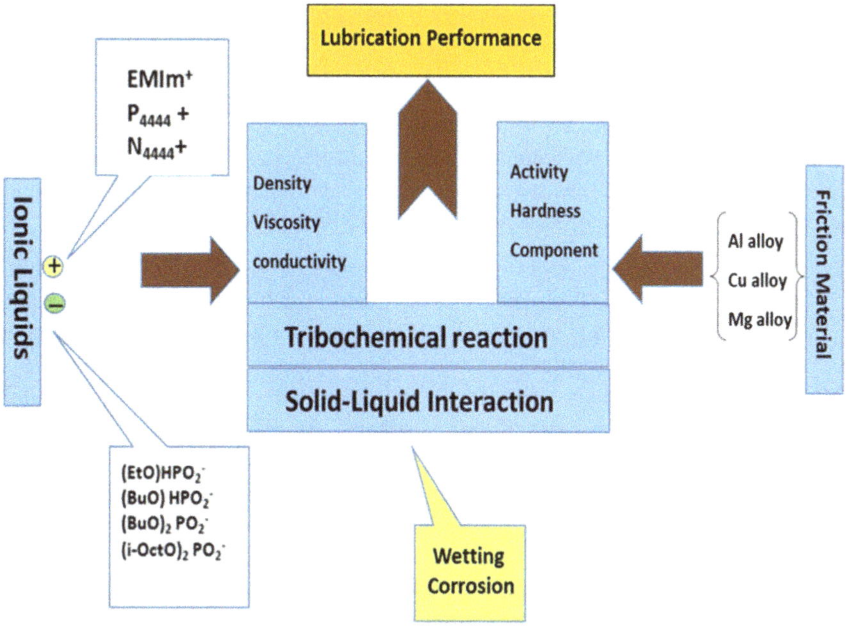


**Fig. 10.6** Effect of phosphate and phosphonate ILs on the lubrication performance of different alloys (Mg, Al, Cu): reproduced from Han et al. (2017)

properties than imidazolium-based ionic liquids having the same anions. Qu et al. also reported friction reduction (upto 35%) and wear reduction (upto 55%) with ammonium based ionic liquids as compared to engine oil (Qu et al. 2009b).

3. **Titanium**: Ti alloys have generally excellent corrosion resistance in aqueous media, under neutral, reducing and oxidising conditions. Recently, imidazolium and ammonium ILs have been investigated as antiwear lubricants for grade 3 titanium (Jimenez and Bermudez 2009). Also, 1-methyl-3-benzylimidazolium chloride and 1-hexyl-3-methylimidazolium hexafluorophosphate showed good tribological results.

Finally, six different ILs containing phosphate and phosphonate moieties were identified as lubricants for three different alloys (Han et al. 2017). For Cu alloy–steel and Mg alloy–steel contacts, the tribological performances were significantly influenced by the IL structures. Figure 10.6 is a schematic representation of these results.

### 10.6.3 ILs as High Temperature Lubricants

Temperature plays a major role in lubricating performance. Most conventional oils except silicones are thermally stable in the range 200–250 °C. However, they are unsuitable for severe environments in any industry, including the glass, chemical, textile processing, aerospace and polymer processing industries. Some IL developments for high temperature lubricants are listed here:

(1) A new variety of dicationic ILs with polyethylene glycol (PEG) functionality and alkyl/polyfluoroalkyl substitutes has emerged as high temperature lubricants (Jin et al. 2006). They are thermally stable up to 300 °C and have very good tribological characteristics.
(2) Three ionic liquids based on imidazolium derivatives, 1-methyl-3-hexylimidazolium hexafluorophosphate (L-P106) and 1-methyl-3-octylimidazolium tetra fluoroborate (L-108) and the quaternary ammonium chloride AMMOENG$^{TM}$ 101(AM-101) have been studied as lubricants for Inconel 600 against AISI 52100 steel. Their tribological performance was compared with that of mineral oil. They concluded that imidazolium derivatives are the best lubricants at higher temperature because of their higher thermal stability upto 300 °C (Jimenez et al. 2009).
(3) Song et al. studied (2014) lithium-containing ILs with sym-triazine and cyclotriphosphazene functionality as high temperature lubricants for $Si_3N_4$–steel contacts. They found remarkable improvements in antifriction and anti-wear proprieties compared with conventional alkylimidazolium ILs at 200 °C.
(4) Recently Wu et al. (2016) studied halogen-free ionic liquids (hf-ILs) as lubricant for poly (ether-ether-ketone)-AISI304 stainless steel (PEEK-304ss) surfaces at higher temperatures. The excellent anti-corrosion and antiwear properties were reported with hf-ILs with short alkyl chains.
(5) The Guanidinium based ionic liquids (GILs) are more thermally stable than perfluoropolyether (PFPE) and 1-hexyl-3-methylimidazolium bis (trifluoromethylsulfonyl) imide (LF106). GILs also exhibited better antifriction and antiwear behaviour than (LF106) and (PFPE) in the range from room temperature to 300 °C (Huang et al. 2017).

The comparisons of antifriction properties amongst PFPE, LF106 and some guanidinium-based ILs are shown in Fig. 10.7 for $Si_3N_4$–steel contacts.

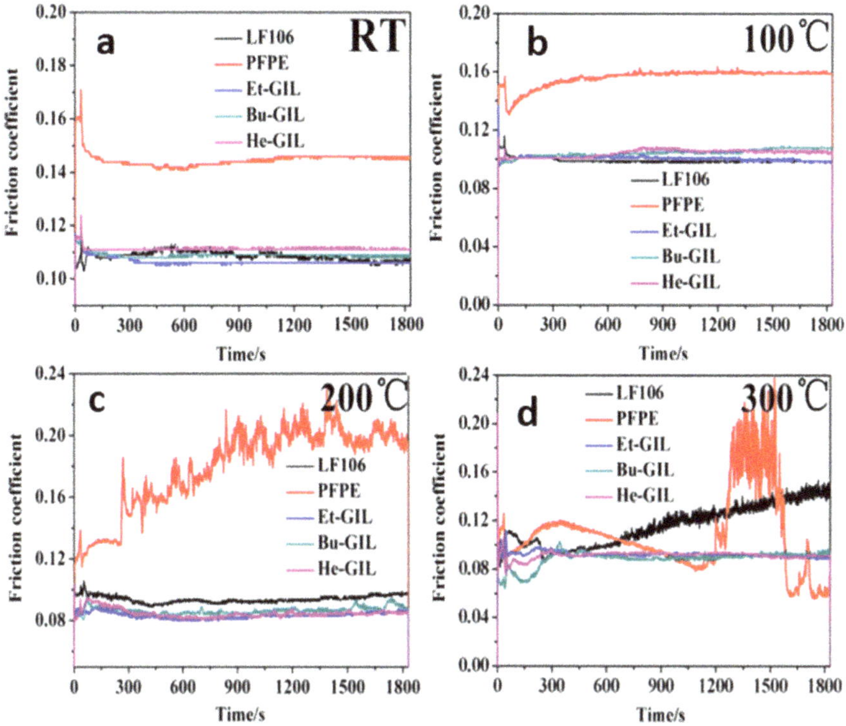


**Fig. 10.7** Evaluation of the coefficient of friction for $Si_3N_4$–steel contacts lubricated by LF106, PFPE, Et-GIL, Bu-GIL and He-GIL at different temperatures: reproduced from Huang et al. (2017)

### *10.6.4 ILs for General Use in Aerospace Applications*

Most aerospace vehicles use liquid lubricants in their mechanical systems. The advantages and disadvantages of conventional lubricants for spacecraft applications are listed in Table 10.5. In comparison, ionic liquids have very low volatility and vapour pressure, and perfect antiwear and antifriction properties, capable of transferring large loads. Also, their property changes are minimal when changing from atmospheric air to a vacuum environment. ILs had better viscosity-temperature characteristics and showed less friction, lower wear rate, and higher load-carrying capacity under high vacuum conditions as compared to conventional lubricants (Kaadonski and Wojdyna 2011). ILs would therefore appear to be ideal lubricating substances for aerospace applications. The borane group containing ionic liquids showed friction factors two times less than the one for PFPE.

**Table 10.5** Advantages and disadvantages of conventional lubricants used in aerospace

| Lubricant | Advantages | Disadvantages |
|---|---|---|
| Mineral oil | Very good in boundary lubrication (BL) and elastohyrodynamic lubrication (EHL) | Unsuitable for quick lubrication |
| Silicone oil | Very low vapour pressure and excellent low temp characteristics | Degrades quickly under BL conditions |
| Perfluoropolyethers | 1. Low freezing point and vapour pressure<br>2. Low surface tension and high viscosity index | Not suitable for BL; degrades under X-ray radiation |
| Polyalphaolephins | 1. Vapour pressure less than that for mineral oils<br>2. High viscosity index | Freezing point is low |
| Multiplyalkylated-cyclopentane (MAC) | Usable over wide temperature range | Less thermally stable and inert than silicones and perfluoropolyethers |
| Silahydrocarbon | Low volatility and very wide viscosity range; suited to BL conditions | Not commercially available (yet) |

## *10.6.5 ILs as Green Engineering Fluids*

The lubrication market is focussing more on 'green' engineering fluids or eco-friendly lubricants, owing to several critical issues like depletion of natural reserves, price increases in oil disposal of petroleum-based products, and lubricant losses due to evaporation and emissions into the environment. The current research in lubrication area is to develop environmentally green lubricants. Initially, many attempts were made to replace petroleum oils with bio-based oils. However, these have high pour points and poor thermal and oxidation resistance, hence suitable alternatives are required.

Ionic liquids are highly biodegradable and can be tailor-made to be non-toxic, as mentioned earlier in this chapter. Benzoic acid and salicylic acid (preservatives) and artificial sweeteners (e.g. saccharin) are selected as anions, with tetraalkylphosphonium salts as cations. The tribological performance of one tetra alkyl phosphonium cation in combination with several carboxylate anions has been investigated and the tribological results were very encouraging as compare to petroleum based lubricants and base oil (Reeves et al. 2012).

### 10.6.6 ILs as Recycling (Life Extension) Liquids in Lubricants

The ILs strihexyltetradecylphosphoniumbis(2,4,4-tri-methylpentyl)phosphinate and trihexyltetradecylphosphoniumbis (2-ethylhexyl) phosphate are used as additives to base oils. Wang et al. (2017) have studied the tribological performance of these phosphonium based ILs as additives to already used engine lubricants. The lubricants regained their good tribological performance and could be re-used. The service life extension of used lubricants gives benefits in terms of fuel economy, engine reliability, and drainage and oil consumption. Anand et al. (2015) observed that after multiple recycling the overall cost of ILs was reduced and the friction and wear behaviour were also improved.

### 10.6.7 ILs in Spacecraft Robotic Applications

This subsection is more specific than Sect. 10.6.4. Near-vacuum and low temperature working conditions are very complex for robotics. Conventional lubricants contaminate under vacuum and ultra-low temperature conditions. A solid lubricant like $MoS_2$ is also unsuitable because of poor thermal stability. However, lubricants based on ionic liquids reduce the possibility of damage to crucial space-based mechanical parts, and pyridinium cation based ILs have been used for this purpose. The polar nature of ILs is beneficial in efficiently coating substrate surfaces via Vander Waals attraction, and this mechanism is helpful in reducing the friction and wear observed for metal–metal, and metal–ceramic parts under mild sliding conditions (David et al. 2014). NASA SBIR/STTR technologies have developed efficient high-performance IL-based lubricants (Iono Glide) for NASA robotic operations (David et al. http://www.innosense.us/documents/ionoglide_brochure.pdf). Figure 10.8 illustrates the use of pyridinium-based complex ILs for a Mars Lander.

### 10.6.8 Roles of ILs as Additives

Following on from the last part of Sect. 10.6.5, additions of small amounts of ILs to water, mineral oils, polyethylene glycol, and synthetic oils are very beneficial to the friction and wear behaviours of contacting materials. The excellent tribological properties of ILs as additives are due to formation of physically absorbed antiwear boundary films during friction.

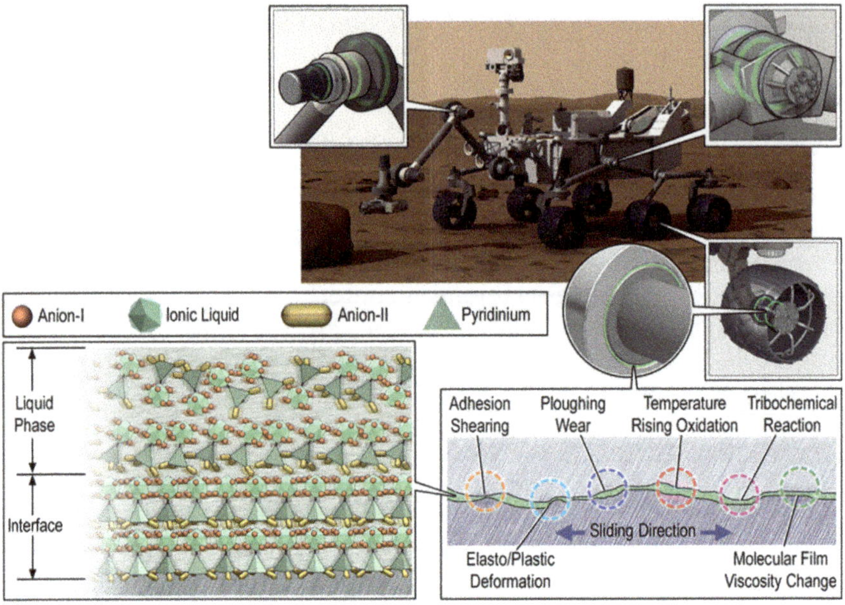


**Fig. 10.8** Modified IL-based high performance lubricants: reproduced from reference (David et al. http://www.innosense.us/documents/ionoglide_brochure.pdf)

#### 10.6.8.1 Additives to Water

ILs can be used as additives to both water and lubricating oils, owing to their miscibility with most solvents:

- Ye et al. (2001b) reported that addition of a small amount of water ($\leq$5 wt%) to ILs could help to improve the antiwear behaviour for some metal–metal contacts by formation of a preventive film of IL molecules.
- The additive behaviour of room temperature IL1 ([BMIM]$PF_6$) in water has been studied (Ge et al. 2007).
- Ibuprofen-based halogen-free ILs, namely 1-hexyl-3-methylimidazolium ibuprofen (L-Ibu106) and 1-octyl-3-methylimidazolium ibuprofen (L-Ibu108), have recently been studied as water-based lubricant additives (Wang et al. 2017). Unlike conventional fluoro-containing ILs, they showed anticorrosion and hydrolytic stable behaviour for steel and cast iron. These water-based novel additives have better extreme-pressure, antiwear and antifriction properties than commercially available antiwear water-based Hostagliss L4 additives. Another significant feature of the ibuprofen-based halogen-free ILs is that they can serve as 'green' water-based lubricants because they do not contain sulphur-, halogen- and phosphorus compounds. Figure 10.9 shows the antifriction behaviour of these green ionic liquids.

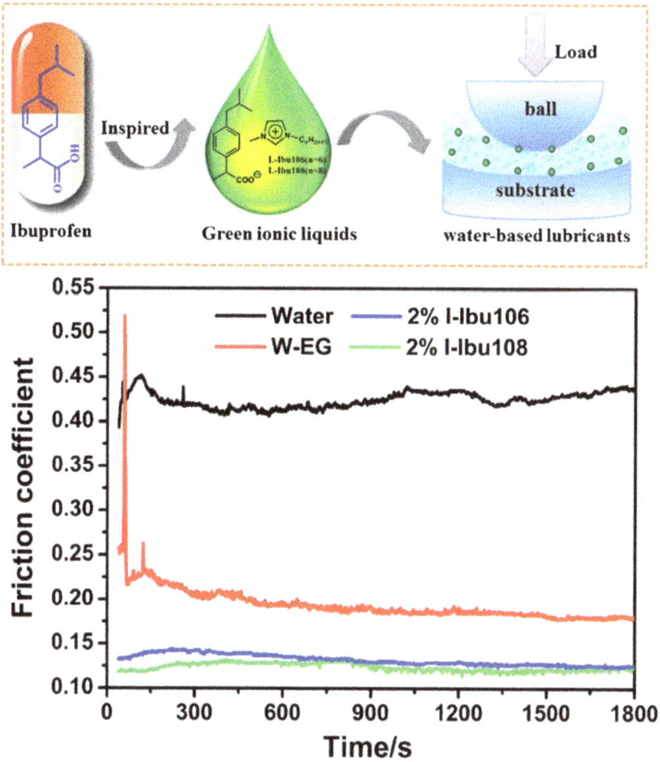


**Fig. 10.9** Ibuprofen-Based ILs as additives for enhancing the lubricity and antiwear of a water–ethylene glycol liquid: reproduced from Wang et al. (2017)

#### 10.6.8.2 Additives to Mineral Oils

- 1-*n*-Alkyl-3-methylimidazolium ILs, with different side chain lengths on the cations, and different fluorine-containing anions have been studied as mineral oil additives (Jimenez et al. 2006). These additives have shown good lubricating performance and significant reductions in wear rate.
- Fluorine incorporated ionic liquids (ILs), namely $[C_8H_{17}]_3NH$. $NTf_2$ and IL C10mim·$NTf_2$, were examined as base oil and as additives to 15W40 engine oil and showed good lubricating performance. A great reduction in wear rate and good increase in lubricity was achieved by adding 5% ILs. One composition even resulted in 9% friction reduction and 34% wear reduction (Qu et al. 2009a).
- The addition of (ILs) to polar oils has also been reported (Zhang et al. 2017). Environmentally friendly poly (isobutylene)-based ionic liquid (PIBIL) was selected. A hybrid film formed by PIBIL added to zinc dialkyldithiophosphate (ZDDP) oil reduced the Wear Scar Diameter (WSD).The formation mechanism of the hybrid film is shown in Fig. 10.10.

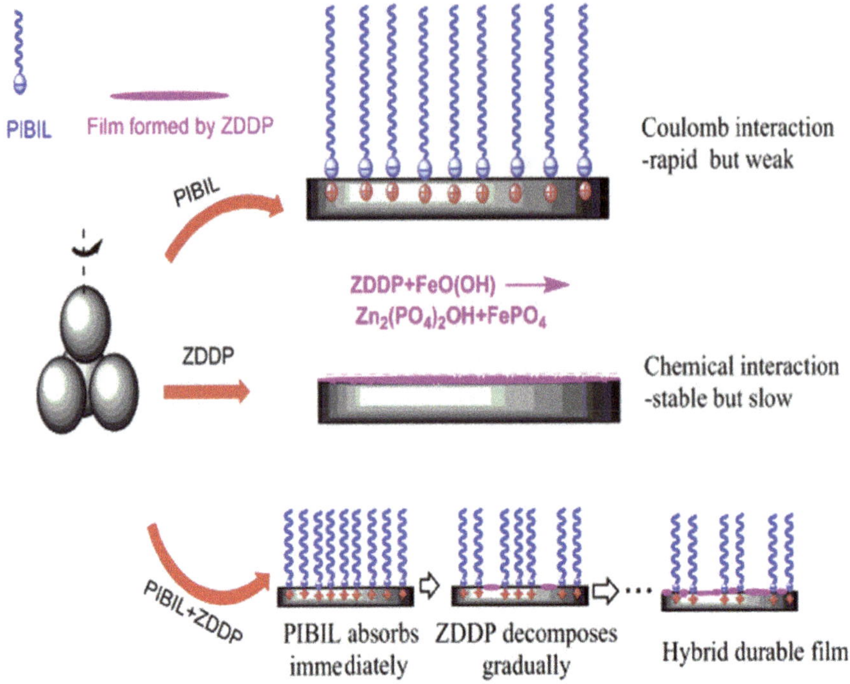


**Fig. 10.10** Environmentally friendly polyisobutylene-based IL containing chelated orthoborate as lubricant additive to ZDDP, resulting in synthesis, synergistic interactions, and improved tribological properties: reproduced from Zhang et al. (2017)

#### 10.6.8.3 Additives to Polyethylene Glycol (PEG)

ILs containing phosphonate anions, e.g. 1-butyl-3-methylimidazolium O-butyl phosphonate and two other ILs are used as additives to PEG for steel–steel contacts (Han et al. 2016). These ILs enhance the tribological capability of the base oil because of competitive adsorption between the base oil and the additives. The tribological mechanism of performance enhancement on steel surfaces is shown in Fig. 10.11 competitive adsorption, going from (a) to (b), changes the steel surface condition.

#### 10.6.8.4 Additives to Natural Oils

Natural oils are being reconsidered as 'green' lubricants. Reeves et al. (2014) did much work on the use of ILs in natural oils. They used phosphonium- and imidazoliums-based ionic liquids as additives to natural oils and concluded that these enhanced the tribological properties. In another study Grace et al. (2015) studied coffee bean oil (CB) with an IL additive for steel–steel contacts. The ionic

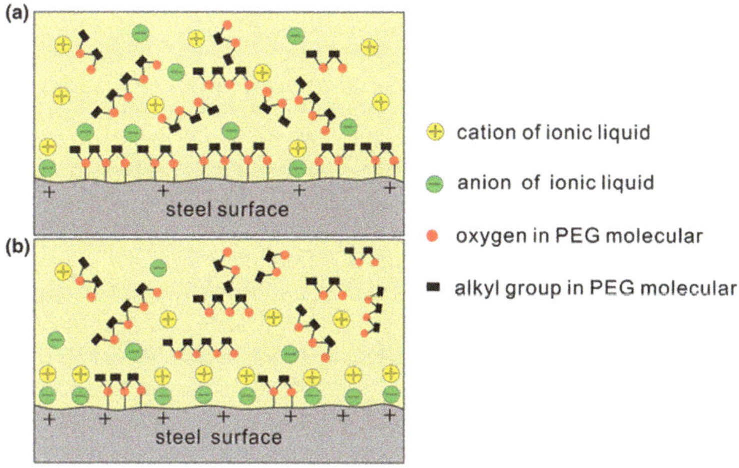


**Fig. 10.11** Influence of competitive adsorption on lubricating property of phosphonate IL additives in PEG: reproduced from Han et al. (2016)

liquid [THTDP][$NTf_2$] minimized the friction coefficient, and the wear volume was reduced. The comparative tribological behaviour is shown in Fig. 10.12.

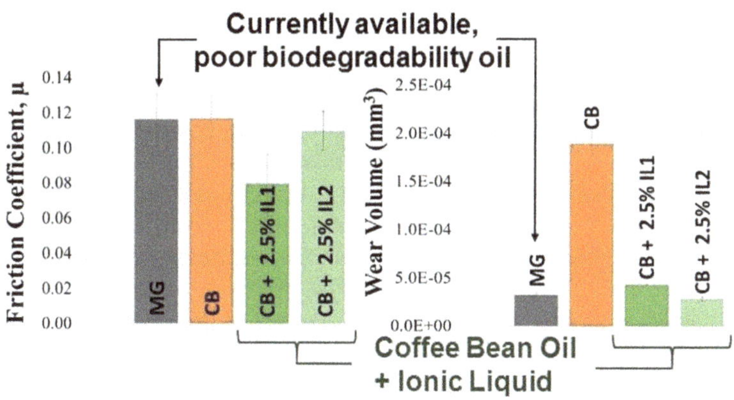


**Fig. 10.12** Tribological behavior of two phosphonium-based ionic liquids (ILs) as additives to coffee bean oil (CB) in steel–steel contact: reproduced from Grace et al. (2015)

#### 10.6.8.5 Additives to Synthetic Oils

The tribological performance of synthetic oils (esters) can also be upgraded by ILs:

- The synthetic ester of Propylene glycol dioleate has been used as base oil for lubrication of steel–aluminium contacts. Imidazolium-based ILs having hexafluorophosphate [$PF_6^-$], tetrafluoroborate [$BF_4^-$] and trifluoromethanesulfonate [$CF_3SO_3^-$] anions added as additives (1%) to this ester minimised friction and wear in the base oil at 100 °C (Jimenez and Bermudez 2008).
- For room temperature application dicationic 2-substituted imidazolium ILs were investigated as additives for polyethylene glycols acting on steel–steel contacts (Yao et al. 2009). The ILs reduced riction and wear. Dicationic-based ILs having different anions like [$PF_6^-$], [$BF_4^-$] and [NT $f_2^-$] showed good tribological behaviour at a concentration of about 3 wt%.
- Recently Gutierrez et al. (2016) reported a novel phosphonium ion based IL additive for wind turbine gearbox oils. An IL added at a concentration of 5 wt% to two commercial lubricants, PAO-40 and mobilgear SHCXMP 320 (MG), resulted in a wear scar reduction of 23% for a steel disc.
- Jia et al. (2010) have studied borate esters containing ILs soluble in poly alfaolefins (PAO) as lubricants for diamond-like carbon coatings (DLCs) sliding against steel and another DLC coating. They concluded that PAO + IL combinations showed enhanced antiwear and antifriction behaviour as compared to PAO + ZDDP.
- Fu et al. (2015) reported on the oil-soluble quaternary ammonium phosphites-based ionic liquids POPA and PTPA added to PAO. The results were very encouraging, and again the PAO + IL combinations showed enhanced antiwear and antifriction behaviour as compared to PAO + ZDDP.
- Qu and Viola (2013) compared the anti-scuffing/antiwear performance of IL[P66614][DEHP] and ZDDP under similar conditions. They added 1.0 wt% (separately) of these additives to PAO 4 base oil and studied the behaviour of both combinations at room temperature (RT) and 100 °C. In the RT tests the IL and ZDDP additives performed equally well in base oil, but at 100 °C the behaviour was entirely different. At higher temperatures the base oil becomes less viscous, and ZDDP was unable to stop scuffing during the early part of testing, resulting in high friction and wear. On the other hand, the IL additive resulted in a lower friction coefficient and a low wear rate, as may be seen in Fig. 10.13. Based on these results, Qu and Viola suggested that the IL additive would perform better in comparison to ZDDP and have the capability to enhance the fuel economy.

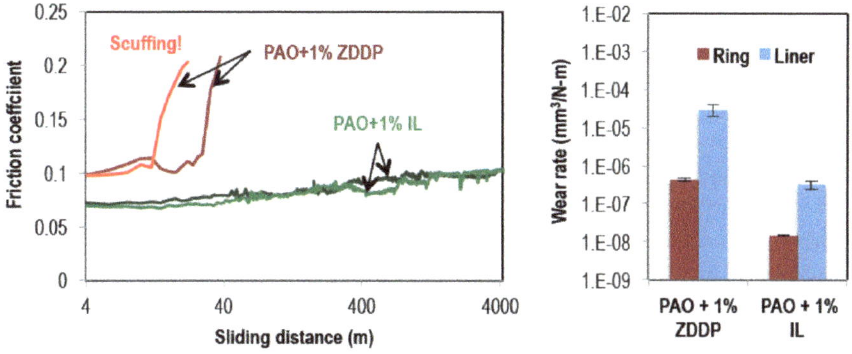


**Fig. 10.13** Friction and wear behavior of PAO with an IL and ZDDP at 100 °C: reproduced from Qu and Viola (2013)

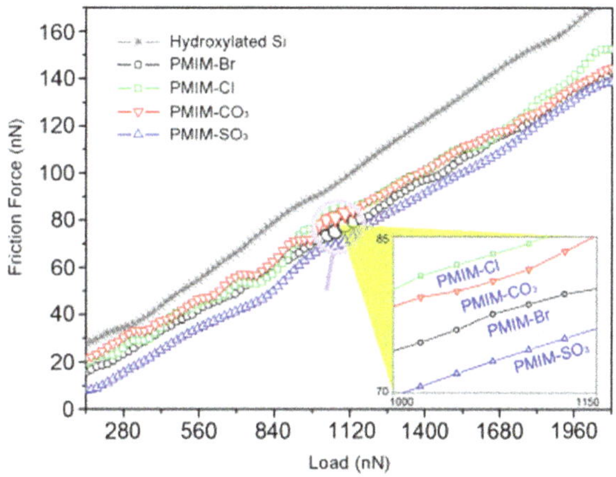


**Fig. 10.14** Plot of friction and load for surfaces of hydroxylated Si, PMIM-Br, $PMIM\text{-}CO_3$, PMIM-Cl and $PMIM\text{-}SO_3$ at a relative humidity (RH) of 15%: reproduced from Mo et al. (2011)

## *10.6.9 ILs for MEMS and NEMS Lubrication*

Micro-electro-mechanical-system (MEMS) and nano-electro-mechanical-systems (NEMS) devices are very sophisticated and have complicated attachments, such that lubrication is very challenging. Some investigations are summarised here:

- The tribological characteristics of octylsulphate and hexafluorophosphate anion-based 1-butyl-3methylimidazolium ILs were studied on Si(100) wafers. The ILs gave enhanced protection of the silicon wafers owing to better film formation (Palacio and Bushan 2008).

- New ILs were produced by joining imidazoliums-, ammonium- and methylpyridinium-containing cations and sulphur-containing anions, namely (bis(2-ethyl-1-hexyl) sulphosuccinate and ethylsulphate, [$C_2SO_4$]) (Cosme et al. 2016). Investigation of their tribological behaviour on silicon and steel was carried out, and their performance was compared with 1-ethyl 3-methylimidazolium cation-based ionic liquids [$C_2$MIM]. It was concluded that the anion [$C_2SO_4$] was responsible for lower friction coefficients on silicon, demonstrating that these new ILs have the potential for lubrication in MEMS and nano-electro-mechanical-systems (NEMS).
- The tribological behaviour of ultrathin films of functionalized imidazolium ILs for MEMS/NEMS devices was studied by Mo et al. (2011). Figure 10.14 describes the correlation between friction and load for different IL films on the hydroxylated Si surface. In detail it is seen that the friction for PMIM-$SO_3$was less than forPMIM-$CO_3$, PMIM-Br and PMIM-Cl.

## 10.7 Study of ILs at DMSRDE, Kanpur, in Collaboration with the Institute of Science, Banaras Hindu University, Varanasi

Lubricants are essential in common and complicated mechanical components/ systems like internal combustion engines, bearings, gearboxes, compressors, and hydraulics. The Directorate of Fuel and Lubricants (DFL) of DMSRDE, Kanpur, is one of the first laboratories in India to undertake R&D in the area of petroleum products and has acquired expertise in this area over a period of nearly 30 years. The laboratory is equipped with all modern analytical test and evaluation equipment, and offers excellent scope for research on lubricants, additives, synthesis, basic tribology and rheology. Delivery of many particular grade of lubricant/fuels for Services has been done. The current main objective is to design and develop some new ILs as high-performance lubricants. The research on this topic done in collaboration with Banaras Hindu University (BHU), Varanasi. Some significant work in the field of ionic liquids (Iwata et al. 2007; Panja et al. 2015, 2018; Panja and Saha 2017; Saha et al. 2017) is highlighted here.

New and unreported room temperature (RT) ILs employing nicotinamide cations are being studied, since nicotinamide cations are expected to be more easily biodegradable than the more popular and prototype alkyl imidazolium cations. Several n-alkylnicotinamide cation-based ILs with various common anions ($BF_4^-$,

$PF_6^-$, $NTf_2^-$) have been developed, including alkylnicotinamide cations with various alkyl chain lengths (n = 4–10).

Table 10.6 gives a list of some of these ILs having $NTf_2^-$ anions. Tables 10.7 and 10.8 show some of the physical properties of pyridinium and piperidium cation-based ILs that are also being studied. All then-alkylnicotinamide cation-based ILs are viscous pale yellow liquids at ambient temperature. The pyridinium and piperidinium cation-based ILs listed in Tables 10.7 and 10.8 are mostly light brown or yellow viscous liquids at room temperature, but three are solid until elevated temperatures are reached.

The tribological properties of these novel ILs are currently being investigated. Also, since recent publications have suggested the greater potentials of ILs with halogen-free anions (Barnhill et al. 2014; Gonzalez et al. 2016), the DMSRDE and BHU have plans to develop various non-halogen anions, mainly borate-based.

**Table 10.6** Nicotinamide-based ILs with a systematic increase in chain length

| Common name of IL | Code | Structure | Viscosity (cP) |
|---|---|---|---|
| N-butylnicotinamidetiflamide | $C_4CPyNTf_2$ | 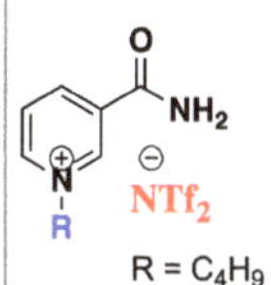  | 102 |
| N-hexylnicotinamidetiflamide | $C_6CPyNTf_2$ | 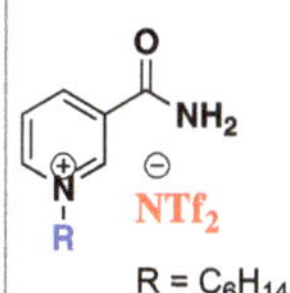  | 152 |
| N-octylnicotinamidetiflamide | $C_8CPyNTf_2$ | 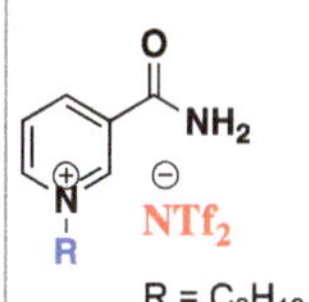  | 96 |
| N-decylnicotinamidetiflamide | $C_{10}CPyNTf_2$ | 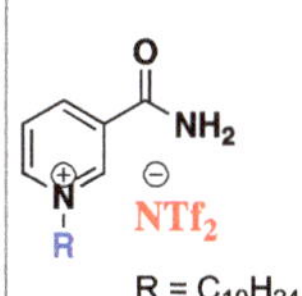  | 200 |

**Table 10.7** Physical properties of pyridinium cation-based ILs

| Code name | Chemical structure | Yield (%) | State/melting point (°C) | Viscosity (cP) | |
|---|---|---|---|---|---|
| | | | | 40 °C | 90 °C |
| PybCl | 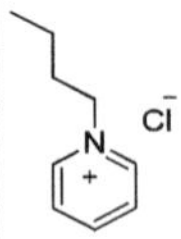 | 26 | Light brown viscous liquid | Unavailable | Unavailable |
| $PybBF_4$ | 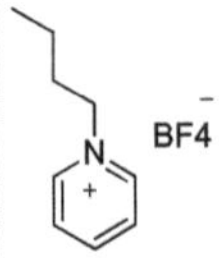 | 42 | Light brown viscous liquid | 34.6 | 8.3 |
| $PyPF_6$ | 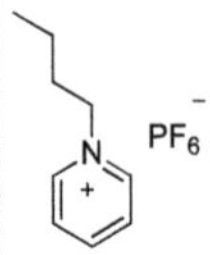 | 62 | White solid MP = 73–76 | Not applicable | Not applicable |
| $PybNTf_2$ | 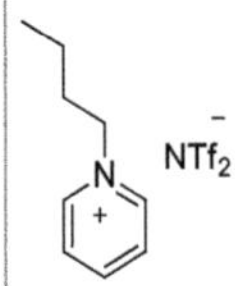 | 74 | Light brown viscous liquid | 25.6 | 6.8 |

## 10.8 Summary

Ionic liquids (ILs) belong to a group of new high-performance functional lubricants that will be very useful for both commercial and high-end technological applications where traditional lubricants are unsuitable or in adequate. ILs have shown promising results in either well-known systems like steel–steel surfaces and challenging systems such as steel–aluminium contacts. ILs also have significant advantages for use in environments such as high vacuum, extreme temperatures and high pressures.

To promote the use of Ils, a larger number should be tested under variable conditions. However, although promising, the use of Ils as neat lubricants is currently not cost-effective with respect to traditional lubricants. Thus their neat lubricant usage should be specific to special applications where small quantities of lubricants are required, such as in MEMS applications. On the other hand, ILs have shown promise as lubricant additives, and in such applications they could see more widespread use. For these applications significant directed research needs to be undertaken to design and develop ILs, specifically to ensure that they are miscible in base oils.

**Table 10.8** Physical properties of N-Alkyl-piperidinium cation-based ILs

| Code name | Chemical structure | Yield (%) | State/melting point (°C) | Viscosity (cP) | |
|---|---|---|---|---|---|
| | | | | 40 °C | 90 °C |
| $PiP_{14}Cl$ | 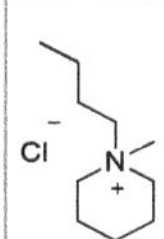 | 31 | Yellow solid | Not applicable | Not applicable |
| $PiP_{14}BF_4$ | 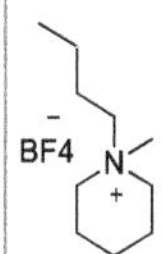 | 14 | Yellow solid MP = 138–140 | Not applicable | Not applicable |
| $PiP_{14}NTf_2$ | 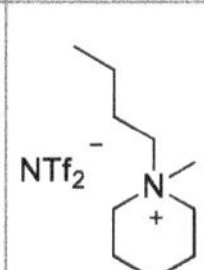 | 89 | Yellow viscous liquid | 50.8 | 9.5 |
| $PiP_{17}NTf_2$ | 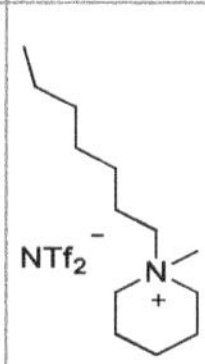 | 72 | Yellow viscous liquid | 128.4 | 20.7 |

**Acknowledgements** The authors are grateful to DRDO for the IL research funding. The contributions of Mr. Ashok Kr. Gautam, Jyoti Prakash Singh, Ms. Pratibha K.B. Pathak, and several colleagues of Fuels and Lubricants division are gratefully acknowledged. The authors are indebted to Dr. R.J.H. Wanhill for a comprehensive review and many suggestions to improve the text.

# References

Anand, M., Hadfield, M., Viesca, J.L., Thomas, B.A., Battez, H., Austen, S.: Ionic liquid as tribological performance improving additive for in-service and used fully formulated diesel engine lubricants. Wear **334–335**, 67–74 (2015)

Barnhill, W.C., Qu, J., Luo, H., Meyer, H.M., Ma, C., Chi, M., Papke, B.L.: Phosphonium-organophosphate ionic liquids as lubricant additives. Effects of cation structure on physicochemical and tribological characteristics. ACS Appl. Mater. Interfaces **6**, 22585–22593 (2014)

Bermudez, M.D., Jimenez, A.E., Sanes, J., Carrion, F.J.: Ionic liquids as advanced lubricant fluids. Molecules **14**, 2888–2908 (2009a)

Bermudez, M.D., Jim Enez, A.E., Sanes, J., Carriun, F.J.: Tribological performance of phosphonium based ionic liquids for an aluminium-on-steel systems and open ions on lubrication mechanism. Wear **10**, 1174–1179 (2009b)

Cabeza, O.: Properties and green aspects of ionic liquids. In: Rios, A.P., Jose, F., Fernandez, H. (eds.) Ionic Liquids in Separation Technology, Ist, pp. 1–93. Elsevier, Netherlands (2014)
Canter, N.: Evaluating ionic liquids as potential lubricants. Tribol. Lubr. Technol. **61**, 15–17 (2005)
Cosme, J., Pedro, D.A. Bastos, Catela, I., Silva, D., Colaço, R., Branco, L.C., Saramago, B.: Task specific ionic liquids based on sulfur for tribological applications. Chem. Select. **1**(13), 3612–3617 (2016)
David, M.H., Selman, C., Winter, R., Kisholoy, G.: Modified Ionic liquids-based high performance lubricants for robotic operations. In: 44th International Conference on Environmental Systems, 1–17 July, 2014, Tuscon, Arizona (2014)
David, M.H.: NASA SBIR/STTR Technologies (Unpublished work): Source http://www.innosense.us/documents/ionoglide_brochure.pdf
Doerr, N., Ille, H.G., Donat, H., Wanzenboeck, A.E., Pauschitz, A., Franek, F.: Evaluation of ionic liquids as lubricants. J. Microeng. Nanoelectron. **1**, 29–34 (2010)
Espinosa, T., Jimenez, Martínez Nicolas, G.E., Sanes, J., Bermudez, M.D.: Abrasion resistance of magnesium alloys with surface films generated from phosphonateimidazolium ionic liquids. Appl. Surf. Sci. **320**, 267–273 (2014)
Forsyth, M., Neil, W.C., Howlett, P.C., Macfarlane, D.R., Hinton, B.R., Rocher, W.N., Kemp, T.F., Smith, M.E.: New insights into the fundamental chemical nature of ionic liquid film formation on magnesium alloy surfaces. ACS Appl. Mater. Interfaces **1**, 1045–1052 (2009)
Fu, X., Sun, L., Zhou, X., Li, Z., Ren, T.: Tribological study of oil-miscible quaternary ammonium phosphites ionic liquids as lubricant additives in PAO. Tribol. Lett. **60**(22), 1–12 (2015)
Ge, L.L., Chen, L.P., Guo, R.: Microstructure and lubrication properties of lamellar liquid crystal in Brij30/[Bmim]PF6/$H_2O$ system. Tribol. Lett. **28**, 123–130 (2007)
Gonzalez, R., Bartolome, M., Blanco, D., Viesca, J.L., Fernandez-Gonzalez, A.: Effectiveness of phosphonium cation-based ionic liquids as lubricant additive. Tribol. Int. **98**, 82–93 (2016)
Grace, J., Vysochanska, S., Lodge, V., Iglesias, P.: Ionic liquids as additives of coffee bean oil in steel-steel contacts. Lubricants **3**, 637–649 (2015)
Gutierrez, M.A., Haselkorn, M., Iglesias, P.: The lubrication ability of ionic liquids as additives for wind turbine gearboxes oils. Lubricants **4**(2), 14 (2016)
Han, Y., Qiao, D., Guo, Y., Feng, D., Shie, L.: Influence of competitive adsorption on lubricating property of phosphonate ionic liquid additives in PEG. Tribol. Lett. **64**, 22 (2016)
Han, Y., Qiao, D., Zhang, S., Dapeng, F.: Influence of phosphate and ionic liquid structures on lubrication for different alloys (Mg, Al, Cu). Tribol. Int. **114**, 4469–4477 (2017)
Hernandez, B.A., Bartolome, M., Blanco, D., Viesca, J.L., Fernandez, G.A., Gonzalez, R.: Phosphoniumcation-based ionic liquids as neat lubricants: physicochemical and tribological performance. Tribol. Int. **95**, 118–131 (2016)
Huang, G., Yu, Q., Cai, M., Zhou, F., Liu, W.: Investigation of the lubricity and antiwear behavior of guinidinium ionic liquids at high temperature. Tribol. Int. **114**, 65–76 (2017)
Iwata, K., Okajima, H., Saha, S., Hamaguchi, H.O.: Local structure formation in alkyl-imidazolium-based ionic liquids as revealed by linear and nonlinear Raman spectroscopy. Acc. Chem. Res. **40**, 1174–1181 (2007)
Jia, Z.F., Xia, Y.Q., Li, J.L.: Frictional and wear behavior of diamond like carbon coating on plasma nitride steel under boundry lubrication. Tribol. Int. **43**, 474–482 (2010)
Jimenez, A.E., Bermudez, M.D.: Imidazolium ionic liquids as additives of the synthetic ester propylene glycol dioleate in aluminum-steel lubrication. Wear **265**, 787–798 (2008)
Jimenez, A.N.E., Bermudez, M.D.: Ionic liquids as lubricants of titanium-steel contact. Tribol. Lett. **33**, 111–126 (2009)
Jimenez, A.E., Bermudez, M.D., Carrion, F.J., Martinez-Nicolas, G.: Room temperature ionic liquids as lubricant additives in steel-aluminium contacts: influence of sliding velocity, normal load and temperature. Wear **261**, 347–359 (2006)
Jimenez, A.E., Bermudez, M.D., Iglesias, P.: Lubrication of INCONEL 600 with Ionic liquids at high temperature. Tribol. Int. **42**, 1744–1751 (2009)

Jin, C.M., Ye, C.F., Phillips, B.S., Zabinski, J.S., Liu, X.Q., Liu, W.M., Shreeve, J.M.: Polyethylene glycol functionalized dicationic ionic liquids with alkyl or polyfluoroalkyl substituents as high temperature lubricants. J. Mater. Chem. **16**, 1529–1535 (2006)

Kaadonski, T., Wojdyna, P.: Liquid lubricants for space engineering and methods for their testing. J. KONES **18**(1), 163–184 (2011)

Mang, T.: Lubicants and their markets. In: Mang, T., Dresel, W. (eds.) Lubricants and Lubrication, 2nd edn., pp. 1–6. Wiley, Germany (2007)

Minami, I.: Ionic liquids in tribology. Molecules **14**, 2286–2305 (2009)

Mo, Y., Huanga, F., Zhaob, F.: Functionalize dimidazolium wear-resistant ionic liquid ultrathin films for MEMS/NEMS applications. Surf Ace Interface Anal. **43**, 1006–1014 (2011)

Mu, Z.G., Zhou, F., Zhang, S.X., Liang, Y.M., Liu, W.M.: Effect of the functional groups in ionic liquid molecules on the friction and wear behavior of aluminum alloy in lubricated aluminum-on steel contact. Tribol. Int. **38**, 725–731 (2005)

Palacio, M., Bhushan, B.: A review of ionic liquids for green molecular lubrication in nanotechnology. Tribol. Lett. **40**, 247–268 (2010)

Palacio, M., Bushan, B.: Ultra thin wear-resistant ionic liquid films for novel MEMS/NEMS applications. Adv. Mater. **10**, 1194–1198 (2008)

Panja, S.K., Saha, S.: Microheterogeneity in imidazolium and piperidiniumcation based ionic liquids 1D and 2D NMR studies. Magn. Reson. Chem. **56**(2), 95–102 (2017)

Panja, S.K., Dwivedi, N., Noothalapati, H., Shigeto, S., Sikder, A.K., Saha, S., Sunkari, S.A., Saha, S.: Significance of weak interactions in imidazolium picrate ionic liquids: spectroscopic and theoretical studies for molecular level understanding. Phys. Chem. **17**, 18167–18177 (2015)

Panja, S.K., Srivastava, N., Srivastava, J., Prasad, N.E., Noothalapati, H., Shigeto, S., Saha, S.: Evidence of C--FP and aromatic π--FP weak interactions in imidazolium ionic liquids and its consequences. Spectrochim. Acta: Part A Mol. Biomol. Spectrosc. **194**, 117–125 (2018)

Qu, J., Truhan, J., Dai, S., Luo, H., Blau, P.: Ionic liquids with ammonium cations as lubricants or additives. Tribol. Lett. **22**, 207–214 (2006)

Qu, J., Blau, P.J., Dai, S., Luo, H., Meyer, III H.M.: Ionic liquids as novel lubricants and additives for diesel engine applications. Tribol. Lett. **35**, 181–189 (2009a)

Qu, J., Blau, P.J., Sheng, D., Huimin, L., Meyer, J.M., Truhan, J.J.: Tribological characteristics of aluminium alloys sliding against steel lubricated by ammonium and imidazolium ionic liquids. Wear **267**, 1226–1231 (2009b)

Qu, J., Viola, M.B.: Ionic liquids as novel lubricants and/or lubricant additives, CRADA final report for CRADA Number NFE-08-1715, U.S. Department of Energy under contract DE-AC05-00OR22725, USA (2013)

Reeves, C.J., Garvey, S.L., Menezes, P.L., Dietz, M.L., Jen, T.C., Lovell, M.R.: Tribological Performance of Environmentally Friendly Ionic Liquid Lubricants Proceedings of the ASME/STLE 2012 International Joint Tribology Conference October 8–10, Denver, Colorado, USA (2012)

Reeves, C.J., Menezes, P.L., Garvey, S.L., Jen, T.C., Dietz, M.L., Lovell, M.R.: The effect of phosphonium and imidazolium based ionic liquids as additives in natural oil: an investigation of tribological performance. In: Proceedings STLE Annual Meeting and Exhibition, 18–21 May, Lake Buena, Vista Florida, USA (2014)

Saha, S., Okajima, H., Homma, O., Hamaguchi, H.: Microscopic solvation environments in a prototype room-temperature ionic liquid as elucidated by resonance Raman spectroscopy of iodine and bromine. Spectrochem. Acta: Part A Mol. Biomol. Spectrosc. **176**, 79–82 (2017)

Silvester, D.S., Compton, R.G.: Electrochemistry in room temperature ionic liquids: a review and some possible applications. Zeitschrift Physikalische Chem. **220**(10), 1247–1274 (2006)

Song, Z., Liang, Y., Fan, M., Zhou, F., Liu, W.: Lithium based ionic liquids functionalized by sym-triazine and cyclotriphosphazene as high temperature lubricants. Tribol. Int. **70**, 136–141 (2014)

Sotres, J., Arnebrant, T.: Experimental investigations of biological lubrication at the nanoscale: the cases of synovial joints and the oral cavity. Lubricants **1**, 102–131 (2009)

Sowmiah, S., Srinivasadesikan, V., Tseng, M.C., Chu, Y.: On the chemical stabilities of ionic liquids. Molecules **14**, 3780–381311 (2009)

Street, K.W., Jr., Morales, W., Koch, V.R., Valco, D.J., Richard, R.M., Hanks, N.: Evaluation of vapor pressure and ultra-high vacuum tribological properties of ionic liquids. Tribol. Trans. **54**, 911–919 (2011)

Totolin, V., Gabler, C., Dorr, N., Minami, I.: Halogen-free borate ionic liquids as novel lubricants for tribological applications. Tribol. Int. **67**, 191–198 (2013)

Vekariya, R.L.: A review of ionic liquids: applications towards catalytic organic transformations. J. Mol. Liq. **227**, 44–60 (2017)

Wang, Y., Yu, Q., Cai, M., Shi, L., Zhou, F., Liu, W.: Ibuprofen-based ionic liquids as additives for enhancing the lubricity and antiwear of water-ethylene glycol liquid. Tribol. Lett. **65**, 55 (2017)

Wilkes, J.S., Zaworotko, M.J.: Air and water stable 1-ethyl-3-methylimidazolium based ionic liquids. J. Chem. Soc. Chem. Commun. 965–967 (1992)

Wu, J., Lua, X., Fenga, X., Shib, Y.: Halogen-free ionic liquids as excellent lubricants for PEEK-stainless steel contacts at elevated temperatures. Tribol. Int. **104**, 1–9 (2016)

Yao, M., Liang, Y., Xia, Y., Zhou, F.: Bisimidazolium ionic liquids as the high performance anti wear additives in poly (ethylene glycol) for steel−steel contacts. ACS Appl. Mater. Interfaces **1**(2), 467–471 (2009)

Ye, C., Liu, W., Chen, Y., Yu, L.: Room-temperature ionic liquids: a novel versatile lubricant. Chem. Commun. **21**, 2244–2245 (2001a)

Ye, C., Liu, W.M., Chen, Y.X., Yu, L.G.: Room-temperature ionic liquids a novel versatile lubricant. Chem. Commun. **7**(21), 2244–2245 (2001b)

Zhang, Y., Cai, T., Shang, W., Sun, L., Liu, D., Tong, D., Liu, D.: Environmental friendly polyisobutylene-based ionic liquid containing chelated orthoborate as lubricant additive: synthesis, tribological properties and synergestic interactions with ZDDP in hydrocarbon oils. Tribol. Int. **115**, 297–306 (2017)

Zhou, F., Liang, Y., Liu, W.: Ionic liquid lubricants: designed chemistry for engineering applications. Chem. Soc. Rev. **38**, 2590–2599 (2009)

Zhou, M., Zhang, J., Wang, S., Bover, D., Guo, H., Li, W., Wu: Laterally substituted ionic liquid crystals and the resulting rheological behavior. Soft Matters **8**(30), 7945–7951 (2012)

GPSR Compliance
The European Union's (EU) General Product Safety Regulation (GPSR) is a set of rules that requires consumer products to be safe and our obligations to ensure this.

If you have any concerns about our products, you can contact us on

ProductSafety@springernature.com

In case Publisher is established outside the EU, the EU authorized representative is:

Springer Nature Customer Service Center GmbH
Europaplatz 3
69115 Heidelberg, Germany

www.ingramcontent.com/pod-product-compliance
Ingram Content Group UK Ltd.
Pitfield, Milton Keynes, MK11 3LW, UK
UKHW022215051026
15624UKWH00001B/163
*9789819997909*